THE DIRECT OBSERVATION
OF DISLOCATIONS

SOLID STATE PHYSICS

Advances in
Research and Applications

Editors

FREDERICK SEITZ

Department of Physics
University of Illinois
Urbana, Illinois

DAVID TURNBULL

Division of Engineering
and Applied Physics
Harvard University
Cambridge, Massachusetts

The following monographs are published within the framework of the series:

1. T. P. Das and E. L. Hahn, *Nuclear Quadrupole Resonance Spectroscopy*, 1958
2. William Low, *Paramagnetic Resonance in Solids*, 1960
3. A. A. Maradudin, E. W. Montroll, and G. H. Weiss, *Theory of Lattice Dynamics in the Harmonic Approximation*, 1963
4. Albert C. Beer, *Galvanomagnetic Effects in Semiconductors*, 1963
5. Robert S. Knox, *Theory of Excitons*, 1963
6. S. Amelinckx, *The Direct Observation of Dislocations*, 1964

ACADEMIC PRESS • New York and London • 1964

THE DIRECT
OBSERVATION OF
DISLOCATIONS

S. AMELINCKX

Departement Fysika Vaste Stof
Studiecentrum voor Kernenergie
Mol-Donk, Belgium

ACADEMIC PRESS • New York and London • 1964

ACADEMIC PRESS INC.
111 Fifth Avenue
New York, N. Y. 10003

United Kingdom Edition
Published by
ACADEMIC PRESS INC. (London) Ltd.
Berkeley Square House, London W. 1

Library of Congress Catalog Card Number: 63-21399

PRINTED BY THE ST CATHERINE PRESS LTD., BRUGES, BELGIUM.

Preface

At the time I agreed to write a review paper on "the direct observation of dislocations" for *Solid State Physics*, I expected that this could be done very easily in about forty pages using perhaps ten additional pages of photographic illustrations. At that time the role of dislocations in crystal growth had been recognized, etching had become a well-established tool, and decoration techniques had produced some quite striking results. However, the pioneering observations of dislocations in the electron microscope had only just started.

From the fact that this turns out to be a book of reasonable size rather than a review article, one can readily deduce that important developments must have taken place in the meantime. The most important development is the use of the electron microscope. Its success has been such that in a period of three or four years the amount of material published is so enormous that it is almost impossible to cover it completely. It is also practically impossible to include all references. A choice had therefore to be made, which inevitably reflects the author's interest.

Although the number of papers published on the subject is still maintaining the same level, the number of significant contributions seems to have slowed down during the last year or so, and it is therefore felt that this is a suitable time to present a survey of what has been done up till now.

The field of transmission electron microscopy owes a great deal to the work of the Cambridge group, which not only introduced most of the experimental techniques but also deserves most of the credit for the development of an adequate theory of contrast formation.

Parallel to the use of the electron microscope, very refined X-ray techniques have been discovered. The difference in resolution and scale between the two methods makes them complementary.

The combined use of all these methods had led to a detailed verification of large chapters of dislocation theory. Moreover, several new and unexpected facts have been discovered which have led to a revision of some parts of the existing theory and to the development of new theory.

It is the purpose of this book to discuss all these methods and to illustrate their use by specific examples. Although emphasis is on methods and observations, it has been found necessary to include some of the underlying theory of dislocations in order to make clear the significance of the observations. However, this has been kept to a minimum and it does not go beyond the qualitative stage unless quantitative results are required. Some knowledge of dislocation theory, as can be found in the standard texts on this subject,* is assumed.

It was considered worth while to develop also in some detail the theory on which the interpretation of electron micrographs is based. As the theory of image formation becomes better understood, more and more detailed information can be obtained by combining direct observation with electron diffraction; for the reader who wishes to work in this field it is therefore of interest to be acquainted with this theoretical background.

A number of people have collaborated a great deal in producing this book. In the first place I would like to thank Prof. Dr. R. Gevers for reading the manuscript thoroughly and for having suggested a large number of improvements especially in the theoretical treatment of contrast effects. He also kindly gave permission to use some of his unpublished work. Many thanks are due to Dr. Delavignette for the use of many photographs and to Dr. Siems for a critical reading of the manuscript and for making useful suggestions. I had the privilege of receiving a large number of photographs from different sources with the permission to use them as illustrations. In this connection I would like to thank: Drs. A. Art, R. S. Barnes, G. Bassett, A. Berghezan, H. Betghe, H. Blank, W. L. Bond, S. Borrmann, W. Carrington, R. Cotterill, W. C. Dash, A. Fourdeux, J. Gilman, A. Glossop, R. Guard, P. B. Hirsch, A. Howie, V. Indenbohm, W. G. Johnston, D. Jones, A. R. Lang, J. Linde, J. Low, M. Marcinkowski, D. Miller, J. W. Menter, J. W. Mitchell, J. Newkirk, D. Pashley, P. B. Price, J. Silcox, R. Smallman, P. Swann, G. Thomas, A. Tweet, W. W. Webb, M. J. Whelan, and F. W. Young.

---

* J. Friedel, "Les dislocations." Gauthiers-Villars, Paris, 1956; W. T. Read, "Dislocations in Crystals." McGraw-Hill, New York, 1955; A. H. Cottrell, "Dislocations and Plastic Flow in Crystals." Oxford Univ. Press, London and New York, 1953.

Thanks are due to Mr. Van Roy for typing the manuscript, to Mr. Van Summeren for preparing the line drawings, and to Mr. Beyens for making the prints of many photographs.

Last but not least, I would like to thank the Directors of C.E.N.-S.C.K. (Centre d'Étude de l'Énergie Nucléaire—Studiecentrum voor Kernenergie) for permission to spend the necessary time for preparing the manuscript. Their understanding and appreciation of fundamental studies in the different fields encompassed by the department has allowed to pursue in particular the research resulting in the writing of this book.

April 1964 S. AMELINCKX

Contents

Introduction

In recent years a number of methods for the direct observation of dislocations has been developed. This has made possible not only a detailed verification of large chapters of dislocation theory, but the discovery of several new and unexpected facts, which have led to a revision of the existing and to the development of a new theory.

It is the author's purpose to review these methods and to illustrate their use by specific examples. It is difficult to include all applications of some of the methods that are discussed because the volume of publication has become enormous in just a few years. It is also practically impossible to include all references. Therefore a choice has to be made, which inevitably reflects the author's interest. Although emphasis is on methods and observations it has been found necessary to include some of the underlying theory of dislocations in order to make clear the significance of the observations. However, this has been kept to a minimum and it does not go beyond the qualitative stage unless quantitative results are required.

Also, it was considered worth while to develop in some detail the theory on which the interpretation of electron micrographs is based. As the theory of image formation becomes better understood, more and more detailed information can be obtained by combining direct observation with electron diffraction. For the reader who wishes to work in this field it is therefore of interest to be acquainted with this theoretical background.

I. Surface Methods

1. CRYSTAL GROWTH

a. Growth Spirals

(1) *Formation mechanism.* The first phenomenon which was proved to be unambiguously related to the presence of dislocations was crystal growth on close-packed faces and at low supersaturations.[1-3]

A growing crystal is soon bounded by its slowest growing faces, since the others eliminate themselves in the process. The slowest growing faces are in general the close-packed ones. The nucleation of a new close-packed layer on a perfect crystal requires the formation of a two-dimensional nucleus, which is a difficult process. Once the nucleus is formed the crystal face is rapidly covered with a new layer. The lateral displacement of growth fronts is governed by the concentration of "kink sites" in the steps. These kink sites are formed as a consequence of thermal fluctuations, and under normal growth conditions, even in the close-packed fronts, a sufficient concentration of them is present to ensure growth. However, according to theory a large supersaturation is required (50%) in order to obtain an observable growth rate, on the assumption that two-dimensional nucleation is the rate determining step. In practice, growth is observed at supersaturations of 1% and less.

Frank[2] therefore suggested that growth might be catalyzed by the presence of dislocations having a Burgers vector which has a component perpendicular to the crystal face. For geometrical reasons a step is attached to the emergence point of such a dislocation (Fig. 1a). During growth, building units are attached to this step which tends as a consequence to be displaced parallel to itself. However, since it is anchored at the emergence point P it will wind up into a spiral

---

[1] W. K. Burton, N. Cabrera, and F. C. Frank, *Phil. Trans. Roy. Soc.* **A243**, 299 (1951).

[2] F. C. Frank, *Discussions Faraday Soc.* **5**, 48 (1949).

[3] N. Cabrera and W. K. Burton, *Discussions Faraday Soc.* **5**, 40 (1949).

(Fig. 1b). During further growth each part of the spiral moves outward and as a result the spiral apparently rotates around its center, maintaining a stationary shape. Direct evidence for the occurrence of this process was obtained first by Griffin[4] on surfaces of beryl crystals, and subsequently by a number of authors for a large number of substances.[5-7] Figure 2b,c shows examples as observed on the basal plane of silicon carbide.[8,9]

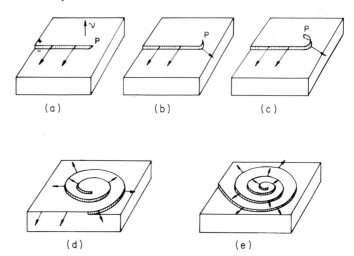

(a)   (b)   (c)

(d)   (e)

FIG. 1. (a) Step attached to the emergence point P of a dislocation with a Burgers vector which is not parallel to the surface. The step height $h = \bar{b} \cdot \bar{\nu}$, where $\bar{\nu}$ is the unit normal on the surface. (b) Under the influence of a supersaturation the step in (a) winds up into a spiral centered on P.

(2) *Shape of growth spirals.* The shape of the growth spiral is given approximately by the following equation[1] in polar coordinates (ρ, ϑ):

$$\rho = 2\rho_c \vartheta \qquad (1.1)$$

[4] L. J. Griffin, *Phil. Mag.* [7] **41**, 196 (1950); **42**, 775, 1337 (1951); **43**, 651 (1952).

[5] W. Dekeyser and S. Amelinckx, "Les dislocations et la croissance des cristaux." Masson, Paris, 1955.

[6] A. R. Verma, "Crystal Growth and Dislocations." Butterworths, London, 1953.

[7] A. J. Forty, *Advan. Phys. (Phil. Mag. Suppl.)* **3**, 1 (1954).

[8] S. Amelinckx, *Nature* **167**, 939 (1951); **168**, 431 (1951); *J. Chim. Phys.* **48**, 1 (1951); **49**, 411 (1952); **50**, 45 (1953); S. Amelinckx and G. Strumane, "Silicon Carbide." Pergamon Press, New York, 1960.

[9] A. R. Verma, *Nature* **167**, 939 (1951); **168**, 431, 783 (1951); **169**, 540 (1952); *Phil. Mag.* [7] **42**, 1005 (1951); **43**, 441 (1952).

where ρ_c is the radius of the critical nucleus. The speed of rotation ω of the spiral during growth is given by

$$\omega = v_\infty/2\rho_c \qquad (1.2)$$

where v_∞ is the displacement velocity of a straight growth front.

FIG. 2. (a) Spiral depressions on the c face of crystals of salol grown from a carbon disulfide solution. [After S. Amelinckx, *J. Chim. Phys.* **50**, 218 (1953).] (b) and (c) Growth spirals on the basal plane, $c(0001)$ of silicon carbide. [After S. Amelinckx, *J. Chem. Phys.* **49**, 411 (1952).]

To a first approximation the spiral is Archimedean, the distance between successive turns being constant and equal to $4\pi\rho_c$.

In reality the displacement rate of a growth front depends on its radius of curvature ρ according to the relation

$$v(\rho) = v_\infty \left(1 - \frac{\rho_c}{\rho}\right) \tag{1.3}$$

and as a consequence of this, the outer turns of the spiral will grow outward more rapidly than the inner ones. A more accurate calculation (1.1) shows that the spiral has an equation of the form

$$\vartheta = a\rho + b \ln \rho \tag{1.4}$$

where $b \ll a$. The slow increase in spacing for the outer turns is evident on most micrographs showing rounded spirals. In many cases the spirals are polygonal having the same symmetry as the face on which they are formed; the spacing is then more uniform.

(3) *Interactions between spirals.* Complications arise when several dislocations emerge in one face. Consider a simple case of the interaction between two dislocations of the same Burgers vector but of opposite sign. Successive stages in the growth process are shown in Fig. 3. It is clear that closed fronts will be formed provided the

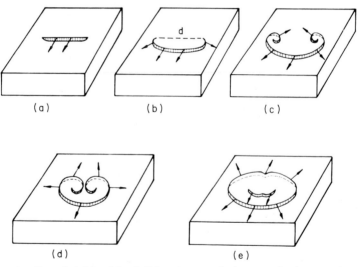

FIG. 3.   Growth initiated by 2 dislocations producing steps in the opposite sense. The emergence points of the 2 dislocations are connected by a step. In (d) the 2 spiral arms meet and a closed growth front is formed, at the same time the original segment is again available for the formation of a new growth front.

distance d between the emergence points is large enough to insure that the closed loop, once it is formed, is larger than the critical two-dimensional nucleus ($2\rho_c$) under the growth conditions considered. This imposes roughly the condition $d < 2\rho_c$. Figure 4 shows a growth pattern of this type.

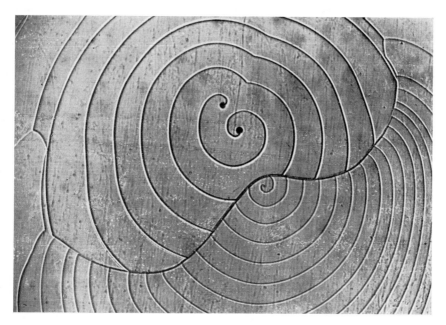

FIG. 4.   Growth pattern due to 2 spirals of opposite sense on the c face of silicon carbide. [After S. Amelinckx, *J. Chem. Phys.* **49**, 411 (1962).]

If the two or more dislocations have Burgers vectors producing steps of the same sign, cooperating multiple spirals may result as, e.g., in Fig. 5.

More complicated interactions are discussed in Amelinckx[8] and Verma.[9]

(4) *Information obtainable from growth spirals.*   It is clear that there is now a means of locating the emergence points of those dislocations having a Burgers vector which has a component perpendicular to the crystal phase. One can moreover determine this perpendicular component since it is equal to the step height of the spiral.

If the emergence point of such a dislocation moves, it leaves a step in its wake. On intersecting previously formed growth fronts the latter are deformed in a characteristic, recognizable way. One therefore has a means of studying dislocation motion (see Fig. 6). The first evidence for movement was obtained in this way.[4] The mutual annihilation of dislocations of opposite sign was also demonstrated

FIG. 5.   Growth promoted by a number of spirals of the same sense originating in a group of dislocations with components perpendicular to the *c* face of the sign. The step heights are — 13 A [After S. Amelinckx and G. Strumane, in "Silicon Carbide" (J. R. O'Connor and J. Smittens, eds.), p. 16?. Pergamon Press, New York, 1960.]

first by the use of growth features. Two dislocations of opposite sign give rise to closed growth fronts as shown in Fig. 7a. Mutual annihilation of the two dislocations just before cessation of growth leads to the features of Fig. 7b, i.e. to a sequence of closed growth fronts without a visible center.

Betghe and Schmidt[10] have worked out a method to mark the emergence points of dislocations by means of small growth hills. On a fresh cleavage face of sodium chloride, kept at 300°C, they evaporate in high vacuum a 1000–10,000 A thick layer of sodium chloride at a rate of 100–300 A per second. Without breaking the vacuum

[10] H. Betghe and V. Schmidt, *Z. Naturforsch.* **14a**, 307 (1959).

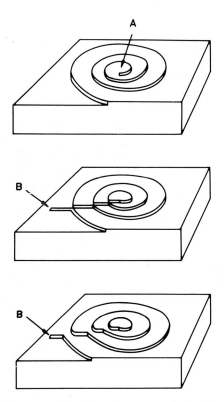

FIG. 6. Surface features produced by the movement of a dislocation from A to B. The dislocation in A originally gave rise to a spiral.

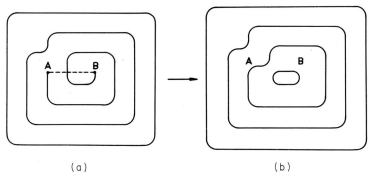

FIG. 7. The mutual annihilation of 2 dislocations of opposite sign emerging in A and B, as reflected in the pattern of surface steps. (a) During growth. (b) After mutual annihilation of the dislocations in A and B.

a carbon film is deposited. On dissolving the crystal the carbon replica is liberated and subsequently shadowed with ThF_4. Small well-defined growth hills mark emergence points of dislocations; it is not possible to distinguish between screws and edges.

b. Spiral Depressions

(1) *Formation mechanism.*[11,12] Crystals which grow while floating on their nutrient solution develop spirally terraced depressions on their top faces, so-called "hopper" faces. This growth feature is also directly related to the presence of screws, although the growth mechanism is quite different; it is pictured in Fig. 8. Suppose that

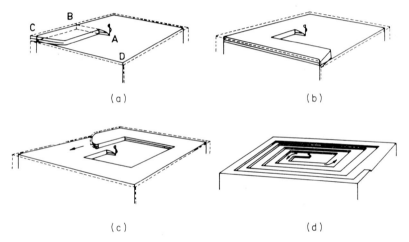

FIG. 8. Formation mechanism for spiral depressions. The crystal grows while floating on its solution. Growth is only possible along the periphery of the top face.

the step associated with the screw (or screws) emerging in A, intersects the periphery of the growing crystal in B. Since growth is only possible along the periphery a narrow strip of material BC will be deposited. This strip grows all along the periphery and it would form some kind of "box" if it were not that in the mean time the crystal has also grown laterally. As a consequence each turn is displaced outward with respect to the previous one and a spiral depression is formed.

[11] S. Amelinckx, *Phil. Mag.* [7] **44**, 337 (1953).
[12] S. Amelinckx, *J. Chim. Phys.* **50**, 218 (1953).

(2) *Shape of spirals.* It is easy to show[12] that the spiral is of the logarithmic type since the spacing between successive turns is proportional to the instantaneous size of the crystal. These features are visible in Fig. 2a. If the crystal face were circular the equation of the spiral in polar coordinates (ρ, ϑ) would be the solution of the differential equation

$$d\rho = \frac{v}{V} \rho d\vartheta. \tag{1.5}$$

This equation describes the path followed by a point that travels along the periphery of a circle at a speed V, the displacement rate of the growth front, while the circle grows radially at the speed v, the lateral growth speed of the crystal. The solution evidently is

$$\rho = a \exp\left(\frac{v}{V} \vartheta\right). \tag{1.6}$$

The conclusion still holds for a polygonal crystal: the distance between successive turns increases exponentially. However, the shape is that of the crystal face on which it is formed. The spiral is turned with respect to the periphery of the crystal face over an angle

$$\alpha = \text{arc tg} \frac{v}{V} \tag{1.7}$$

Once formed, the inner branches of the spiral no longer move in contrast to the spirals of the Frank type.

Photographs of successive stages in the formation of the spiral depressions have been photographed confirming the described process.[12]

(3) *Information obtainable from spiral depressions.* From the formation mechanism it follows that the step height of the spiral is equal to the projection on the normal on the top face of the algebraic sum of the Burgers vectors of all the dislocations emerging in that face.

Glide may change the character of the growth features. If a slip step reaches the periphery it will in turn give rise to the formation of a spiral branch. If the slip step has the same sign as the central spiral, cooperating multiple spirals result. If on the contrary the slip step is of opposite sign and of larger height than the central spiral, the sense of winding of the spiral is inverted as illustrated by Figs. 9 and 10.

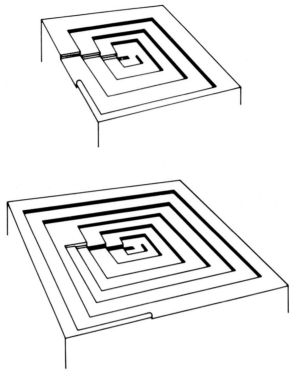

FIG. 9. The sense of winding of a spiral depression is inverted if a higher slip step of opposite sign is formed before the cessation of growth.

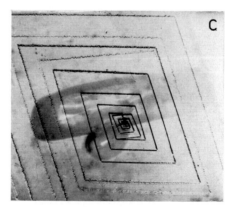

FIG. 10. Observed growth pattern on the *c* face of salol illustrating the process represented in Fig. 9. [After S. Amelinckx, *J. Chim. Phys.* **50**, 218 (1953).]

The center of the spiral depression usually corresponds to the center of a Frank spiral on the bottom surface. This was proved directly in the case of potassium ferrocyanide.[13]

2. EVAPORATION

The reverse phenomenon, i.e., evaporation under conditions of low undersaturation also produces spiral steps. These are depressions instead of hills, i.e., their center is the lowest point instead of the highest. It is clear that this phenomenon can equally well be used to locate intersection points of dislocations with the surface and to measure the component of the Burgers vector perpendicular to that surface.

The method was first applied to alkali halides[14] using optical microscopy. Recently Sella et al.[15] and Bassett[16] have developed an electron optical method for making visible monoatomic steps on alkali halides such as those formed on evaporation or growth. The procedure consists in evaporating in vacuum a very thin layer of gold onto the surface of the crystal which is heated at some 150–200°C. Under these circumstances the gold does not form a continuous layer, but instead small isolated gold particles nucleate epitaxially, preferentially along surface steps. A layer of carbon is deposited on top of the gold layer. The carbon replica is then taken off by dissolving the crystal in water. The gold particles remain imbedded in the carbon film which is then examined in the electron microscope. Recently Betghe and co-workers[17-19] have shown how detailed information can be obtained from surface studies by the use of this technique. Figure 11 shows one of their micrographs; single as well as double spirals are observed. The double spirals are attributed to the dissociation of a dislocation with a Burgers vector a [100] into two dislocations with vectors $(a/2)$ [110] and $(a/2)$ [1$\bar{1}$0]. If such a dislocation emerges

[13] S. Amelinckx and E. Votava, Nature 172, 538 (1953).

[14] S. Amelinckx and E. Votava, Naturwissenschaften 41, 422 (1954).

[15] C. Sella, P. Conjeaud, and J. J. Trillat, Compt. Rend. 249, 1978 (1959).

[16] G. A. Bassett, Phil. Mag. [8] 3, 1042 (1958).

[17] H. Betghe and W. Keller, Z. Naturforsch. 15a, 271 (1960).

[18] H. Betghe and W. Keller, Proc. European Reg. Conf. on Electron Microscopy, Delft, 1960, Vol. 1, p. 266 (1961).

[19] H. Betghe, Proc. 4th Intern. Conf. on Electron Microscopy, Berlin, 1958 1, 409 (1960).

in the (100)face two monoatomic steps are formed. If no dissociation
takes place single spirals with step height equal to 2.8 Å result; they are
in general squared instead of rounded. This difference in morphology
can be understood on the basis of Stranski's theory. Figure 12 shows
a number of evaporation spirals centered on a subboundary.

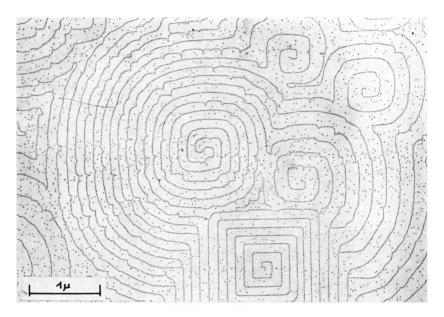

FIG. 11.   Evaporation spirals observed on the cube plane of sodium chloride. The
steps are decorated by means of small gold particles. Note that the step height of the
square spiral is twice that of the rounded double spiral. This is concluded from the
way the steps fuse. (Courtesy of Betghe and co-workers.)

Votava *et al.*[20,21] have used evaporation spirals as a means of studying
the geometry of coherent twin boundaries in copper. They found
that evaporation spirals are centered on the noncoherent parts of the
twin interface and none on the coherent parts. The noncoherent parts
occur as kinks in the coherent boundary; it is found that the sense
of winding of the spirals is uniquely related to the sign of the kinks.
This is interpreted as the result of the presence of twin dislocations
in the noncoherent parts.

[20] E. Votava, A. Berghezan, and R. H. Gillette, *Naturw.* **44**, 372 (1957).
[21] E. Votava and A. Berghezan, *Acta Met.* **7**, 6 (1959).

3. ETCHING

a. Introduction

The first direct proof that dislocations can be revealed by means of etching was given independently by Gevers *et al.*,[22] Gevers,[23,24] and Horn.[25] These authors showed that etch pits are developed at the

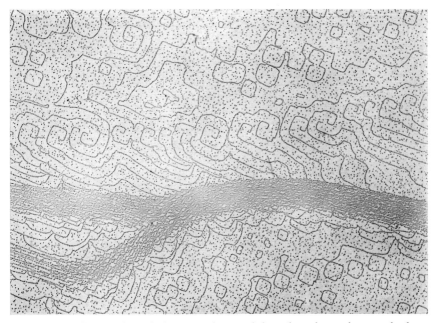

FIG. 12. Evaporation spirals, centered on a sub-boundary observed on a cube face of sodium chloride. (Courtesy of Betghe and co-workers.)

centers of growth spirals. The experiments were performed on silicon carbide crystals which were etched in fused borax at 900°C for 1 hr. In these experiments etch pits which were not located at the centers of spirals also were found and it was concluded that these correspond to edge dislocations. Figure 13 shows some more recent observations on the same system. The two kinds of pits can be recognized. At present etching has become the most widely used method

[22] R. Gevers, S. Amelinckx, and W. Dekeyser, *Naturwissenschaften* **39**, 448 (1952).

[23] R. Gevers, *J. Chim. Phys.* **50**, 321 (1953).

[24] R. Gevers, *Nature* **171**, 171 (1953).

[25] F. H. Horn, *Phil. Mag.* [7] **43**, 1210 (1952).

for measuring dislocation densities. An excellent review on the etching of nonmetallic crystals has been presented by Johnston.[26a]

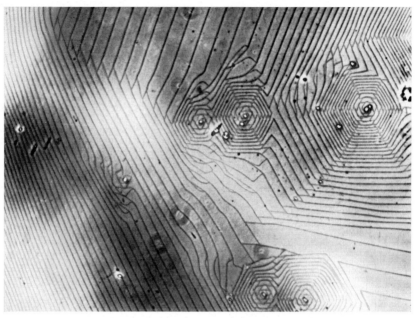

FIG. 13. Etch pattern on the *c* face of silicon carbide. Pits located at the centers of growth spirals, as well as pits located elsewhere, can be seen. The growth spirals have become visible as a consequence of preferential attack along the growth steps. Single steps are only 7 A high.

b. Principle of the Method

The method consists simply in immersing the crystal in a suitable medium, e.g., a liquid, a solution, or a gaseous chemical reagent. It is found that small pits are developed at the emergence points of dislocations.

Different methods of etching have been succesful.

(1) *Thermal etching.* For instance, heating of silver in an atmosphere containing some oxygen produces small pits at the dislocations.[26b]

[26a] W. G. Johnston, Dislocation etchpits in non-metallic crystals. "Progress in Ceramic Science" (J. E. Burke, ed.), Vol. 2. Pergamon Press, New York, 1962.

[26b] A. A. Hendrickson and E. S. Machlin, *Acta Met.* **3**, 64 (1955).

(2) *Chemical etching.* The reaction between molten borax and silicon carbide is of this type.[22,23] The etching of calcium fluoride with concentrated sulfuric acid is also a chemical attack.[27]

(3) *Solution etching.* Sodium chloride can be etched by dipping it for 1 sec in anhydrous methylalcohol.[28] Lithium fluoride is etched by means of an aqueous solution of ferric fluoride.[29]

(4) *Preferential oxidation.* In a sense this can be considered as an etching method. For example, Young has shown[30] that copper, for example, is oxidized preferentially at dislocation sites.

(5) *Electrolytic etching.* This has for instance been used by Jacquet[31] on specimens of α-brass.

(6) *Cathodic sputtering.* Cathodic sputtering in mercury vapor has been tried as a method for revealing emergence points of dislocations in germanium.[32] However the ion bombardment revealed features which did not correspond to the etch pits produced by CP_4 etching.

A similar approach was made by Meckel and Swalin [33]; they bombarded the surface with argon ions after CP_4 etching and find, confirming Wehner's findings, that the chemically produced etch pits disappear. The authors found spirally terraced hillocks which they interpret as the intersection of screw dislocations with the surface.

Sirotenko and Spivak[34] repeated the experiments and found that with a critical choice of the sputtering conditions (a short-time bombardment with a high-density ionic current) and the use of krypton, reproducible etch patterns can be obtained.

c. *Theory of Etching*

Etching and evaporation are phenomena which are essentially the reverse of growth. It is therefore quite natural that the same concepts that play a role in growth do so in evaporation or dissolution.

[27] W. Bontinck, *Phil. Mag.* [8] **2**, 516 (1957).

[28] S. Amelinckx, *Acta Met.* **2**, 848 (1954).

[29] J. J. Gilman and W. G. Johnston, *in* "Dislocations and Mechanical Properties of Crystals" (J. C. Fisher *et al.*, eds.), p. 116. Wiley, New York, 1957.

[30] F. W. Young and A. T. Gwathmey, *J. Appl. Phys.* **31**, 225 (1960).

[31] P. A. Jacquet, *Acta Met.* **2**, 725, 770 (1954).

[32] G. K. Wehner, *J. Appl. Phys.* **29**, 217 (1958).

[33] B. B. Meckel and R. A. Swalin, *J. Appl. Phys.* **30**, 89 (1959).

[34] I. G. Sirotenko and G. V. Spivak, *Soviet Phys.—Cryst.* (*English Transl.*) **6**, 274 (1961).

Evaporation presumably takes place, at least under conditions of low undersaturation, by the migration of surface steps. Further the active sites in the steps are the "kinks" which according to theories by Frenkel and by Burton, Cabrera, and Frank are easily formed as a consequence of thermal fluctuations. Normal surface steps, not attached to dislocations, will disappear in the evaporation process. However steps attached to a screw dislocation wind up in spiral shape. The center of the spiral is now the lowest point. Evaporation spirals have been discussed in the previous section.

Etch pits will be visible only if successive turns of the evaporation or dissolution spiral are close enough so that the slope of the sides of the etch pit is sufficiently large. According to the theory by Cabrera and Levine[35,36] this distance d is about 19 ρ_c where ρ_c is the radius of the critical two-dimensional dissolution nucleus. This radius is given by

$$\rho_c = \frac{\gamma \Omega}{kT \ln (c/c_0)}$$

where γ is the surface energy of the interface crystal-solution, Ω is the molecular volume, c_0 is the equilibrium concentration of the crystalline substance in the solvent, or the equilibrium vapor pressure in the case of evaporation, while c is the actual concentration or the actual vapor pressure. Provided the undersaturation is large enough, i.e., ρ_c small enough, a visible pit will result.

This theory has to be modified for etching at edge dislocations. Two-dimensional nucleation is now necessary and one has to account for the preferential nucleation of pits at the emergence points of edge dislocations. This follows because the activation energy for two-dimensional nucleation is locally decreased since strain energy is gained by dissolving deformed material in the immediate vicinity of the dislocation.

It is clear that this factor is of importance for screws as well as for edges. The speculations presented above for etching at screws have therefore to be amended by allowing for preferential nucleation also at screws.

d. Role of Impurities in the Crystal

The reason why emergence points of dislocations are attacked preferentially are probably more complex than suggested in Section 3c

[35] N. Cabrera and M. M. Levine, *Phil. Mag.* [8] 1, 450 (1956).
[36] N. Cabrera, M. M. Levine, and J. Plaskett, *Phys. Rev.* 96, 1153 (1954).

and are not well understood at present. However it seems established that in metals some impurity segregation or Cottrell atmosphere formation is necessary before dislocations can be reliably etched.[37] This is probably due to the inevitable presence of a surface film on metal specimens; in fact it is this film which is attacked. Dislocation sites in this film are probably differentiated as a consequence of the impurities.

In ionic crystals the presence of impurities does not seem to play an important role. On clean cleavage faces of such crystals fresh dislocations, (i.e., newly introduced) as well as old ones (i.e., grown-in) can be revealed. In some cases the aspects of the pits may differ for fresh and for grown-in dislocations.[28,29] One can even develop etches that will reveal only fresh dislocations.[38] This is particularly useful if one wishes to study the movement of dislocations.

From the fact that fresh dislocations are better attacked one may conclude that the strain around dislocations is responsible for the increased reactivity. When an atmosphere of impurities forms, the strain is relieved, and hence the reactivity decreases. On the other hand, an atmosphere may locally increase the solubility and hence increase the etching rate. Which of the two effects actually predominates will depend on the particular etchant used.

e. The Effect of Impurities in the Etchant

The cross section of a pit is essentially determined by the ratio v_l/v_n where v_l is the lateral displacement velocity of surface steps and v_n is the normal dissolution rate, i.e., the rate at which the pit deepens. Well-defined pits result only if the ratio v_l/v_n is sufficiently small. A good etchant is one for which v_l/v_n is at most 10. The quality of an etchant can be improved by changing either v_n or v_l. The usual method of decreasing v_l is to add impurities to the solvent, which poison the kink sites in the surface steps. For LiF, Gilman et al.[39] have shown that v_l depends very much on the iron concentration of a dilute solution of iron fluoride (FeF_3). In many cases the active poison does not seem to be very specific, which makes it easier to develop an etchant.

[37] G. Wyon and P. Lacombe, *Rept. Conf. Defects in Crystalline Solids, London, 1954* p. 187 (1955).

[38] J. J. Gilman and W. G. Johnston, *J. Appl. Phys.* **31**, 687 (1960).

[39] J. J. Gilman, W. G. Johnston, and G. W. Sears, *J. Appl. Phys.* **29**, 747 (1958).

According to Sears[40] and Vermilyea[41] the role of the impurities is simply to slow down the dissolution rate so that a high undersaturation can be maintained.

f. Effect of the Surface Orientation

Usually one finds a pronounced effect of the crystallographic orientation of a surface on its etching behavior. As a rule the close-packed planes are more easily etched than the others; deviations from this orientation may result in nonpreferential etching for some specific etchant. A proper choice of the etchant may allow etching of higher index faces.

We indicate in Table I[27,42,43] a few examples of specific dislocation etchants and the corresponding tolerances in the orientation of the face to be attacked.

TABLE I

Substance	Face	Tolerance	Reference[a]
LiF	(100)	10°	(1)
	(111)	5°	(1)
Cu	(111)	3°	(2)
AgBr	(100)	5°	(1)
Si	All orientations		
CaF$_2$	(111)	?	(3)

[a] References: (1) W.G. Johnston, Unpublished data, 1959; (2) J.D. Livingston, *J. Appl. Phys.* **31**, 1071 (1960); (3) W. Bontinck, *Phil. Mag.* [8] **2**, 516 (1957).

Using germanium single crystal spheres Holmes[44] has made a systematic study of the orientation dependence of the appearance of dislocation etch pits for various etchants. For some etchants pits are produced only over a limited range of orientations near to close-packed planes. For other etchants dislocation pits are produced over the complete surface of the sphere.

The effect is not well understood. Possible explanations are (a) the slope of the pit sides with respect to the close-packed planes is smaller

[40] G. W. Sears, *J. Chem. Phys.* **32**, 1317 (1960).
[41] D. A. Vermilyea, *Acta Met.* **6**, 381 (1958).
[42] W. G. Johnston, Unpublished data, 1959.
[43] J. D. Livingston, *J. Appl. Phys.* **31**, 1071 (1960).
[44] P. J. Holmes, *Acta Met.* **7**, 283 (1959).

than the deviation of the etched plane with respect to the close-packed orientation. In this case no well-defined pits are produced for a trivial, purely geometrical reason. Etchants that produce steeper pits should be effective on a wider range of planes. (b) It is also possible that on high index faces enough nucleation sites are available, preventing in this way the building up of a sufficient undersaturation to allow the nucleation of the pits at dislocations.

Etchants that are based on impurity segregation at the dislocation are in general less sensitive to surface orientation.

g. *Reliability of an Etching Method*

If one could make a crystal with a predetermined dislocation density the easiest check on the one-to-one correspondence etch pit dislocation would consist in comparing calculated and observed dislocation densities. The simplest method of introducing a known number of dislocations into the crystal consists in bending it to a given radius of curvature R. The "excess" density of dislocations n of a given sign is then

$$1 \mid n = R\, b \cos \phi \tag{3.1}$$

if the bending axis is chosen perpendicular to $\bar{b}$; ϕ is the angle between $\bar{b}$ and the neutral plane of the specimen. Usually it is found that after bending more etch pits are present than predicted by relation (3.1). After a slight anneal, however, polygonization takes place and the density of pits now corresponds to that predicted by (3.1). Moreover the pits are arranged in a way such that they reveal the polygonized structure of the crystal; they are on lines perpendicular to the active glide planes[28,45,46].

In order to make sure that all dislocations are etched it is desirable to perform a few simple tests. If an annealed and slowly cooled specimen can be succesfully etched, it is demonstrated that dislocations with an atmosphere or with precipitates can be revealed. If the same etchant also reveals dislocations in freshly formed slip traces it is reasonable to conclude that the impurities in the crystal are not important. This was, e.g., shown for NaCl etched with anhydrous methyl alcohol[28] and for lithium fluoride.[29]

By a proper choice of slip traces one can also select emergence points of screws and edges separately and check whether both types

[45] F. L. Vogel, *Acta Met.* **3**, 95 (1955).
[46] W. R. Hibbard and C. G. Dunn, *Acta Met.* **4**, 306 (1956).

of dislocations are etched. This is again especially simple for cleaved specimens of alkali halides. The 45° traces are due to edge dislocations while the bands parallel to the crystal edges contain screws, inclined 45° with respect to the surface. These circumstances then permit, at the same time, investigation of the effect of the inclination of dislocations with respect to the etched face. Usually inclined dislocations produce asymmetrical pits, but they can still be revealed (Fig. 14).

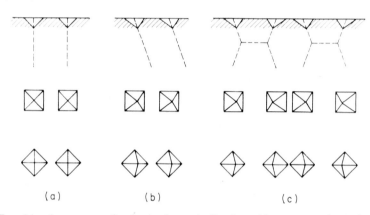

(a)   (b)   (c)

FIG. 14. Asymmetry of etch pits due to inclination with respect to the surface of the dislocation lines giving rise to them. (a) Parallel lines perpendicular to the surface produce symmetrical pits. (b) Parallel lines inclined with respect to the surface, as in a tilt boundary, produce asymmetrical pits all oriented the same way. (c) Pits formed at the emergence points of an hexagonal grid of dislocation; the pits are asymmetrical and successive pits are oriented differently.

It is difficult to prove that all pits correspond to dislocations. There is in fact some evidence that certain pits do not. For instance precipitates may cause etching. Gilman[47] has shown conclusively that precipitates in LiF cause the formation of terraced pits. Fission tracks can also be etched.[48]

Intermittent etching can still be due to dislocations. In deformed LiF and MgO intermittent etching due to dislocation dipoles, broken up in elongated loops, has been found.[49]

When cleaved specimens are used, as, e.g., in the ionic solids, the easiest check on reproducibility and reliability consists in comparing

[47] J. J. Gilman, *J. Appl. Phys.* **31**, 936 (1960).
[48] D. A. Young, *Nature* **182**, 375 (1958).
[49] W. G. Johnston and J. J. Gilman, *J. Appl. Phys.* **31**, 632 (1960).

the 2 specimen halves. The etch patterns should be mirror images. For example Fig. 15 shows the etch patterns of corresponding halves of a LiF crystal. Small deviations may occur, e.g., due to branching of the dislocations in the cleavage plane.[50] In the latter case 1 pit on 1 face corresponds to 2 closely spaced pits on the second face. When metal specimens are used, successive attacks, alternated with electropolishing, should reveal consistently the same pattern.

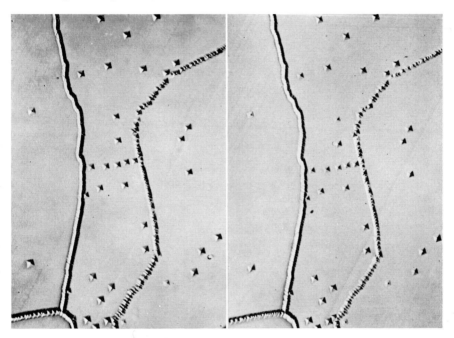

FIG. 15. The etch patterns on 2 matching halves of a cleaved lithium fluoride crystal. The patterns are mirror images, but on printing 1 image has been reversed so that they become identical. (Courtesy of Gilman and Johnston.)

A comparison with an independent method, e.g., decoration, also can be used. This has been done by Dash[51] for silicon. He has shown that a decorated dislocation line corresponds with every etch pit. The same kind of evidence was obtained for NaCl[28] and for silver bromide.[52] A direct check has also been obtained for germanium.

[50] S. Amelinckx, *Phil. Mag.* [8] **1**, 269 (1956).
[51] W. C. Dash, *in* "Dislocations and Mechanical Properties of Crystals" (J. C. Fisher *et al.*, eds.), p. 57. Wiley, New York, 1957.
[52] D. A. Jones and J. W. Mitchell, *Phil. Mag.* [8] **2**, 1047 (1957).

Vogel et al.[53] demonstrated that the distance between etch pits in a tilt boundary corresponds, within the experimental error, to the distance between dislocations D as given by the Burgers model $D = b/\vartheta$. The misorientation angle ϑ was determined with a refined X-ray method. In this way they proved both the correctness of the Burgers model and the reliability of the etchant.

It is possible to make use of similar evidence without measuring the small angle ϑ, which is difficult. It is clear that a relation exists between the dislocation densities ρ_i in 3 intersecting tilt boundaries; one simply has

$$\sum_i \rho_i/(\cos \phi_i + \sin \phi_i) = 0 \tag{3.2}$$

where the ρ_i are the densities of dislocations as measured from the etch pattern and ϕ_i are the angles between the directions of the boundaries and the directions of the symmetrical tilt boundaries. This relation was verified for sodium chloride,[28] germanium,[54] and silicon iron.[55] Thus demonstrating the validity of the dislocation model for asymmetrical tilt boundaries and giving confidence in the etching method. Figure 16 represents a sub-boundary triple junction in sodium chloride. The insets show the dislocation densities in the three branches of the junction; they are in accord with relation (3.2).

h. Information Obtainable from Etch Pits

Etch pits essentially reveal emergence points in the surface and they therefore give a direct measure of dislocation densities. Since the etch pits have a certain depth they may also give some indication concerning the general direction of the dislocation lines. If the line intersects the surface perpendicularly, a symmetrical pit results; if the line is oblique the pits become slightly asymmetrical and from this asymmetry the inclination of the line can be deduced, as shown in Fig. 14.

Gradual removal of the surface layers alternated with etching permits the exploration of dislocation configurations in space. For example, this technique was applied by Gilman and Johnston[29] for

[53] F. L. Vogel, W. G. Pfann, H. E. Corey, and E. E. Thomas, *Phys. Rev.* **90**, 489 (1953).

[54] W. G. Pfann and C. C. Lovell, *Acta Met.* **3**, 512 (1955); J. Okada, *J. Phys. Soc. Japan* **10**, 1018 (1955).

[55] C. G. Dunn and W. R. Hibbard, *Acta Met.* **3**, 409 (1955).

mapping dislocation loops. If the dislocation line is perpendicular to the surface and has a helical shape, the etch pit acquires the form of a conical spiral.[56] Direct evidence for this in silicon was obtained by Dash[57] by the use of the decoration technique. Helical dislocations

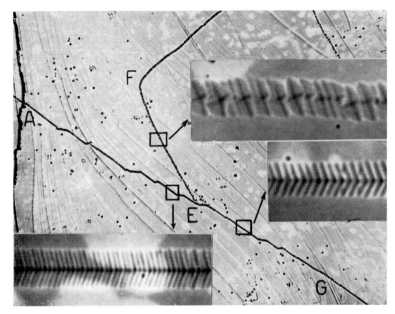

FIG. 16. Junction of 3 sub-boundaries in sodium chloride. The insets allow measuring the dislocation densities in the 3 branches and to verify the relation between them. [After S. Amelinckx, *Acta Met.* **2**, 848 (1954).]

were shown to produce spiral etch pits; the reverse is not necessarily true.

A flat-bottomed pit results if a dislocation moves out of a pit produced by a first etching; further etching then only develops the pit laterally, but no longer in depth. Jerky motion of a dislocation during etching would result in a configuration as shown in Fig. 17. The flat-bottomed pits represent intermediate positions of the dislocation, the centered pit represents the final position. Since etch pits do not pin the dislocation very much dislocation movement can be studied.[29]

[56] S. Amelinckx, W. Bontinck, and W. Dekeyser, *Phil. Mag.* [8] **2**, 48 (1957).

[57] W. C. Dash, *in* "Properties of Elemental and Compound Semiconductors, AIME Conf. (H. C. Gatos, ed.), p. 195. Wiley (Interscience), New York, 1960.

A dislocation half-loop is represented as 2 coupled pits, which on continued etching approach one another and finally coalesce. A helix intersected along its axis is in fact equivalent to a sequence of such half-loops; therefore helices can easily be recognized in an etch pattern.

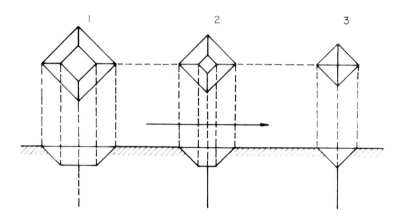

FIG. 17.   Etch pits due to moving dislocation. The flat-bottomed pits (1) and (2) correspond to intermediate positions of the dislocation. The final position of the dislocation is marked by the centered pit (3).

Figure 18a shows the pattern obtained when a half-loop in LiF is etched under stress.

It has been shown that not only dislocations but also tracks due to fission fragments can be etched; also "debris" left by moving dislocations etches, but discontinuously. The "debris" may consist either of agglomerates of point defects or of dislocation "dipoles" broken up in small elongated loops.

In exceptional cases dislocations practically parallel to and close to the surface produce grooves along their length, especially if they have acquired an impurity atmosphere, or if decorated first.[58,59] In pure floating zone grown crystals of silicon such grooves have also been found.[60] Apparently the strain around the dislocation is sufficient to cause enhanced dissolution. Figure 19 shows such grooves observed in silicon.

[58] J. R. Low and R. W. Guard, *Acta Met.* **7**, 171 (1959).
[59] W. W. Tyler, and W. C. Dash, *J. Appl. Phys.* **28**, 1221 (1957).
[60] W. C. Dash, *J. Appl. Phys.* **29**, 705 (1958).

i. Specimen Preparation

In order to obtain reliable etch patterns, surface preparation is of primary importance. Whenever possible, cleavage is to be preferred if this produces a nondeformed specimen. In this case clean as well as aged dislocations can be etched. In metal specimens electrolytic

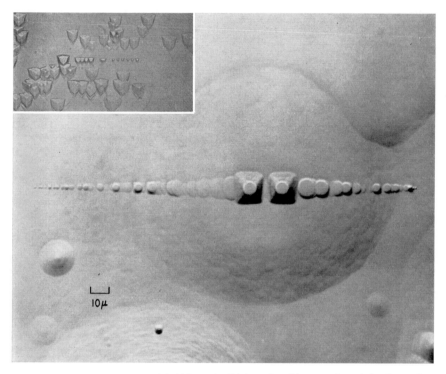

FIG. 18. (a) Extension of half-loop in lithium fluoride as reflected in the etch pattern. The crystal was etched during application of the stress. (Courtesy of Gilman and Johnston.) (b) Movement of a dislocation in copper deformed during etching. (Courtesy of Young.)

polishing is the usual procedure; this, however, produces in general a surface covered by some oxide film. Thus it is often necessary to age the specimen in order to segregate impurities along the dislocations and produce "sensitive" emergence points in this surface film. In many cases intentional impurity addition may be required, e.g., cadmium to zinc,[61] iron to aluminum,[37] tellurium to copper.[62] The deformed surface of mechanically polished specimens has first to be removed chemically before representative etch patterns can be obtained. For more detailed information we refer to Faust.[63]

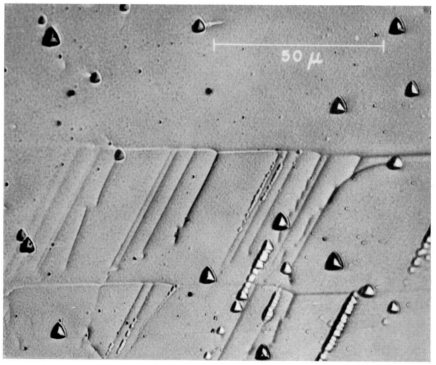

FIG. 19.   Etch grooves in a (111) plane of silicon. Notice the cusps in some of the dislocations, as well as the trails connected to them. [Courtesy of W. C. Dash, *J. Appl. Phys.* **29**, 705 (1958).]

j. Application of Etching to Dislocation Problems

We now review a few typical examples of the application of etching techniques to dislocation problems.

(1) *Stress-velocity relation for individual dislocations.* One of the most elaborate problems that has been solved successfully by an

[61] J. J. Gilman, *Trans. A.I.M.E.*, **206**, 998 (1956).

[62] F. W. Young, *J. Appl. Phys.* **29**, 760 (1958); F. W. Young and N. Cabrera, *ibid.* **28**, 787 (1957).

[63] J. W. Faust, *in* "Methods of Experimental Physics" (K. Lark-Horovitz and V. A. Johnson, eds.), Vol. 6, Part A, p. 147. Academic Press, New York, 1959.

etching procedure probably is the experimental determination of the stress velocity relationship for individual dislocations by Gilman and Johnston.[64] Grown-in dislocations appear to be immobile as a consequence of pinning by impurities. Therefore it is necessary to introduce first a fresh dislocation half-loop in a lithium fluoride

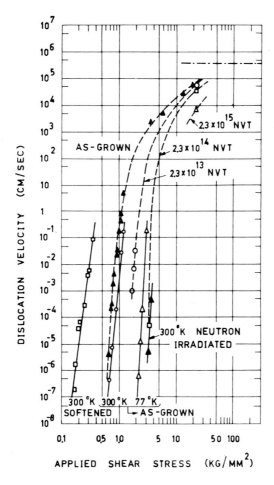

FIG. 20.   Dislocation velocity as a function of the applied stress in as-grown and in neutron-irradiated lithium fluoride. [After W. G. Johnston and J. J. Gilman, *J. Appl. Phys.* **30**, 129 (1959).]

[64] W. G. Johnston and J. J. Gilman, *J. Appl. Phys.* **30**, 129 (1959).

cleavage prism. This is done by producing a small indentation rosette and then polishing away all but one of the half-loops so formed. A given stress is now applied during a known period of time and the distance traveled by the dislocation of the loop is deduced from the difference in the etch patterns before and after application of the stress. An example of the patterns obtained is shown in Fig. 18a. The velocity turns out to be an extremely sensitive function of the applied stress; for a certain critical stress it rises rapidly over several decades, and then approaches some limiting value as shown in Fig. 20. Similar observations have been made for silicon iron.[65] The results are qualitatively the same.

(2) *Deformation patterns.* The distribution of dislocations in slip traces and deformation bands can easily be revealed by etching. From the patterns obtained therefore one can make deductions concerning the deformation process, glide planes, dislocation density, distribution, etc.

(a) Pileups. Pileups against boundaries or even sub-boundaries as well as at intersection points of glide traces are often observed. For example, they have been found in sodium chloride,[66] lithium fluoride,[29] and magnesium oxide,[67] in metals, e.g., in α-brass[31] and in silicon[68] and in silicon carbide.[69,70] In the case of α-brass the obstacles are presumably Lomer-Cottrell barriers.[31] A detailed study of pileups in α-brass has been made by Meakin and Wilsdorf.[71a,b]

The distribution of dislocations in pileups has been calculated by Frank *et al.*[72] and by Leibfried.[73] A comparison between theory and experiment is available for a number of crystals.[31,68,70] According

[65] D. Stein and J. Low, *J. Appl. Phys.* **31**, 362 (1960).

[66] S. Amelinckx, Thesis, Ghent University, 1955.

[67] Y. T. Chou and R. W. Whitmore, *J. Appl. Phys.* **32**, 1920 (1961); R. J. Stokes, T. L. Johnston, and C. H. Li, *Phil. Mag.* [8] **3**, 718 (1958); J. Washburn, A. E. Gorum, and E. R. Parker, *Trans. A.I.M.E.* **215**, 230 (1959).

[68] J. Patel, *J. Appl. Phys.* **29**, 170 (1958).

[69] S. Amelinckx and G. Strumane, *in* "Silicon Carbide" (J. R. O'Connor and J. Smiltens, eds.), p. 162. Pergamon Press, New York, 1960.

[70] S. Amelinckx, G. Strumane, and W. W. Webb, *J. Appl. Phys.* **31**, 1359 (1960).

[71a] J. D. Meakin and H. G. F. Wilsdorf, *Trans. A.I.M.E.* **218**, 737 (1960).

[71b] J. D. Meakin and H. G. F. Wilsdorf, *Trans. A.I.M.E.* **218**, 745 (1960).

[72] J. D. Eshelby, F. C. Frank, and F. R. N. Nabarro, *Phil. Mag.* [7] **42**, 351 (1951).

[73] G. Leibfried, *Z. Physik.* **130**, 214 (1951).

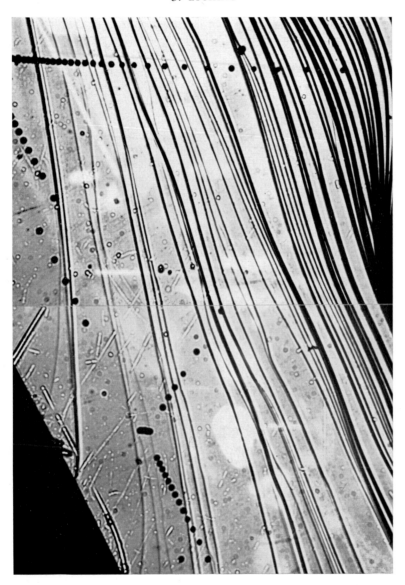

FIG. 21.   Pileup of dislocations as observed by means of etch pits on the *c* plane of a silicon carbide crystal. The obstacles are large grown-in dislocations on which growth spirals were centered (not visible on the photograph since no phase contrast was used). [After S. Amelinckx and G. Strumane, "Silicon Carbide." Pergamon Press, New York, 1960.]

to the calculations by Eshelby *et al.*[72] one should have the following relation:

$$\left(\frac{\pi i}{2}\right)^2 = 2nx\sigma_0/A \qquad (3.3)$$

where i is the etch pit number ($i = 0$ at the locked dislocation); σ_0 is the applied shear stress, n is the number of dislocations in the slip plane, x is the etch pit distance from the origin, and finally, $A = \mu b/(2\pi(1-v))$. A plot of the square root of the distance versus the dislocation index should yield a straight line.

An example of an observed pileup is shown in Fig. 21; the crystal is silicon carbide; the obstacle is a grown-in dislocation which has a growth spiral centered on it. Figure 22 shows the correspondence between calculated and observed spacings, for pileups in silicon carbide.[70]

(b) Dislocation multiplication and movement. From etch pits observations on the growth of glide bands in deformed lithium fluoride Gilman and Johnston[74] concluded that the multiplication of dislocations takes place by the double cross slip mechanism

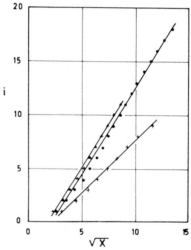

FIG. 22. Correspondence between the observed and calculated positions of dislocations in pileups in silicon carbide. [After S. Amelinckx and G. Strumane, "Silicon Carbide." Pergamon Press, New York, 1960.]

[74] W. G. Johnston and J. J. Gilman, *J. Appl. Phys.* **31**, 632 (1960).

originally proposed by Koehler[75] and Orowan.[76] By comparing macroscopic measurements of the flow stress and velocity measurements of individual dislocations [see Section 10a (1)] it was shown that the flow stress is determined by the resistance of the dislocations to motion. A detailed discussion of these observations and theories is given in another contribution to this series.[77]

Because of the possibility of etching preferentially newly formed dislocations in lithium fluoride it becomes feasible to study the behavior of individual dislocations in strain-hardened crystals.[78] After the crystal has been deformed up to the desired level it is heat-treated at 300°C and slowly cooled so as to pin the dislocations with impurities. New dislocations can then be distinguished from old ones and their motion studied. It is found that the stress to move dislocations is larger in the strain-hardened crystals than in the undeformed ones.

Reliable etch pitting techniques for copper have been developed by Young,[79,80] Livingston,[81,82] and Lovell and Wernick[83] and applied to a number of problems. Young and Noggle[80] found evidence for dislocation reactions at the intersection points of slip traces in copper, which was radiation-hardened prior to bending. By etching before and after bending Young[84] was able to observe the movement of individual dislocations and hence to deduce the minimum stress at which dislocations move in copper (Fig. 18b). It is found that grown-in dislocations move at a resolved stress of about 4 gm/mm² while the multiplication of dislocations is observed at a resolved stress of about 18 gm/mm². Dislocations are found to move back over some distance when the stress is removed.

Although the technique developed by Livingstone[84a] allows to observe the motion of single dislocations no stress-velocity relationship could be established, because the motion appears to be too irregular. Beautiful examples of cooperative motion of dislocations

[75] J. S. Koehler, *Phys. Rev.* **86**, 52 (1952).

[76] E. Orowan, "Dislocations in Metals," p. 103 A.I.M.E., New York, 1954.

[77] J. J. Gilman, and W. G. Johnston, *Solid State Phys.* **13**, 148 (1962).

[78] J. J. Gilman and W. G. Johnston, *J. Appl. Phys.* **31**, 687 (1960).

[79] F. W. Young, *J. Appl. Phys.* **32**, 192 (1961).

[80] F. W. Young and T. Noggle, *J. Appl. Phys.* **31**, 3 (1960).

[81] J. D. Livingston, *J. Appl. Phys.* **31**, 1071 (1960).

[82] J. D. Livingston, *Acta Met.* **10**, 229 (1962).

[83] L. C. Lovell and L. H. Wernick, *J. Appl. Phys.* **30**, 590 (1959).

[84] F. W. Young, *J. Appl. Phys.* **32**, 1815 (1961).

[84a] J. D. Livingstone, G. E. Report 62-RL-3173M (1962).

in walls were produced. Such motion already takes place at a resolved shear stress of 15 gm/mm². Small angle boundaries are found to be obstacles for dislocation motion as deduced from the formation of pileups.

(3) *Origin of dislocations in as-grown crystals.* Etch pit studies, especially in germanium and silicon, have contributed considerably in elucidating the origin of dislocations in melt-grown crystals.[85-87] It was demonstrated that the major part of the dislocations result, not from the condensation of point defects, but from plastic deformation due to thermal stresses. A (111) section perpendicular to the pulling axis of a germanium crystal usually exhibits an etch pattern in the form of a star as a result of deformation on the intersecting {111} planes.[85,86]

Methods eliminating these dislocations have been derived from systematic etch pits and decoration studies.[88,89] The dependence of crystal perfection on the growth direction has been demonstrated convincingly by Rosi.[90] This can be understood in terms of Dash's findings that dislocations in silicon tend to adopt certain preferred orientations.[88] It is clear that crystals grown along such directions will retain many more dislocations than if grown along another direction. Dash[88,89] also demonstrated that dislocations are inherited from the seed; they are often introduced as a consequence of bad contact between seed and melt. Once dislocations are present they propagate and multiply as a consequence of thermal stresses. In order to produce dislocation-free crystals it is therefore necessary to start with a perfect seed. This is done by using a pointed ⟨111⟩ (or ⟨100⟩) seed and growing a certain length of a filamentary crystal on it. This filament has a large chance of being perfect at its lower end, because dislocations are eliminated from it progressively by climb. This filament is then allowed to thicken and a perfect crystal usually results. If the crystal is initially dislocation-free, thermal stresses are no longer of importance as sources of dislocations.

[85] E. Billig, *Proc. Roy. Soc.*, **A235**, 37 (1956).

[86] P. Penning, *in* "Halbleiter und Phosphore," (M. Schön and H. Welker, eds.), p. 482. Vieweg, Braunschweig, 1958.

[87] E. Billig, *Brit. J. Appl. Phys.* **7**, 375 (1956).

[88] W. C. Dash, *J. Appl. Phys.* **30**, 459 (1959).

[89] W. C. Dash, *J. Appl. Phys.* **29**, 736 (1958).

[90] F. D. Rosi, R. C. A. Rev. **19**, 349 (1958); *Trans. A.I.M.E.* **209**, 76 (1957).

(4) *Polygonization.* If a crystal is bent normal to a Burgers vector and subsequently annealed, pure symmetrical tilt boundaries form. The plane of the boundaries is perpendicular to the glide planes. The formation of these boundaries is known as polygonization since curved lattice planes become polygonal during this process, as shown schematically in Fig. 23. The phenomenon essentially consists in a re-ar-

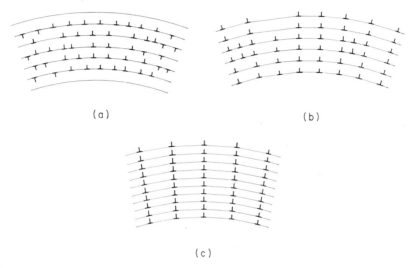

(a)   (b)

(c)

FIG. 23.   Dislocation rearrangement during polygonization: (a) Dislocations are distributed at random in their glide planes. (b) Dislocations rearrange themselves in their glide planes so as to adopt positions in walls perpendicular to active glide planes. (c) Dislocations climb and walls develop further. The distance between dislocations in the walls becomes smaller than the distance between active glide planes.

rangement of dislocations into an equilibrium configuration, which is such that the edge dislocations arrange themselves "one on top of the other." During the first stage, which may even occur during glide, the dislocations rearrange themselves in their glide plane. In a further stage, after a heat treatment, tilt boundaries build up. This implies the occurrence of climb, since the distance between dislocations in the walls is smaller than the distance between active glide planes. Etching and decoration[91,92] have illustrated convincingly the

[91] C. G. Dunn and F. W. Daniels, *Trans. A.I.M.E.* **191**, 147 (1951).
[92] S. Amelinckx and R. Strumane, *Acta Met.* **8**, 312 (1960).

correctness of this picture which is due to Cottrell.[93] Figure 24 shows, e.g., a polygonized area of an etched silicon carbide crystal. In a few places, indicated by arrows, characteristic configurations called "doublets" are observed; they are stable with respect to glide, but not with respect to climb.

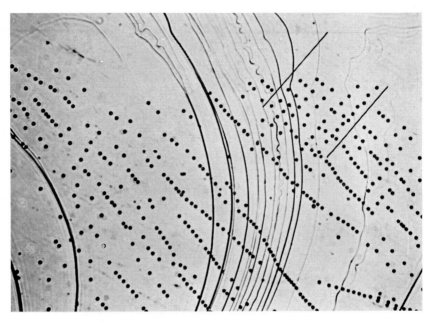

FIG. 24. Polygonized region in a deformed crystal of silicon carbide. The dislocations are marked by etch pits. [After S. Amelinckx and G. Strumane, "Silicon Carbide." Pergamon Press, New York, 1960.] The direction of the active glide planes is indicated.

A striking and characteristic feature of polygonized crystals is the frequent occurence of Y junctions, i.e., 2 boundaries with tilt angles ϑ_1 and ϑ_2 fusing into a single one with tilt angle $\vartheta_1 + \vartheta_2$. By this process energy is gained; it supplies, therefore, at least partly, the driving force for polygonization.

Polygon walls have been revealed by means of etch pits, in sodium chloride[28] and in other alkali halides.[94,95] Polygonization in zinc has

[93] A. H. Cottrell, "Dislocations and Plastic Flow in Crystals," p. 181. Oxford Univ. Press, London and New York, 1953.

[94] M. Sakamoto and S. Kobayashi, *J. Phys. Soc. Japan* **13**, 800 (1958).

[95] A. A. Urusovskaia, *Kristallografiya* **3**, 726 (1958).

been studied by Gilman[96] and more recently by Sinha and Beck.[97] The latter authors demonstrated in particular that no climb is required for the formation of small angle tilt boundaries in bent single crystals of high-purity zinc. They were able to do this because they developed an etching solution which is capable of etching clean dislocations. Gilman[96] on the other hand had to age his crystals, which contained cadmium, before etching.

Polygonization in copper was studied by Young and Cabrera.[98] Rearrangement of dislocations into vertical walls, but without leaving their glideplane (so called "glide polygonization"), was noticed in copper by Livingston.[81]

(5) *Climb.* The displacement of dislocations in a direction perpendicular to their glideplane was demonstrated[99] by etching before and after heat treatment. Crawford and Young[100] etched sodium chloride before and after exposure to gamma rays from Co^{60}. *No* observable climb of dislocations was found with the optical microscope, which suggests that color center production is not accompanied by climb. Similar results were obtained by Schüller[101] on lithium fluoride, by the electron microscopic observation of replicas.

(6) *Radiation hardening.* It is well-known that alkali halides, when irradiated with X-rays or other ionizing radiation, harden considerably.[102] The hardening can be followed by measuring the cross section of indentations produced by a Vickers pyramid on cleavage planes. A more sensitive method of measuring the effect consists in comparing the distance traveled by dislocations under the influence of a constant load. The dislocation pattern associated with an indentation has a characteristic rosette shape for crystals with the NaCl structure and indented on the cube plane. It can easily be revealed by etching. The diameter of the wings is a convenient measure for the hardening; it has perhaps also a more direct physical meaning than the diameter of an indentation. Moreover it is possible to compare the effect of irradiation on edges and screws separately. There seems

[96] J. J. Gilman, *Trans. A.I.M.E.* **206**, 998 (1956).
[97] P. Sinha and P. Beck, *J. Appl. Phys.* **32**, 1222 (1961).
[98] F. W. Young and W. Cabrera, *J. Appl. Phys.* **28**, 787 (1957).
[99] J. J. Gilman and W. G. Johnston, *J. Appl. Phys.* **27**, 1018 (1956).
[100] J. H. Crawford and F. W. Young, *J. Appl. Phys.* **31**, 1688 (1960).
[101] E. Schüller, Private communication, 1959.
[102] M. N. Podachewsky, *Physik. Z. Sowjetunion* **8**, 82 (1935).

to be a linear relationship between the inverse of the diameter of the wings and the diameter of the indentation.

The experiments on LiF[103] and NaCl[104] lead to the conclusion that radiation hardening consists at least partly in friction hardening since the distance traveled by dislocations under the same load decreases rather rapidly on irradiation.

Gilman and Johnston[105] have measured the speed of propagation of dislocations in neutron-irradiated lithium fluoride by using the method they developed for unirradiated crystals [see Section 10a (1)]. They find that the slope of the curve relating velocity to applied stress is the same in irradiated and unirradiated material apart from a parallel displacement towards higher stresses (Fig. 20).

(7) *Fracture and dislocations.* The relation between the initiation of cracks and pileups of dislocations in MgO has been studied extensively by means of etch pits. Stokes *et al.*[106],[107] have demonstrated that small cracks are nucleated in the glide bands at their intersection points. In accord with Stroh's theory of crack formation they attributed these to the coalescence of piled-up dislocations. The cracks are actually small slits situated in the (110) plane perpendicular to the glide plane of the active pileup. The slits extend through the whole crystal in the direction parallel to the intersection line of the active glide planes. Once nucleated the cracks may switch over to the normal cleavage plane, which is the cube plane.

(8) *Special features.* (a) Spiral pits. Although it has been shown that helical dislocations produce spiral pits on etching[56],[57] the reverse is not necessarily true. On slowly etched germanium and silicon surfaces one sometimes observes large numbers of spirals pits which have apparently no relation with helical dislocations. No satisfactory explanation of this phenomenon has been given.

(b) Dislocations in polar crystals. In polar crystals, i.e., crystals with a polar axis, having for instance the zincblende structure, positive and negative dislocations are distinguishable.

In terms of supplementary half-planes one can say that, for instance

[103] W. H. Vaughan and J. W. Davisson, *Acta Met.* **6**, 554 (1958); A. D. Whapham, *Phil. Mag.* [8] **3**, 103 (1958).

[104] E. Aerts, S. Amelinckx, and W. Dekeyser, *Acta Met.* **7**, 29 (1959).

[105] J. J. Gilman and W. G. Johnston, *J. Appl. Phys.* **29**, 877 (1958).

[106] R. J. Stokes, T. L. Johnston, and C. H. Li, *Phil. Mag.* [8] **3**, 31 (1958).

[107] R. J. Stokes, T. L. Johnston, and C. H. Li, *Phil. Mag.* [8] **4**, 137 (1959).

in indium antimonide, dislocations of one sign have a supplementary half plane ending at indium atoms, whereas the dislocations of the opposite sign would have a supplementary half plane ending at antimony atoms. It has been shown[108-110] that the 2 types of dislocations behave differently with respect to etching on {110} planes. The 2 types of dislocations appear also to behave differently in their influence on the electrical properties of indiumantimonide: the indium dislocations make the sample more p type whereas the antimony dislocations make it more n type.

The (0001) and (000$\bar{1}$) planes or the (111) and ($\bar{1}\bar{1}\bar{1}$) planes of crystals with the wurtzite and the zincblende structure respectively behave quite differently with respect to the same etching solution. Usually well-defined pits are developed only on 1 of these planes. The 2 kinds of planes can be identified by means of X-ray diffraction, as zinc or sulfur planes.[111] Silicon carbide crystals show this behavior in a pronounced way[70].

(c) Positive and negative dislocations. Livingston[82] made the astonishing discovery that positive and negative dislocations in copper can be distinguished by etching on (111) planes with a proper etchant. The 2 kinds of dislocations produce pits of different depths on the same crystal face. They appear as "light" and "dark" pits on the same optical micrograph. It is even possible to conclude which side of the slip plane is the compression side. The origin of the difference in etching rate is not well understood at present.

(d) Dislocations accompanying electrical breakdown. Dislocation loops are nucleated during electrical breakdown of lithium fluoride[112] and magnesium oxide.[113]

(e) Surface pits due to vacancy condensation. When a pure polished aluminium specimen is cooled from an elevated temperature it is found that the surface is pitted[113a,b]. The formation of these pits was attributed to the condensation of vacancies, which were in

[108] H. C. Gatos, M. C. Finn, and M. C. Lavine, *J. Appl. Phys.* **32**, 1174 (1961).
[109] M. C. Lavine, A. J. Rosenberg, and H. C. Gatos, *J. Appl. Phys.* **29**, 1131 (1958).
[110] J. Venables and R. Broudy, *J. Appl. Phys.* **29**, 1025 (1958).
[111] D. Coster, R. S. Knol, and J. A. Prins, *Z. Physik* **63**, 345 (1930).
[112] J. J. Gilman and D. Stauff, *J. Appl. Phys.* **29**, 120 (1958).
[113] E. Mendel and S. Weinig, *J Appl. Phys.* **31**, 738 (1960).
[113a] P. E. Doherty and R. S. Davis, *Acta Met.* **7**, 118 (1959).
[113b] O. P. Arora and M. Metzger, *Acta Met.* **8**, 49 (1960).

TABLE II ETCHANTS.

Crystals	Etchant	Remarks	References
Aluminum	(1) 47 % HNO_3, 50 % HCl, 3 % HF temp: room ref.: (1)	With Fe impurity	(1) S. Amelinckx, *Phil. Mag.* [7] **44**, 1048 (1953).
	(2) HF temp: room ref.: (2)	After decoration in vacuum at 350°C	(2) P.B. Hirsch, R.W. Horne, and M.J. Whelan, *Phil. Mag.* [8] **1**, 677 (1956).
Antimony	(1) CP 4 temp: room; time: 2-3 sec plane: (111) ref.: (3)	Cleaved surface	(3) J.H. Wernick *et al.*, *J. Appl. Phys.* **29**, 1013 (1958).
	(2) 1 pt HF, 1 pt Superoxal, 4 pt H_2O temp: room; time: 1 sec plane (111) ref.: (3)		
Bismuth	1 % I_2 in methanol temp: room; time: 15 sec plane: (11) ref.: (4)	After prepolishing	(4) I.C. Lovell and J.H. Wernick, *J. Appl. Phys.* **30**, 234 (1959).
Brass	0.2 % $Na_2S_2O_4$ temp: room; time: 60 sec plane: (111) ref.: (5)	Electrolytic: 0.1 A cm^{-2}	(5) P.A. Jacquet, *Compt. Rend.* **237**, 1248 (1953).

Material	Procedure		Reference
Cadmium	(1) 2 pt orthophosphoric acid, 2 pt glycerine, 1 pt water time: 20-40 sec ref.: (6)	Electrolytic: 0.9-10 v after polishing in same bath at 2.1-2.2 v for 9-12 min	(6) A.A. Predvohitiv and N.A. Tiapunia, Phys. Metals Metallog (U.S.S.R.) (English Transl.) 7, No. 6, 55 (1959).
	(2) Sat. sol. of picric acid in acetone time: 2 min ref.: (7)		(7) J. George, Nature 189, 743 (1961).
Copper	(1) Impure: 60 % H_3PO_4 in H_2O temp: room; time: 10 sec ref.: (8)	Electrolytic	(8) F.W. Young and N. Cabrera, J. Appl. Phys. 28, 339 (1957).
	(2) Pure: 4 pt sat. aqueous $FeCl_3$, 4 pt HCl, 1 pt acetic acid plus few drops bromine temp: room; time: 15-30 sec ref.: (9)	Rinse in NH_4OH, prepolish necessary	(9) L.C. Lovell and J.H. Wernick, J. Appl. Phys. 30, 590 (1959).
	(3) Several etchants planes: (111) (110) (100) ref.: (10)	Distinguish fresh dislocations and dislocations with Cottrell atmosphere	(10) F.W. Young, J. Appl. Phys. 32, 192 (1961).
Germanium	(1) CP4 time: 1 min; planes: (111) (100) ref.: (11)		(11) W.G. Johnston, Gen. Elec. Rept. 61-RL-2649M (1961); Progr. Ceramic Sci. 2, 1 (1962); R.D. Heidenreich, U.S. Patent 2, 619, 414 (1952); F.L. Vogel et al., Phys. Rev. 90, 489 (1953).
	(2) 40 ml HF, 20 ml HNO_3, 40 ml H_2O, 2 gm $AgNO_3$ time: 1 min; planes: (111) (110) ref.: (11)		

TABLE II. ETCHANTS (continued)

Crystals	Etchant	Remarks	References
Iron	(1) 2 % Nital containing 2 % saturated picral time: 15 min ref.: (12) (2) Saturated picral time: 4 min ref.: (12)	Anneal at 1750°F to decorate dislocations.	(12) L.C. Lovell, F.L. Vogel, and J.H. Wernick, *Metal Progr.* **75**, No. 5, 96 (1959).
Silicon iron	(1) 133 ml acetic acid 25 gm CrO_3, 7 ml H_2O temp: room time: 5–20 min ref.: (13)	Electrolytic: 30 mA/cm^2 after predecoration	(13) W.R. Hibbard and C.G. Dunn, *Acta Met.* **4**, 306 (1956).
	(2) Dilute phosphoric acid (200 gm/liter H_3PO_4) temp: 20–25°C; time: 4–12 min ref.: (14)	Electrolytic: 0.15 A/cm^2	(14) S. Felin and G. Castro, *Acta Met.* **10**, 543 (1962).
Nickel-manganese	100 cc H_3PO_4, 100 cc ethanol temp: 40°C; time: 2 min ref.: (15)	Electrolytic: 2 A/cm^2 Cu-cathode	(15) T. Taoko and S. Aoyagi, *J. Phys. Soc. Japan* **11**, 522 (1956).
Niobium	(1) 10 cc H_2O, 10 cc H_2SO_4, 10 cc HF, few drops superoxil temp: room ref.: (16)	Agitate specimen in solution	(16) A.B. Michael and F.J. Huegel, *Acta Met.* **5**, 339 (1957).
	(2) H_2SO_4(95 %), HNO_3(70 %), HF(48 %) in ratio 5 : 2 : 2 ref.: (17)	Strongly orientation-dependent	(17) R. Bakish, *Trans. A.I.M.E.* **212**, 818 (1958).

Material	Etchant	Remarks	Reference
Silicon	(1) 1 HF, 3 HNO$_3$, 10 CH$_3$COOH time: 10 min to "overnight" ref.: (18)		(18) W.C. Dash, *J. Appl. Phys.* **27**, 1193 (1956).
	(2) 10 cc hydrofluoric acid (37-38%), 10 cc HNO$_3$, 10 cc glacial acetic acid temp: 30-35°C; time: 2-3 min ref.: (19)	Adding 15 cc double distilled water reduces time to 1.5-2 min	(19) N.N. Sirota and A.A. Tonoyan, *Proc. Acad. Sci. U.S.S.R., Phys. Chem. Sect.* (*English Transl.*) **134**, 987 (1960).
Tantalum	5 pt H$_2$SO$_4$, 2 pt HNO$_3$, 2 pt HF temp: room; plane: (112) ref.: (20)	After predecoration	(20) R. Backish, *Acta Met.* **6**, 120 (1958).
Tellurium	(1) 3 pt HF, 5 pt HNO$_3$, 6 pt acetic acid time: 1 min; temp.: room plane: (10$\bar{1}$0) ref.: (21)	Cleaved surface	(21) L.C. Lovell, J.H. Wernick, and K.E. Benson, *Acta Met.* **6**, 716 (1958).
	(2) Conc. HNO$_3$ ref.: (22)		(22) A.I. Blum, *Soviet Phys.—Solid State* (*English Transl.*) **2**, 1509 (1961).
	(3) 43 gm H$_3$PO$_4$(d = 1.55), 1 cc conc. H$_2$SO$_4$, 5 gm CrO$_3$ time: 1-3 min temp: 90-100°C ref.: (22)	Another etchant with nearly the same composition (84.4 gm H$_3$PO$_4$, 2 cc H$_2$SO$_4$, 4.4 gm CrO$_3$) works at 150-160°C	
Tungsten	(1) 2 pt CuSO$_4$(25%), 1 pt conc. NH$_4$OH ref.: (23)		(23) U.E. Wolff, *Acta Met.* **6**, 559 (1958).
	(2) 32.7 gm K$_3$Fe(CN)$_6$, 4.78 gm NaOH, 107 ml H$_2$O plane: (110) ref.: (24)	No guarantee that all pits correspond to dislocations	(24) I. Berlec, *J. Appl. Phys.* **33**, 197 (1962).

TABLE II. ETCHANTS (*continued*)

Crystals	Etchant	Remarks	References
Uranium	170 cc H_2SO_4(98%), 100 cc H_2O(dist.), 5 gm CrO_3, 25 cc dehydrated glycerol time: 60 sec ref.: (25)	Electrolytic:2 A/cm^2 similar etchants with slightly different composition are given in same reference	(25) A. Bassi, S. Granata, and G. Imarisio, *Energie Nucl.* **8**, 744 (1961).
Zinc	(1) 160 gm chromic acid, 50 gm hydrated sodium sulfate, 500 ml H_2O ref.: (26) (2) 2 gm $NH_4(NO_3)_2$, 10 cc NH_4OH, 50 cc deionized water time: 10 sec; temp: room ref.: (27) for other etchants: see ref.: (28)	Rinsed in water in methyl alcohol and dried in a stream of air	(26) J.J. Gilman, *Gen. Elec. Rept.* 56-RL-1575 (1956); *J. Metals* **8**, 998 (1958). (27) P.P. Sinka and P.A. Beck, *J. Appl. Phys.* **33**, 625 (1962). (28) V.M. Kosevich and V.P. Soldatov, *Soviet Phys.—Cryst.* (*English Transl.*) **6**, 347 (1961).
		Nonmetals	
$CaCO_3$	HCl (10%) time: 10-60 sec; plane: (010) cleavage ref.: (29) other etchants: see ref.: (30) (31) (32)		(29) R.C. Stanley, *Nature* **183**, 1548 (1959). (30) G.B. Rais, *Dokl. Akad. Nauk S.S.S.R.* **117**, 419 (1957). (31) H. Watts, *Nature* **183**, 314 (1959). (32) R.E. Keith and J.J. Gilman, *Acta Met* **8**, 1 (1960).

CaF$_2$	Conc. H$_2$SO$_4$ time: 10-30 min; plane: (111) ref.: (33) other etchants: see ref.: (34)	(33) W. Bontinck, *Phil. Mag.* [8] **2**, 561 (1957). (34) W.G. Johnston, unpublished data, cited in ref. (26).
CdS	Vapor of conc. HCl time: 5-10 sec; plane: (0001) ref.: (35) other etchants: see ref.: (36) (37) (38) (39)	(35) D.C. Reynolds and L.C. Greene, *J. Appl. Phys.* **29**, 559 (1958). (36) J. Nishimura, *J. Phys. Soc. Japan* **15**, 232 (1960). (37) J. Woods, *Brit. J. Appl. Phys.* **11**, 296 (1960). (38) A.J. Eland, *Philips Tech. Rev.* **22**, 288 (1961). (39) Zh. G. Pizarenko and M.K. Sheinkman, *Soviet Phys.—Solid State* (English *Transl.*) **3**, 838 (1961).
GaAs	2 HCl, 1 HNO$_3$, 2H$_2$O time: 10 min; plane: (1̄1̄1̄) ref.: (40) other etchants: see ref.: (41) (42) (43)	(40) J.G. White and W.C. Roth, *J. Appl. Phys.* 30, 946 (1959). (41) H.A. Shell, *Z. Metallk.* **48**, 158 (1957). (42) J.L. Richards, *J. Appl. Phys.* **31**, 600 (1960). (43) J.L. Richards and A.J. Crocker, *J. Appl. Phys.* **31**, 611 (1960).
InAs	0.4 N Fe^{++} in conc. HCl time: 3 min; temp: 25°C; plane: (1̄11) ref.: (44) other etchants: see ref.: (45) (46)	(44) H.C. Gatos and M.C. Lavine, *J. Electrochem. Soc.* **107**, 427 (1960). (45) J.W. Faust and A. Sagar, *J. Appl. Phys.* **31**, 331 (1960). (46) E.P. Warekois and P.H. Metzger, *J. Appl. Phys.* **30**, 960 (1959).

TABLE II. ETCHANTS (continued)

Crystals	Etchant	Remarks	References
InSb	25 HNO$_3$, 20 CH$_3$COOH, 10 HF, 1 Br$_2$ time: 5 sec; plane: ($1\bar{1}1$) ref.: (47) other etchants: see ref.: (44) (45) (48) (49) (50) (51) (52)	Stop etching by adding H$_2$O	(47) W. Bardsley and R.L. Bell, *J. Electron. Control* **3**, 103 (1957). (48) J.W. Allen, *Phil. Mag.* [8] **2**, 1475 (1957). (49) R.E. Maringer, *J. Appl. Phys.* **29**, 1261 (1958). (50) J.D. Venables and R.M. Broudy, *J. Appl. Phys.* **29**, 1025 (1958). (51) H.C. Gatos and M. Lavine, *J. Appl. Phys.* **31**, 743 (1960). (52) H.C. Gatos and M. Lavine, *J. Electrochem. Soc.* **107**, 433 (1960).
KBr	Glacial acetic acid time: 3 sec; plane: (100) ref.: (53) other etchants: see ref.: (54)	Rinse in CCl$_4$	(53) P.R. Moran, *J. Appl. Phys.* **29**, 1768 (1958). (54) J.S. Cooks, *J. Appl. Phys.* **32**, 2492 (1961).
KCl	Conc. sol. polyvinyl butyrol in buryl or ethyl time: 5-7 sec; plane: (100) ref.: (55) other etchants: see ref.: (53) (54) (56) (57) (58) (59)		(55) M.P. Shaskolskaya, W. Yang-Wen and K. Shu-Chao, *Soviet Phys.—Cryst.* (*English Transl.*) **6**, 220 (1961). (56) S.A. Slack, *Phys. Rev.* **105**, 832 (1957). (57) M. Sakamoto and S. Kobayashi, *J. Phys. Soc. Japan* **13**, 800 (1958). (58) L.W. Barr et al., *Trans. Faraday Soc.* **56**, 697 (1960). (59) J.J. Gilman, W.G. Johnston, and R. Sears, *J. Appl. Phys.* **29**, 747 (1958).

	Etchant and conditions	Remarks	References
KI	Isopropyl alcohol time: 25 sec; plane (100) ref.: (53) other etchants: see ref.: (53)		
LiF	$H_2O + FeF_3$ ($\sim 2 \times 10^{-6}$ molar) time: 1 min; plane: (100) ref.: (60) other etchants: see ref.: (60) (61) (62) (63) (64) (65) (66)	Distinguishable between aged and fresh dislocations	(60) J.J. Gilman and W.G. Johnston, "Dislocations and Mechanical Properties of Solids," p. 116. Wiley, New York, 1957. (61) J.J. Gilman and W.G. Johnston, *J. Appl. Phys.* **27**, 1018 (1956). (62) L.S. Birks and R.T. Seal, *J. Appl. Phys.* **28**, 541 (1957). (63) A.A. Urusovshaia, *Soviet Phys.—Cryst. (English Transl.)* **3**, 731 (1958). (64) A.P. Kapustin, *Soviet Phys.—Cryst. (English Transl.)* **4**, 247 (1959). (65) M.P. Ives and J.P. Hirth, *J. Chem. Phys.* **33**, 517 (1960). (66) A.R.C. Westwood, H. Opperhauser, and D.L. Goldheim, *Phil. Mag.* [8] **6**, 1475 (1961).
MgO	1 H_2SO_4, 1 H_2O, 5 NH_4Cl sat. solution time: 15 min; plane: (100) ref.: (67) other etchants: see ref.: (34) (68) (69) (70)	Distinguishes between fresh and aged dislocations	(67) R.J. Stokes, T.L. Johnston, and C.H. Li, *Phil. Mag.* [8] **3**, 718 (1958). (68) R.J. Stokes, T.L. Johnston, and C.H. Li, *Phil. Mag.* [8] **4**, 137 (1959). (69) J. Washburn, A.E. Gorum, and E.R. Parker, *Trans. A.I.M.E.* **215**, 718 (1958).

TABLE II. ETCHANTS *(continued)*

Crystals	Etchant	Remarks	References
			(70) T.K. Ghash and F.J.P. Clarke, *Brit. J. Appl. Phys.* **12**, 44 (1961); A.S. Keh, *J. Appl. Phys.* **31**, 1538 (1960).
NaCl	Glacial acetic acid time: 0.1 sec; plane: (100) ref.: (71) other etchants: see ref.: (34) (53) (54) (71) (72) (73) (74)		(71) B. Jessensky, *Nature* **181**, 559 (1958). (72) S. Amelinckx, *Acta Met.* **2**, 848 (1954). (73) S. Mendelson, *J. Appl. Phys.* **32**, 1579 (1961). (74) M.P. Shaskolskaya and S. Jui-Fang, *Soviet Phys.—Cryst. (English Transl.)* **4**, 74 (1960).
NaF	Same as the one mentioned for KCl ref.: (54)		
NiO	Hot nitric acid ref.: (75) other etchant: see ref.: (76)		(75) T. Takeda and H. Kondoh, *J. Phys Soc. Japan* **17**, 1315 (1962). (76) T. Takeda and H. Kondoh, *J. Phys. Soc. Japan* **17**, 1317 (1962).

PbTe	$10\ ml\ H_2O + 5\ gm\ NaOH + 0.2\ gm\ I_2$ time: 5 min.; temp.: 94-98°C plane: (100) cleavage ref.: (77) other etchants: see ref.: (77) (78)	Rinse in water	(77) B.B. Houston and M.K. Norr, *J. Appl. Phys.* **31**, 615 (1960). (78) G.P. Tilly, *Brit. J. Appl. Phys.* **12**, 524 (1961).
SiC	Fused borax temp: 800-1000°C; plane: (0001) ref.: (79) other etchants: see ref.: (80)		(79) R. Gevers, S. Amelinckx, and W. Dekeyser, *Naturwissenschaften* **39**, 448 (1952). (80) F.H. Horn, *Phil. Mag.* [7] **43**, 1210 (1952).
UO$_2$	H_2O_2, H_2O, and conc. H_2SO_4 in proportion 6 : 3 : 1 by volume time: 3 min; temp: 20°C; plane (111) (100) ref.: (81) other etchant: see ref.: (81)		(81) A. Briggs, Harwell Report AERE M-859 (1961).
ZnS	Aqueous H_2O_2 solutions of 7.5 % to 30 % time: 10 min-1 hr temp: 60-80°C ref.: (82)	Wash with KCN solution, water, acetic acid, water, dry at 100°C	(82) P. Goldberg, *J. Appl. Phys.* **32**, 1520 (1961).

thermodynamic equilibrium at the high temperature. On cooling they diffuse to the surface and form the small pits. The same phenomenon was recently observed in a more pronounced way in a copper—4.8% Si alloy when cooled in argon[113c]. The observation was made at the temperature of formation, by the use of a hot stage. The phenomenon is reversible: on heating the pits disappear again, on renewed cooling many pits reappear at the same site. This suggests that the nucleation sites may be the emergence points of the dislocations.

k. Etching Procedure

No well-defined rule can be formulated for developing an etchant for a given crystal. Etchants are usually discovered on purely empirical grounds. Fortunately etchants are not specific; many etchants will attack a given crystal, and a given etchant may be applicable to many crystals. Table II lists etchants that have been found satisfactory for revealing dislocations.

4. CLEAVAGE STEPS

A limited amount of information on the geometry of dislocations can be obtained by a study of the cleavage steps. If cleavage proceeds along a plane with unit normal $\bar{\nu}$, and which is intersected by a stationary dislocation with Burgers vector $\bar{b}$, a cleavage step of height $h = \bar{b} \cdot \bar{\nu}$ is formed in the wake of the emergence point. This was first shown for an individual screw dislocation of large Burgers vector in mica[114] using multiple beam interferometry and has since been extended to dislocations in boundaries.[115]

In principle the step height gives the component of the Burgers vector perpendicular to the cleavage plane, while the sign of the step gives the sign of the dislocation. However cleavage steps 1 unit cell high are difficult to detect and therefore the method is not practical.

Nevertheless the phenomenon may be used to determine the character of a sub-boundary with a relatively large angle of misorientation. If a cleavage front passes a tilt boundary, as in Fig. 25a,

[113c] M. B. Kasen and D. H. Polonis, *Acta Met.* **10**, 821 (1962).
[114] S. Amelinckx, *Nature* **169**, 580 (1952).
[115] S. Amelinckx and E. Votava, *Naturwissenschaften* **41**, 422 (1954).

no cleavage step will be produced, but if on the contrary a twist boundary is intersected a sequence of steps will result as in Fig. 25b.

Gilman[116] demonstrated that if cleavage proceeds slower than a critical speed it is accompanied by plastic flow, ahead of the crack tip. Cleavage steps will then be produced by intersected screws, which were formed just before the crack passes. The half-loops formed during this process were revealed by means of etch pits.

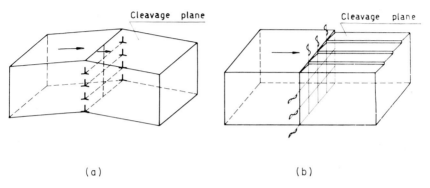

Cleavage plane   Cleavage plane

(a)   (b)

FIG. 25. Production of surface steps on cleavage. (a) If the cleavage front passes a dislocation for which the Burgers vector is parallel to the surface, no step results. (b) If the cleavage front passes a dislocation for which the Burgers vector has a component perpendicular to the cleavage plane, a step results. The step height is equal to this vertical component.

5. SLIP STEPS

We will only be concerned here with slip steps caused by the movement of one or a small number of dislocations.

As already mentioned in the previous paragraph the movement of the emergence point of a dislocation with Burgers vector $\bar{b}$ leaves a trace of height $\bar{b} \cdot \bar{v}$ if $\bar{v}$ is the unit normal on the observation face. Such monosteps were observed for the first time on faces of silver crystals prepared by sublimation and decorated by chance.[117] Similar monosteps have been observed by Suzuki[118] in copper, heated to very near the melting point. After cooling, the steps were decorated with oxide particles and could be observed in optical microscopy. The

[116] J. J. Gilman, *Trans. A.I.M.E.* **209**, 449 (1957).

[117] A. J. Forty and F. C. Frank, *Proc. Roy. Soc.* **A217**, 262 (1953).

[118] H. Suzuki, *Sci. Repts. Res. Inst., Tohoku Univ.* **A6**, 573 (1954); *J. Phys. Soc. Japan* **10**, 11 (1955).

oxide particles apparently formed progressively during the movement of the dislocations. The movement of complete boundaries as well as of individual dislocations was continuously recorded in this way.

Bassett[119] revealed slip step traces, left by individual dislocations, in KBr. The steps were decorated with gold particles in the way described in Section 2. He found, in particular, evidence for cross slip from (110) planes onto (100) planes. He could also prove that monoionic steps were revealed, by measuring the offset caused in one slip trace by a known number of dislocations in an intersecting slip system.

A detailed study of elementary slip steps on alkali halides was made recently by Betghe and co-workers[17-18] using the gold step decoration technique. One of their striking photographs is reproduced in Fig. 26.

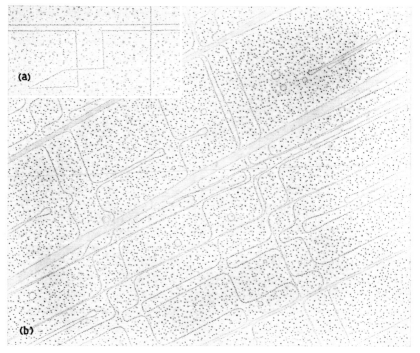

FIG. 26.   Slip steps on sodium chloride cleavage face revealed by gold decoration. (a) Complicated slip path. (b) Intersecting slip steps after some evaporation. (Courtesy of Betghe and co-workers.)

[119] G. Bassett, *Phil. Mag.* [8] **3**, 1042 (1958).

5a. ETCH SPIRALS

Etch spirals have been observed in special circumstances on electrolytically polished surfaces of aluminium[119a] and silicon iron[119b]. The origin of the phenomenon is not clear. According to Marchin and Wyon[119a] the spirals have no direct relation with lattice imperfections whereas Feliu and Castro[119b] have shown that there is a certain correlation between etch pits and the presence of spirals. In any case there is no direct relation with growth or dissolution spirals observed on habit faces. In the latter case the spirals consist in fact of a spirally wound surface step, whereas the spirals on polished surfaces consist of surface grooves.

[119a] J. M. Marchin and G. Wyon, *Acta Met.* **10**, 915 (1962).
[119b] S. Feliu and G. Castro, *Acta Met.* **10**, 543 (1962).

II. Bulk Methods

6. DECORATION TECHNIQUE

a. Principles of the Method

The methods discussed so far reveal only emergence points of dislocations and are therefore restricted in their application. We will now discuss methods which reveal the dislocations along their length, and which are therefore suitable for studying the complete geometry.

The technique consists in heating a crystal in such a way that small but visible particles are formed along the dislocation lines. The particles should be small enough to permit resolution of the dislocations. On the other hand they should be large enough to scatter sufficient light to become visible in the optical microscope. The particles can be observed either in transmitted light, if they are large enough, or in scattered light (ultramicroscopic observation). The latter method is more sensitive but requires longer exposure times for the photographs.

The possibility of producing decoration by the use of a suitable heat treatment is due to the following factors:

(a) Cottrell interaction, which causes an elastic interaction of impurities with dislocations, and possibly also electrical interaction;

(b) enhanced diffusion along dislocations;

(c) preferential nucleation of particles along the dislocations;

(d) the ability of a dislocation to act as a source of vacancies and thus make room for the growing precipitate particles.

From this it will be clear that heat treatment is generally required so that migration of the decorating agents towards the dislocations will occur. The more easily the impurity diffuses, the lower will be the temperature of the heat treatment. It is therefore convenient to use impurities which migrate along interstitial sites, e.g., Cu in Si,

Li in Ge, etc. In such cases impurity diffusion may be appreciable at temperatures where self-diffusion is still small. It is also required that the solubility of the decorating agent be small at room temperature and increases with temperature.

It is possible to treat the crystal in such a way that a chemical transformation takes place, the reaction product being insoluble and therefore precipitating, e.g., formation of CaO in CaF_2.[120] The "impurity" may be a stoichiometric excess of one of the constituents, e.g., Na in NaCl.[50]

The inherent limitations of the methods are obviously the following:

(a) the heat treatment may disturb the dislocation pattern so that only annealed structures can be observed;
(b) the dislocations are pinned by the precipitates; the study of motion is therefore difficult. Whereas dislocations are observed *in vivo* in the electron microscope, they are observed *in vitro* by decoration techniques.

b. Decoration Methods for Specific Crystals

We now discuss different methods that have been used for specific crystals. We only give the practical prescriptions, without going into the details of the decoration mechanisms, which are often not well understood.

(1) *Silver halides*.[121] Exposure at room temperature to light absorbed in the tail of the fundamental absorption band is all that is needed to cause the formation of photolytic silver particles along dislocations in silver bromide and silver chloride.[121] The silver migrates as interstitial ions in the silver halides. The formation process of the silver particles is discussed by Mitchell and Mott[122] and by Seitz.[123] The patterns can be observed in transmission using inactive light. Figure 27 is an example of a pattern revealed by the use of this method. From the work of Gardner Miller, Childs, and Slifkin[124] it follows

[120] W. Bontinck, *Physica* 24, 650 (1958).

[121] J. M. Hedges and J. W. Mitchell, *Phil. Mag.* [7] 44, 223, 357 (1953).

[122] J. W. Mitchell and N. F. Mott, *Phil. Mag.* [8] 2, 1149 (1957).

[123] F. Seitz, *Rev. Mod. Phys.* 23, 328 (1951).

[124] C. Childs, Thesis, University of North Carolina, 1959; C. Childs and L. Slifkin, *Phys. Chem. Solids* 12, 119 (1959); *Phys. Rev. Letters* 5, 502 (1960); M. G. Miller, Thesis, University of North Carolina, 1961.

that dislocations in silver halides can be decorated with silver by sweeping photoelectrons into the crystal. This produces a decoration which extends further into the bulk than the decoration obtained by exposing to light only. The effect seems to be very much dependent on the presence of small amounts of impurities since it depends on the nature of the substrate on which the crystals have been annealed.

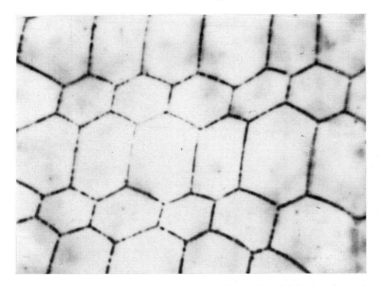

FIG. 27.   Dislocation pattern in silver chloride made visible by the print-out effect (transmitted light). (Courtesy of J. W. Mitchell.)

(2) *Alkali halides.* (a) Additive coloration.[50] Heavy additive coloration of NaCl results in the precipitation of colloidal sodium along dislocations. The procedure consists in making a cavity in a cleavage block, filling this with sodium metal, and closing it, like a box, with another cleavage fragment. The whole assembly is kept together in a steel frame and heated to a temperature not far below the melting point of the alkali halide, for 1 or 2 hr and finally fairly rapidly cooled. Thin cleavage slices are then examined in ultramicroscopic illumination.

(b) Doping with heavy metal halides.[125,126] The dislocations in sodium chloride, potassium chloride, and potassium bromide can be

[125] W. Van der Vorst and W. Dekeyser, *Phil. Mag.* [8] 1, 882 (1956).
[126] S. Amelinckx, *Acta Met.* 6, 34 (1958).

decorated by doping the crystals with a small amount of the corresponding silver halide and annealing them in hydrogen at a temperature not too far below the melting point (for 1 or 2 hours). After this treatment the crystals have to be cooled slowly. The silver halide is added to the melt when growing the crystal (0,5% addition by weight to the melt). The decorating particles are silver, produced by reduction in hydrogen. An example is shown in Fig. 28. Lead or copper halides can also be used as doping agents.

(c) Cavity decoration. In crystals doped with 0,1% of silver nitrate, the decoration procedure is as follows[127]: The crystal is first

Fig. 28.   Dislocation network of a mixed type in potassium chloride (dark field image). The decorating particles are silver. They result from an anneal in hydrogen of a silver-doped crystal. Notice the zig-zag shaped dislocation lines. [After S. Amelinckx, *Acta Met.* **6**, 34 (1958).]

[127] S. Amelinckx, W. Maenhout-Van der Vorst, and W. Dekeyser, *Acta Met.* **7**, 8 (1959).

irradiated with 50 kv X-rays for 2-3 hr and then annealed at some 500-600°C for 1 hr. Small cavities filled with the decomposition gases of the NO_3 group, are then found along the dislocations. Large cavities result on annealing in hydrogen. After the irradiation alone, the crystal is not visibly changed apart from coloration; the only effect is the decomposition of the nitrate group. The anneal apparently agglomerates the decomposition products. Figure 29 shows patterns revealed in this way.

An interesting feature of this method is that dislocations present before the irradiation are decorated much better than those introduced after the irradiation but before the anneal. This applies as well to grown-in, as to freshly introduced dislocations. Apparently the decomposition of the nitrate group takes place preferentially along dislocations possibly because excitons dissipate their energy there.[128a]

In a somewhat similar way it is possible to decorate dislocations in LiF by heavy neutron irradiation ($\sim 10^{18}$ nvt) followed by anneal above 550°C. The dislocations are then decorated by small cavities.[128b]

The latter method allows one to make visible deformation structures since only a heat treatment at a moderate temperature is necessary.

(d) Doping with divalent impurities. In lithium fluoride decoration can be achieved by adding calcium fluoride to the melt while growing the crystal.[129] The calcium fluoride precipitates along the dislocations on cooling. Suitable additives for sodium chloride are calcium and barium chloride.[130]

(e) Diffusion from outside. In undoped alkali halide crystals the decoration can be achieved by diffusing in some impurities from outside. The following method is convenient for KBr and NaCl.[131] The alkali halide crystals, together with a small quantity of the corresponding gold halide, i.e., $HAuCl_4$ or $HAuBr_4$, are sealed in an evacuated quartz capsule and heated for 2 or 3 hr at a temperature about 100°C below the melting point, and subsequently cooled at a

[128a] S. Amelinckx, W. Maenhout-Van der Vorst, and W. Dekeyser, *Proc. Intern. Conf. Solid State Phys. Electron. Telecommun., Brussels, 1958*, p. 909 (1960).

[128b] K. Kubo and Y. Katano, *J. Phys. Soc. Japan*, **16**, 347, (1961).

[129] J. Washburn and J. Nadeau, *Acta Met.* **6**, 665 (1958); J. Washburn, *in* "Growth and Perfection of Crystals" (R. H. Doremus *et al.*, p. 342. Wiley, eds.), New York, 1958.

[130] S. Amelinckx, W. Maenhout-Van der Vorst, R. Gevers, and W. Dekeyser, *Phil. Mag.* [7] **46**, 450 (1955).

[131] D. J. Barber, K. B. Harvey, and J. W. Mitchell, *Phil. Mag.* [8] **2**, 704 (1957).

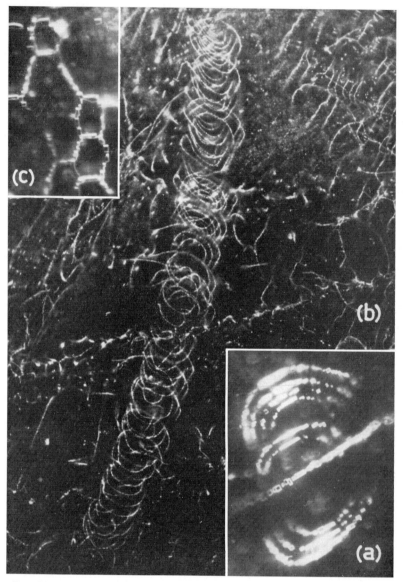

FIG. 29. Dislocations decorated by small cavities in potassium chloride containing silver nitrate. The crystal is first irradiated with X-rays and then annealed. [After S. Amelinckx, *Phil. Mag.* [8] **3**, 307 (1958).] (a) Small glissile loops, due to surface sources. (b) Surface sources along the instantaneous position of a crack front. (c) A few meshes of a network.

moderate rate. The cooling rate is rather critical, especially with small crystals. The decorating particles are gold resulting from the thermal decomposition of the gold halide. The same method can also be applied to cesium bromide.[132] A pattern revealed by the use of this method in cesium bromide is shown in Fig. 30.

(3) *Calcium fluoride.* It is sufficient to heat calcium fluoride[120] in moist air for a short time, e.g., 30 min at 800°C, in order to obtain

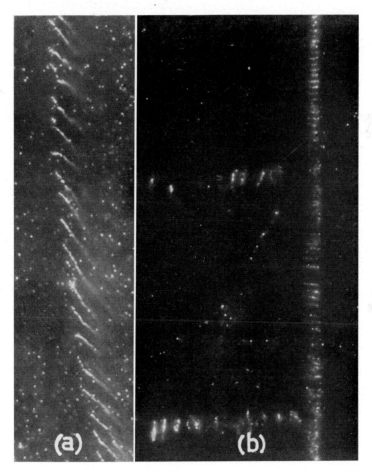

FIG. 30. Tilt boundary n cesium bromide revealed by decoration with gold particles. [After S. Amelinckx, *Phil. Mag.* [8] **3**, 307 (1958).]

[132] S. Amelinckx, *Phil. Mag.* [8] **3**, 307 (1958).

a precipitation of CaO along the dislocations. The phenomenon consists in fact in a partial hydrolysis of calcium fluoride. The temperature is not critical; at lower temperature the annealing time has to be correspondingly longer. Figure 31 represents a dislocation pattern in calcium fluoride revealed in this way.[133,134]

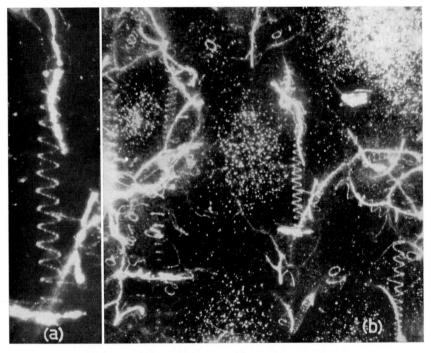

Fig. 31. Decorated helical dislocations in calcium fluoride. The decorating particles are CaO. [After W. Bontinck and S. Amelinckx, *Phil. Mag.* [8] **2**, 1 (1957).]

(4) *Silicon and germanium.* Decoration in silicon[135] is achieved by diffusing in copper at about 900°C. This is done by putting a few drops of a copper nitrate solution on the crystal slab and heating it subsequently in hydrogen. Copper then diffuses through interstitial sites and precipitates in substitutional positions along dislocations, which climb somewhat in order to produce the vacancies required.[136]

133 W. Bontinck and S. Amelinckx, *Phil. Mag.*, [8] **2**, 1, (1957).

134 W. Bontinck, *Phil. Mag.* [8] **2**, 561 (1957).

135 W. C. Dash, *J. Appl. Phys.* **27**, 1193 (1956).

136 F. C. Frank and D. Turnbull, *Phys. Rev.* **104**, 617 (1956).

Since silicon is not transparent to visible light, except in very thin samples, infrared radiation is used to view the crystal slab in transmission. An image convertor, suitable for the near infrared, is used to focus the image, but photographs are made directly on infrared sensitive emulsions. The heat treatment does not seem to disturb very much the dislocation pattern; apparently dislocations are not very mobile at the diffusion temperature. Figure 32 shows an example.

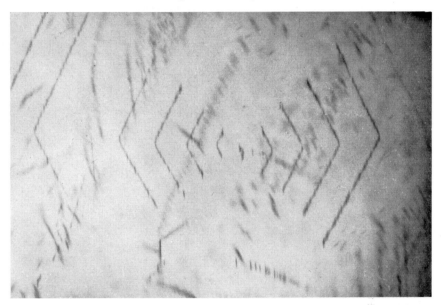

FIG. 32. Single-ended Frank-Read source in silicon deformed by twisting about a $\langle 111 \rangle$ axis. The decorating particles are copper. The nondecorated segments have the $\langle 110 \rangle$ direction and they are probably pure screws. (Courtesy of W. C. Dash.)

In germanium, lithium can be used as a decorating agent.[59] However observations in transmission require the use of more elaborate infrared sensitive detection than for silicon. For this reason decoration etching is used. (see Section 6d).

Annealing silicon crystals containing aluminium at 1200–1300°C results in the formation of fine precipitates visible in optical microscopy (infrared) on structural defects such as dislocations and partly coherent twin boundaries.[136a] The precipitation of aluminium consists

[136a] R. Bullough, R. C. Newman, J. Wakefield, and J. B. Willis, *J. Appl. Phys.* **31**, 707 (1960).

presumably in a chemical reaction involving oxygen. It is found that dislocations introduced by deformation do not act as sites for precipitation on heat treatment at 1250°C, only grown in dislocations do so. This shows that the latter have in their core impurities picked up from the melt, and which act as nuclei.

In heat treated crystals also containing phosphorus the dislocations are surrounded by a cylindrical region of p-material in an n-type matrix. This is shown by etching.

Bond and Harvey[136b] have shown that dislocations in aluminum oxide can be decorated by doping the crystal with 1% (by weight) zirconium oxide while growing it by the Verneuil method. After reannealing in air at 1500°C for 4 hr followed by air quenching the dislocations are decorated throughout the crystal.

c. Applications of the Decoration Techniques

We will now consider a few typical examples where the application of decoration techniques have been particularly successful. At the same time the theory required for a proper understanding of the features will be discussed briefly.

(1) *Geometry of sub-boundaries.* Since the decoration techniques always involve a certain amount of annealing, they are ideally suited to study, e.g., sub-boundaries in an annealed crystal. The geometrical features of sub-boundaries in alkali halides were studied in detail[50,126] and these studies made possible a direct verification of the models predicted by theory.

The Burgers vectors of dislocations in crystals of the sodium chloride structure are of the type $(a/2)$ [110]. With this type of dislocation a variety of sub-boundaries can be formed. They fall into 3 classes.

(a) Boundaries consisting of parallel lines. These are tilt boundaries, i.e., the rotation axis is in the plane of the boundary and all lines are edge dislocations parallel to the rotation axis. In the simplest case all dislocations have the same Burgers vector. The boundary plane is then a symmetry plane for the bicrystal and it is perpendicular to the Burgers vector (Fig. 33). If the boundary plane has a more general orientation, more than one kind of dislocation is required (Fig. 34). Of course, on a decoration pattern no difference is visible.

[136b] H. E. Bond and K. B. Harvey, *J. Appl. Phys.* **34**, 440 (1963).

Examples of tilt boundaries are presented in Fig. 35 and Fig. 30. The first refers to potassium chloride and the decorating particles are silver. The second example is in cesium bromide, the decorating particles being gold. In the latter case the Burgers vectors are of the type a [100] and the symmetrical tilt boundaries are in the cube plane.

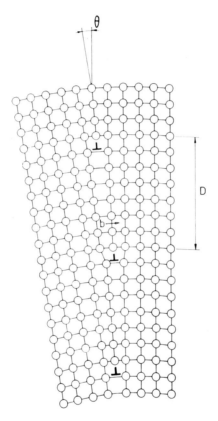

FIG. 33. Symmetrical pure tilt boundary in the primitive lattice. The boundary consists of parallel edge dislocation all of the same kind.

(b) Square networks. The model for a twist boundary having a a [100] rotation axis and a (100) contact plane consists of a square grid of pure screw dislocations having directions [011] and [01$\bar{1}$] and Burgers vectors $(a/2)$ [011] and $(a/2)$ [01$\bar{1}$], respectively. This can be seen intuitively from Fig. 36 for a primitive lattice. If the contact plane deviates from the cube plane, the network acquires lozenge-

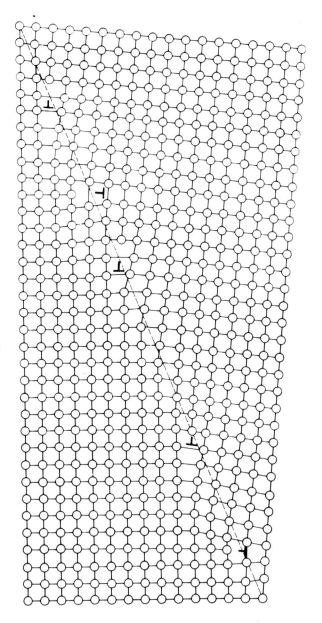

FIG. 34. Asymmetrical pure tilt boundary in the primitive lattice. The boundary consists of two kinds of dislocations with mutually perpendicular Burgers vectors.

shaped meshes; its projection along [100] on the (100) plane is, however, still the same square grid.

Examples of square grids are shown in Fig. 37. The directions of the lines are in agreement with the theoretical predictions. It is obvious that the node points are decorated preferentially. In some cases, where the resolution of the method becomes insufficient, only the node points are decorated. Figure 38 is an array of dots corresponding to a square network in the cube plane of CsBr.

(c) Hexagonal networks. If the contact plane of the boundary is the (111) plane, and if the rotation axis is parallel to the [111] direction,

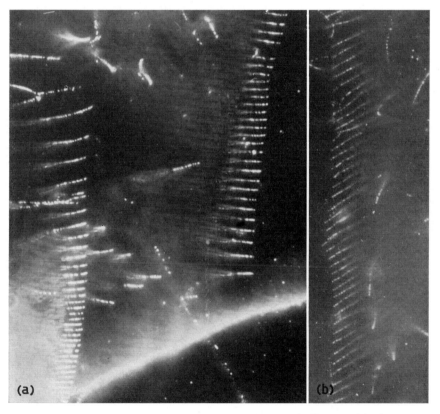

(a)   (b)

Fig. 35. (a) Tilt boundaries revealed by decoration with silver particles in potassium chloride. Notice the gradual increase in spacing towards the end of the wall. [After S. Amelinckx, *Acta Met.* **6**, 34 (1958).] (b) Tilt boundary in sodium chloride decorated with silver.

the model for the boundary consists of a hexagonal grid of screw dislocations having Burgers vectors $(a/2) [1\bar{1}0]$, $(a/2) [\bar{1}01]$, and $(a/2) [01\bar{1}]$. This can be seen intuitively from Fig. 39. Provided the rotation axis remains the same, a change in boundary plane only causes a change in the mesh shape, the pattern becomes such that its projection along [111] on a (111) plane produces the same hexagonal net. An hexagonal grid in potassium chloride is reproduced in Fig. 40; the plane of the net was near to the cube plane; the meshes are therefore elongated hexagons.

In many cases decoration is less or even completely absent along one family of lines and it is found that the segments have a [110] direction, i.e., the direction of a possible Burgers vector. Therefore they are probably screw segments and it can be concluded that the interaction of impurities is presumably weaker with screws.

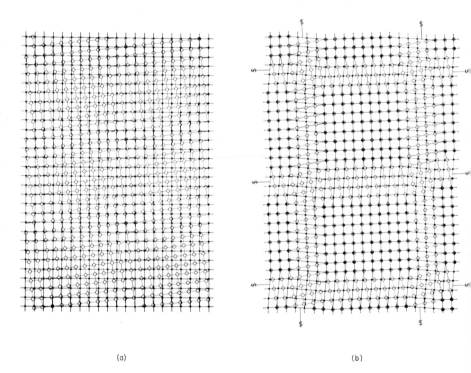

(a)   (b)

FIG. 36. Model for a pure twist boundary in the cube plane of a primitive lattice. (a) The 2 lattice planes on either side of the contact plane are rotated one with respect to the other. (b) The atoms have taken up equilibrium positions.

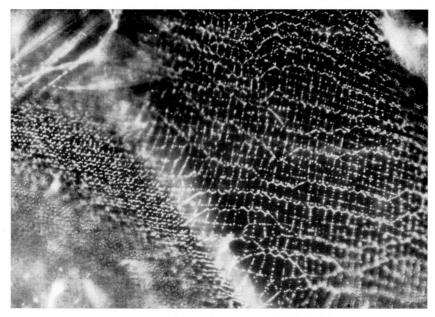

FIG. 37. Square grid of screw dislocations in potassium chloride. The node points are decorated preferentially. Notice the presence of zig-zag shaped singularities. [After S. Amelinckx, *Acta Met.* **6**, 34 (1958).]

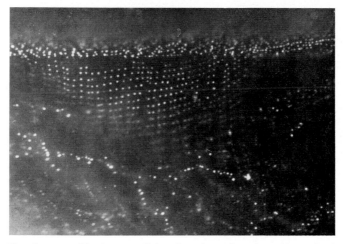

FIG. 38. Square grid of screw dislocations in cesium bromide; only the node points are decorated with gold particles. [After S. Amelinckx, *Phil. Mag.* [8] **3**, 307 (1958).]

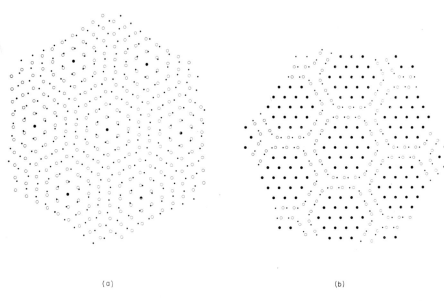

(a)   (b)

Fig. 39.   Model for an hexagonal grid of screw dislocations in the (111) plane of a face-centered cubic crystal. (a) After twisting about the ⟨111⟩ axis. (b) After the atoms have relaxed.

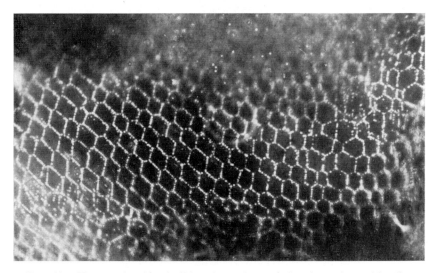

Fig. 40.   Hexagonal grid of dislocations observed by decoration with silver particles in potassium chloride. The plane of the net is not far from the cube plane. Therefore, the meshes are elongated. [After S. Amelinckx, *Acta Met.* **6**, 34 (1958).]

(d) More general patterns. The 3 main types of patterns, described above, can be generated by the intersection of at most 2 families of dislocations. In the case of the hexagonal grid the third family of dislocations is formed at the intersection points of the first 2 families, since this reduces the total energy. In these cases the orientation of the rotation axis is subject to certain restrictions. This is no longer true if 3 families of dislocations having noncoplanar Burgers vectors are allowed.

An arbitrary boundary, i.e., one having an arbitrary rotation axis and an arbitrary contact plane, can now be constructed. Two dislocation boundaries for which the rotation axis is not along a simple direction will contain foreign dislocations of a third kind; these can often be recognized as, e.g., in Fig. 37.

(2) *Dislocation climb.* It is easy to see that a screw dislocation which is subject to a supersaturation or undersaturation of vacancies (or interstitials) will adopt the shape of a helix.[137,138] The reasoning goes as follows: Consider a small segment of dislocation AB of length ds, which is in equilibrium with a supersaturation $(c - c_0)/c_0$ of vacancies. The forces acting on this segment are the line tension and the chemical stress; we neglect externally applied stresses.

The line tension can be represented by the two vectors $\vec{T}_1$ and $\vec{T}_2$ oriented along the tangents in A and B and of equal magnitude.

$$| \vec{T}_1 | = | \vec{T}_2 | = T' = \tfrac{1}{2} \mu b^2. \tag{6.1}$$

The chemical stress has the magnitude $kT/b^2 \ln (c/c_0) \sin \vartheta$ per unit length and it is a vector $\vec{C}$ oriented along $\vec{t} \times \vec{b}$ (see Fig. 41).

In equilibrium we have

$$\vec{T}_1 + \vec{T}_2 + \vec{C} = 0. \tag{6.2}$$

Projecting the equality (6.2) on the direction of $\vec{b}$ leads to

$$T' \cos \vartheta_1 = T' \cos \vartheta_2 \tag{6.3}$$

and hence to

$$\vartheta_1 = \vartheta_2 = \vartheta \tag{6.4}$$

[137] S. Amelinckx, W. Bontinck, W. Dekeyser, and F. Seitz, *Phil. Mag.* [8] **2**, 1 (1957).
[138] J. Weertman, *Phys. Rev.* **107**, 1259 (1957).

where ϑ_1 and ϑ_2 have the significance illustrated in Fig. 41. Projecting the same equality on the plane perpendicular to $\bar{b}$ and expressing the equilibrium between the projected vectors leads to

$$2T' \sin \vartheta \, \frac{d\phi}{2} = \frac{kT}{b^2} \ln \frac{c}{c_0} \, ds \sin \vartheta \qquad (6.5)$$

or taking into account

$$R = \frac{ds}{d\phi} = \frac{\frac{1}{2} \mu \, b^2}{(kT/b^2) \ln (c/c_0)} = \text{constant} \qquad (6.6)$$

where R is the radius of curvature of the projection of the dislocation line, on the plane perpendicular to b. Therefore we can conclude that the curve has the following properties:

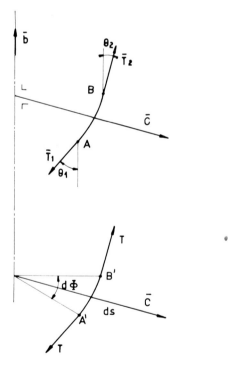

FIG. 41.   Illustrating the derivation of the equilibrium shape of a dislocation under a supersaturation of vacancies or interstitials.

(a) The tangent forms a constant angle with a given direction in space, i.e., with b [from Eq. (6.4)].

(b) The radius of curvature is constant [from Eq. (6.6)], because the projected radius of curvature as well as the angle of projection are constant. The only curve that fits both requirements is the helix with a radius of curvature given by Eq. (6.6).

The phenomenon was first discovered in calcium fluoride[133,134] by the use of decoration techniques in the way described in Section 6f (3). An example of the observed helical dislocations is shown in Fig. 31. The axis of the helix is along [110] in accordance with the assumption that the original dislocation was a screw. In some cases a sequence of loops is observed in the prolongation of the helix. These probably result from the intersection of the helix with itself. Since successive turns of the helix repel each other, and if prismatic glide in the direction of the Burgers vector is possible, loops can be blown off in this direction.

In NaCl polygonized helices were found[139] by means of silver decoration.

The presence of the helices also was demonstrated with an independent method, namely etching. Calcium fluoride cleaves along (111) planes; a cleavage plane therefore may intersect a helix with its axis along [1$\bar{1}$0] and produce a sequence of half-loops. On etching one observes a sequence of paired pits, which on continued etching coalesce and finally become flat-bottomed. This is shown in Fig. 42. In this way it was demonstrated that the helices are not a consequence of the decoration procedure but are only revealed by it.

Electron microscopic observations in quenched metals have confirmed that helices can be generated by a supersaturation of vacancies. [Section 10j (1).]

More recently Dash[140] demonstrated that an undersaturation of vacancies may cause climb as well. The experiment was performed as follows.

First, screw dislocations of a known sign were introduced by twisting a silicon rod about a [111] axis. Then gold was allowed to diffuse into the crystal. Gold is known to diffuse interstitially in silicon but if captured by a vacancy it becomes substitutional and destroys

[139] S. Amelinckx, W. Bontinck, and W. Maenhout-Van der Vorst, *Physica* **23**, 270 (1957).

[140] W. C. Dash, *J. Appl. Phys.* **31**, 2275 (1960).

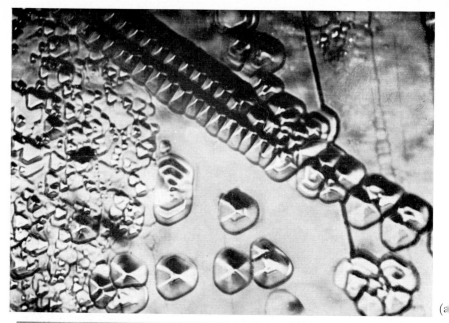

(a)

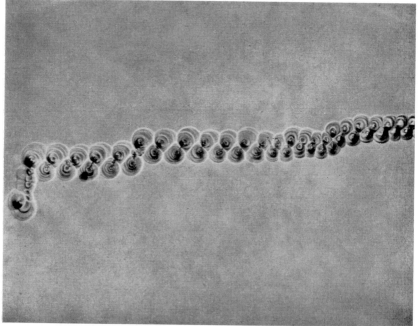

(b)

FIG. 42. (a) Etched section of helical dislocation in calcium fluoride. The helix has been cut lengthwise by a (111) cleavage plane. On etching a sequence of half-loops is revealed. [After W. Bontinck, *Phil. Mag.* [8] **2**, 561 (1957).] (b) Etched helix along [110] direction in germanium single crystal. (Courtesy of Tweet.)

the vacancy. Consequently the dislocations must climb in order to maintain the equilibrium concentration of vacancies, and the screws are thereby driven into helices. The dislocations are subsequently or simultaneously decorated, using copper precipitation. The sign of the helices was deduced by observing them along their axis, moving the focus of the microscope up and down. From the sense of movement of the in focus part of the helix, the sense of winding can be deduced. It was found that left-handed screws were transformed into right-handed helices; this can only be caused by the emission of vacancies (or the absorption of interstitials). Figure 43 shows helices in silicon.

The method for decorating dislocations in sodium chloride by means of barium chloride has been refined considerably by Harvey.[140a] The crystal contains (1% by weight) $BaCl_2$ as grown; i.e., after slow cooling they are opaque. Decoration is achieved by reheating such crystals in air to 650°C during 4 hr and air quenching them to room temperature. Large prismatic loops are observed, apparently resulting from the operation of Bardeen-Herring climb sources. The fact that climb is observed in the presence of such a large amount of divalent impurities is taken as evidence that a high concentration of vacancies is present at 650°C.

(3) *Dislocations in whiskers.* Because of their unusual mechanical properties, there has been much speculation as to the dislocation content and configuration in whiskers. Their great strength has been attributed either to the absence of dislocations or their inability to multiply. On the other hand, the particular habit of these crystals has been explained by assuming that they grew around a single axial screw dislocation.

Decoration methods have shown that at least the particular alkali halide whiskers studied often contain a large number of dislocations. This was confirmed by means of etching. Only in a few exceptional cases is the theoretical picture of a single axial dislocation confirmed.[141]

Figure 44, e.g., shows a whisker containing a single axial dislocation. The whiskers were obtained by growth from a solution containing a small amount (2 mg per liter) of polyvinyl alcohol. This impurity acts as a "poison" which induces the crystals to grow as thin fibers rather than as isometric crystals. In fact this is a form of dendrite growth; the fibers often branch. Figure 45 shows the dislocation configuration

[140a] K. B. Harvey, *Phil. Mag.* **8**, 435 (1963).
[141] S. Amelinckx, *J. Appl. Phys.* **29**, 1610 (1958).

observed in such a branching point. From this observation the
Burgers vector of the axial dislocation can be deduced in the following
way. The angles at the node being approximately equal, the Burgers
vectors must be equal in magnitude, and hence of the form $(a/2)\,[1\bar{1}0]$.
If the dislocations in the 3 branches are further required to emerge
with a screw component in the respective top faces, the only possible
configuration is such that the axial dislocation is a 45° dislocation

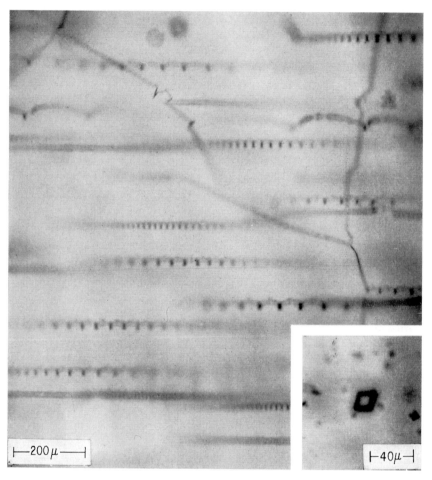

FIG. 43.   Decorated helices in silicon; the helices have $\langle 110 \rangle$ directions; they
are polygonized as can be judged from the end-on view in the inset. (Courtesy W. C.
Dash.)

(Fig. 46). According to Hirth and Frank,[142] the configuration of such a dislocation would *not* be stable.

Decoration in the whiskers was achieved by diffusing in gold from outside. A large number of whiskers were enclosed in an evacuated

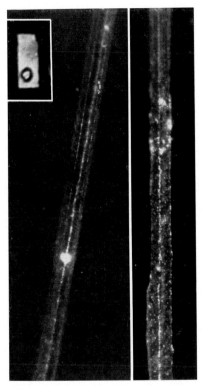

FIG. 44. Sodium chloride whiskers, grown from solution, containing axial dislocations. Decoration was achieved by means of gold diffusion. The inset shows the etched tip of a similar whisker; only 1 etch pit is present.

quartz tube together with a small amount of gold chloride. After heating at 700°C for about 1 hr the crystals were cooled fairly rapidly. For further details of the experimental procedure we refer to Amelinckx.[143]

[142] J. Hirth and F. C. Frank, *Phil. Mag.* [8] 3, 34 (1958).
[143] S. Amelinckx, *in* "Growth and Perfection of Crystals" (R. H. Doremus *et al.*, eds.), p. 139. Wiley, New York, 1958.

(4) *Deformation patterns.* (a) Frank-Read sources.[144] It is now generally accepted that dislocation multiplication is the process responsible for plastic flow. One of the possible mechanisms for dislocation multiplication is the so-called Frank-Read source.

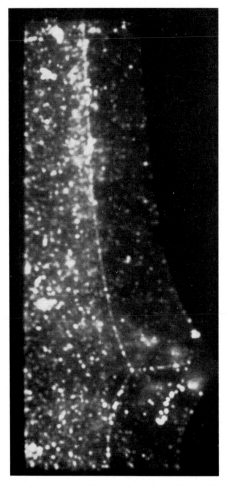

Fig. 45. Dislocations in a branching point of a sodium chloride whisker. This observation allows to deduce that the Burgers vectors of the 3 dislocations is of the type $(a/2)$ [110].

[144] W. C. Dash, *in* "Dislocations and Mechancial Properties of Crystals" (J. C. Fisher *et al.*, eds.), p. 57. Wiley, New York, 1957.

The configuration of dislocations resulting from the operation of a Frank-Read source is as shown in Fig. 47. Dash has twisted silicon crystals at high temperature around a [111] axis, causing glide in (111) planes. The dislocations were then decorated using copper in the way described in Section 6b (4). The result is shown in Figs. 48 and 32.

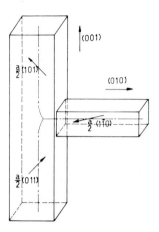

FIG. 46.   Distribution of Burgers vectors in a branching point of the whisker, as deduced from Fig. 45.

It is clear that the observed configuration is of exactly the kind predicted by theory. Since we have to do with a face of threefold symmetry the loops are of course more hexagonal in shall. The non-decorated segments in Fig. 32 are the screw segments. In accordance with the observation of nets in alkali halides the interaction with impurities seems to be less with screws. An alternative reason for the lack of decoration may be that screws do not provide space for precipitates. A single-ended source is shown in Fig. 49. The same pattern exhibits a number of other remarkable features. In a number of places the dislocations present cusps, and behind the cusps linear arrays of precipitates are visible. Similar trails have been observed in potassium chloride decorated with silver.[126] It is believed that these are due to point defects or small loops left by the movement of a jogged dislocation. In view of more recent evidence obtained with the electron microscope in other substances,[145] it now seems more plausible that

[145] J. Washburn, G. W. Groves, A. Kelly, and G. K. Williamson, *Phil. Mag.* [8] 5, 991 (1960).

they represent dislocation "dipoles" caused by large jogs. These dipoles may break up in a sequence of loops as shown by Price[146,147] for zinc. In fact such linear sequences of loops in the (111) plane had been found earlier by Dash, using his copper decoration technique. It now appears that they are related to the trails. Such small loops left in the wake of a moved dislocation have been held responsible for

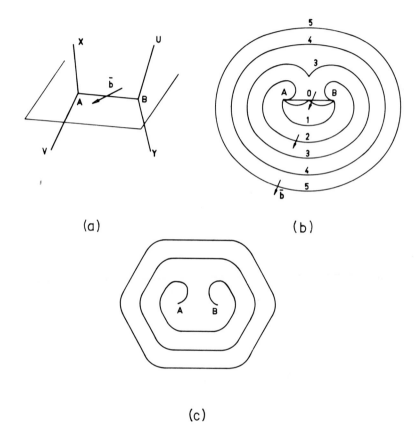

FIG. 47. Configuration of dislocations resulting from the operation of a Frank-Read source of the double-ended type. (a) Configuration of the source AB in space; only the segment AB is in a glide plane. (b) Operation of the source in a primitive lattice. (c) Shape to be expected for dislocation loops in a (111)-type glide plane.

[146] P. B. Price, *Phil. Mag.* [8] **5**, 873 (1960).
[147] P. B. Price, *Phil. Mag.* [8] **6**, 449 (1961).

dislocation multiplication in LiF.[148] More recent observations, obtained with the electron microscope, will be discussed further below. [See Section 10d (5).]

Surface sources of a somewhat different kind are visible in Fig. 29 which represents dislocation patterns in a crystal of KCl doped with silver nitrate and decorated by X-irradiation followed by annealing [see Section 6, b, (2), (c)]. The surface loops are formed as a consequence of plastic deformation, at the tip of a moving crack. Such sequences of half-loops are very unstable and would move out of the crystal on a high-temperature anneal. The picture therefore proves that the decoration technique allows one to reveal deformation patterns, i.e., freshly introduced dislocations.[149]

(b) Prismatic punching. Jones and Mitchell[150] have performed an elegant experiment to simulate the behavior of precipitate particles growing in a matrix. They embedded glass spheres in single crystal

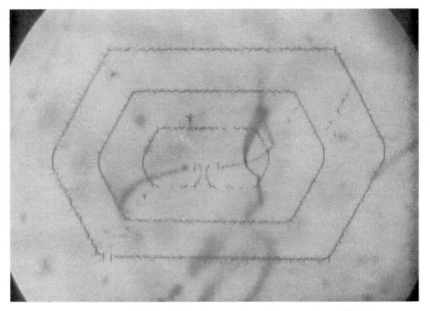

FIG. 48. Double-ended Frank-Read source in silicon made visible by copper decoration. (Courtesy of W. C. Dash.)

[148] S. Amelinckx and W. Dekeyser, *J. Appl. Phys.* **29**, 1000 (1958).
[149] S. Amelinckx, *Phil. Mag.* [8] **3**, 653 (1958).
[150] D. A. Jones and J. W. Mitchell, *Phil. Mag.* [8] **3**, 1 (1958).

sheets of silver chloride. On cooling, the crystal contracts more than the sphere and an inhomogeneous stress field is set up. Stresses are relieved by prismatic glide in a few or sometimes all $\langle 110 \rangle$ directions. The resulting dislocation patterns are revealed by exposing

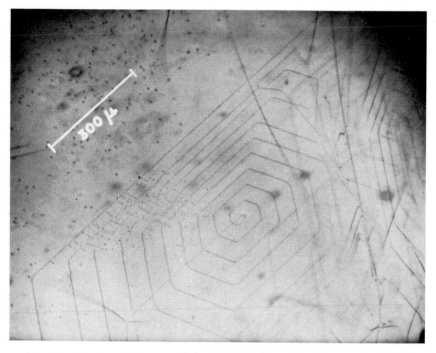

Fig. 49. Single-ended source of the Frank-Read type in silicon, made visible by copper decoration. Notice the cusps in some of the dislocations. Trails due to defects left behind on moving, are visible behind the cusps. (Courtesy of W. C. Dash.)

the silver chloride to light. In some cases helical dislocations are formed. An example of the observed pattern is shown in Fig. 50.

The formation of loops and helical dislocations can be understood as follows. Figure 51 shows a cross section through the stress pattern, for 4 glide directions. It is clear that this pattern as seen along a glide direction in fact has axial symmetry. The stress is a maximum at the surface of the sphere along points lying on the surface of a cylinder which has roughly a diameter equal to $a \sqrt{2}$, where a is the sphere radius. The first step in the generation of the sequence of loops or of the helical dislocation is the nucleation of a small dislocation

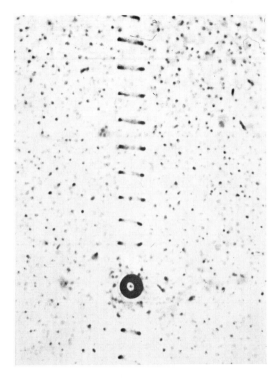

FIG. 50. Prismatic punching around a glass sphere embedded in silver chloride. The dislocations are decorated by exposure to light. (Courtesy of Jones and J. W. Mitchell.)

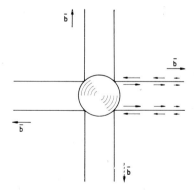

FIG. 51. Schematic representation of the stress pattern around a spherical inclusion. Only stress tunnels in the directions of the Burgers vectors are shown.

loop at the interface glass crystal, as shown in Fig. 52. This loop will evidently be created on the surface of the cylinder since there the stress is maximum. The shear stresses are then such that the edge segment E will move along the cylinder outward, while the screw segments S_1 and S_2 rotate in opposite sense along the cylindrical surface. If no climb is possible the dislocations will remain on the cylinder since they will in general have no pure screw sections. This follows from the simultaneous movement outward and along the cylinder. If the dislocations meet after each has made a half-turn, they annihilate locally and a loop is pinched off, which can glide outward, on the cylinder under the influence of the inhomogeneous shear stress. If on meeting the 2 segments S_1 and S_2 miss one another, they will go on rotating in opposite sense and generate a double helix. It is also possible that only 1 of the screws rotates and generates a helix. If this

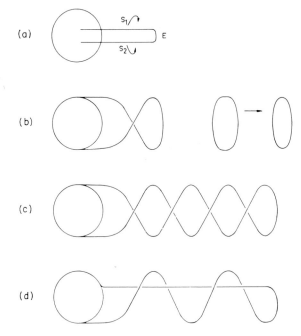

FIG. 52. Illustrating the generation mechanism for helical dislocations and "smokerings" due to prismatic punching. (a) A small loop is nucleated. (b) The 2 screw segments S_1 and S_2 have made one turn around the stress tunnel and pinched off 2 loops. (c) The 2 screws missed each other on traveling around the stress tunnel. (d) Only 1 screw traveled around the stress tunnel.

helix intersects the other screw which was left straight, a sequence of loops can again be pinched off. More variations of the same mechanism are possible; the essential feature is that space is made for the "precipitate" by blowing off a sufficient number of interstitial prismatic loops. The mechanism described here is in fact not very different from one proposed by Seitz.[151] Barber *et al.*[131] made similar observations in KBr and NaCl, decorated by means of gold. In NaCl the prismatic loops also climb and this results in a widening of the outer loop.

(5) *Prismatic loops.* Prismatic loops have been revealed by decoration in sodium chloride and potassium chloride doped with silver.[152] Some examples are visible in Fig. 53a. The loops result from the precipitation, during growth of the crystal, of a layer of vacancy pairs on the cube plane. This would produce a sessile loop with vector $(a/2)$ [100]. In order to remove the high-energy stacking fault of electrostatic origin a shift in the plane over a vector $(a/2)$ [010] takes place and the resulting loop becomes prismatic and has a vector $(a/2)$ [110] (Fig. 54).

Dash[153] has revealed hexagonal prismatic loops on (111) planes in silicon specimens into which gold was diffused. The loops were attributed to the precipitation of silicon interstitials (Fig. 53b). It was verified by more recent electron microscopic observations that the loops contain stacking faults[154]. Therefore they were presumably of the Frank sessile type. The presence of stacking faults in these loops can be understood since the stacking fault energy in silicon is relatively small [see Section 10c (6)]

d. Decoration Etching

The decoration is used to delineate the dislocation by small precipitates. On subsequent etching the dislocation lines which are very nearly parallel and close to the surface will be revealed along the whole or part of their length, either as grooves or as strings of precipitates, if the latter are less soluble than the matrix. The method is

[151] F. Seitz, *Phys. Rev.* **79**, 723 (1950).

[152] S. Amelinckx, *in* "Dislocations and Mechanical Properties of Crystals" (J. C. Fisher *et al.*, eds.), p. 3. Wiley, New York, 1957.

[153] W. Dash, *Phys. Rev. Letters* **1**, 400 (1958).

[154] V. A. Phillips and W. C. Dash, *J. Appl. Phys.* **33**, 568 (1962).

especially useful for revealing glide patterns, since one usually knows *a priori* in which planes to expect them.

The principle was used first by Wilsdorf and Kuhlmann-Wilsdorf.[155] They decorated dislocations in an aluminum copper (4%) alloy; by aging, precipitates of the ϑ phase form along the dislocations.

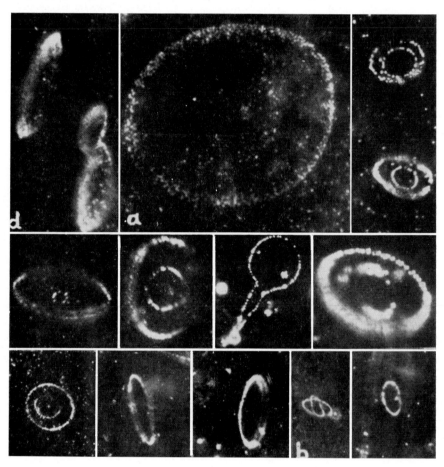

FIG. 53.   (a) Prismatic loops in potassium chloride revealed by decoration with silver; they are presumably of the vacancy type. [After S. Amelinckx, *in* "Dislocations and Mechanical Properties of Crystals" (J. C. Fisher *et al.*, eds.), p. 3. Wiley, New York, 1957.]

[155] H. Wilsdorf and D. Kuhlmann-Wilsdorf, *Rept. Conf. Defects in Crystalline Sdsoil, London, 1954* p. 175 (1955).

Subsequent electropolishing and replication permits parts of concentric dislocation loops to be seen in the electron microscope as quasi-continuous strings of precipitates.

The method has also been used by Gilman.[61] He decorated dislocations in zinc by means of cadmium as an insoluble impurity. He did

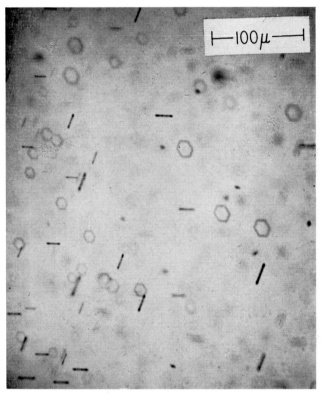

FIG. 53. (b) Prismatic loops in silicon revealed by copper decoration. The loops are presumably due to interstitials caused by diffusing in gold. (Courtesy of W. C. Dash.)

not intentionally slice along specific planes, but occasionally a hexagonal network parallel to the surface was revealed.

Dash[57] examined (111) planes of pure deformed silicon single crystals and found that grooves were produced on etching (111) faces. Although no impurities were added intentionally, the strain around the dislocations was apparently large enough to cause local dissolution. He found evidence for glide loops and for cusps connected with trails. The same features as revealed by infrared microscopy of copper decorated silicon (Section 6c (3) (a)) were found (Fig. 19).

Tyler and Dash[59] used lithium as an impurity to decorate dislocations in germanium deformed by twisting around a [111] axis. Lithium is known to diffuse fairly rapidly in germanium along interstitial sites; it precipitates in substitutional position, behaving like copper in silicon. In germanium the infrared transmission technique is much

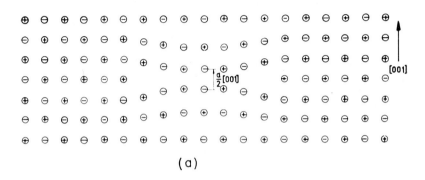

(a)

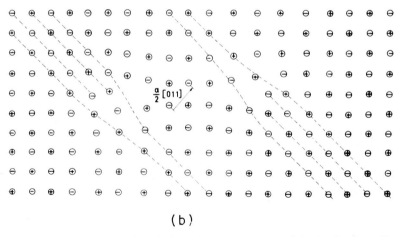

(b)

FIG. 54. Model for a prismatic loop due to vacancy precipitation in the sodium chloride structure. (a) A single layer of vacancies on the cube plane gives rise to a high-energy stacking fault. (b) The stacking fault is eliminated.

more difficult to apply. Frank-Read sources and other features analogous to those found in silicon could be revealed as surface grooves
on polished and etched (111) faces.

In body-centered silicon-iron (31 % Si, 0.005 wt % C) containing a
small quantity of carbon, Low and Guard[58] were able to reveal glide
loops and to obtain information on the geometry of deformation.
The specimens were deformed by bending around an axis normal
to the (112) planes. The crystal was limited by the (112) planes
perpendicular to the bending axis; in this face the edge components
of the dislocations with a Burgers vector $(a/2)$ [111] emerge.

The end faces were parallel to the active (110) glide plane, while
the top faces were perpendicular to (112) and at 45° to the (110)
faces. In the top face screw dislocations emerge. With this well-defined
geometry, emergence points of edges and screws can be observed on
separate faces and moreover the glide pattern usually can be observed
in successive (110) planes by alternating polishing and etching.

Aging the crystals for up to 18 hr at 150°C produces carbon
decoration even on the fresh dislocations produced by deformation.
The aging temperature is sufficiently low so that no appreciable change
in the pattern, due to recovery is to be feared. The specimens are
then etched to produce pits on top and side faces and grooves on the
end faces (Fig. 55). Optical microscopy and 2 stages of carbon
replication, followed by electron microscopy, are the observation
techniques.

The slip traces of the edge dislocations seem to remain very nearly
straight although the linear arrangement of pits occasionally changes
discontinuously over to parallel lines, causing in this way some
broadening, especially in the region of high strain near the surface.
The traces due to screws, on the other hand, appear very wavy,
even curved, especially where meeting traces on a parallel, nearby
plane. The screws can apparently leave the primary glide plane, on
almost any other plane, as a consequence of local stress due to other
dislocations. In the slip plane very elongated loops are found; the
screws move over much shorter distances than the edges; they seem
to leave glide planes rather often in accord with the observations on
the top faces.

From these observations it is concluded that double cross slip
is the mechanism responsible for multiplication of dislocations and
broadening of glide bands. The onset of double cross slip is attributed
to torques locally exerted by stationary dislocations on the moving ones.

A similar method for studying the distribution of dislocations in slip bands was used by Sestak *et al.*[156] However these authors come to a different interpretation of their results. They propose that instead of the relatively uniform distribution of sources found by Low and Guard[58] the dislocations seem to originate from a small number of sources and travel large distances in the same glide plane.

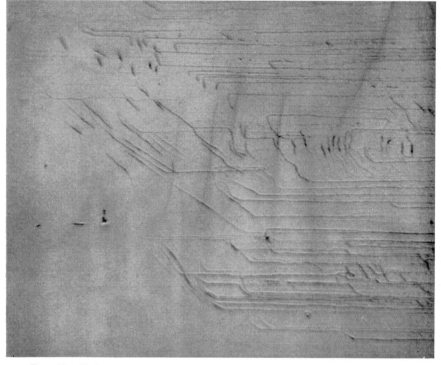

FIG. 55.   Etch grooves in the surface of a deformed silicon-iron crystal observed by electron microscopy. The dislocations were decorated with carbon prior to etching. (Courtesy of Low and Guard.)

7. X-Ray Diffraction Methods

a. Introduction

X-ray diffraction has been used for many years to study imperfections in crystals. Several methods have been developed; all of

[156] B. Sestak, F. Kroupa, and S. Libivicky, *Czech. J. Phys.* in press.

them are based on the principle that small angular differences or eventually small differences in lattice parameter between neighboring crystal blocks are revealed as a fine structure of the diffraction spots. We shall not review these methods since they do not produce a direct image of the dislocation structure. However, since about four years ago a number of methods became available which do reveal images of individual dislocations. They are due to Lang,[157-160] Newkirk[161-163], Borrmann,[164-166] Bonse et al.,[167-169] and others.[170-174a] We will describe the principle of the different methods.

b. Reflection Method

Newkirk used a refined Berg-Barett[163] technique. The crystal is set under the Bragg angle with respect to a beam of parallel filtered X-rays and the diffracted beam is recorded on a photographic plate which is mounted very near to the specimen in order to achieve undeformed mapping. Figure 56a is a schematic view of the experimental arrangements. The photograph obtained in this way represents an unmagnified view, with eventually some distortion in one direction due to the nonparalellism of plate and specimen. The picture has to be studied under a low-power microscope and the use of a high-resolution emulsion is therefore required.

[157] A. R. Lang, *Acta Cryst.* **12**, 249 (1959).

[158] A. R. Lang, *J. Appl. Phys.* **30**, 1748 (1959).

[159] A. R. Lang and G. Meyrick, *Phil. Mag.* [8] **4**, 878 (1959).

[160] A. E. Jenkinson and A. R. Lang, "Direct Observation of Imperfections in Crystals" (J. B. Newkirk and J. H. Wernick, eds.), p. 471. Wiley (Interscience), New York, 1962.

[161] J. Newkirk, *Trans. A.I.M.E.* **215** 483 (1959).

[162] J. Newkirk, *J. Appl. Phys.* **29**, 995 (1958).

[163] J. Newkirk, *Phys. Rev.* **110**, 1465 (1958).

[164] G. Borrmann, *Physik. Bl.* **15**, 508 (1959).

[165] G. Borrmann, W. Hartwig, and H. Irmler, *Z. Naturforsch.* **13a**, 423 (1958).

[166] G. Borrmann and G. Hildebrandt, *Z. Naturforsch.* **11a**, 585 (1956).

[167] U. Bonse, *Z. Physik* **153**, 278 (1958).

[168] U. Bonse and E. Kappler, *Z. Naturforsch.* **13a**, 348 (1958).

[169] U. Bonse, E. Kappler, and F. J. Simon *Z. Naturforsch.* **14a**, 1079 (1959).

[170] H. Barth and R. Hosemann, *Z. Naturforsch.* **13a**, 792 (1958).

[171] A. Authier, *J. Phys. Radium* **21**, 655 (1960).

[172] A. Authier, Thesis, University of Paris, 1961.

[173] K. A. Carlson and R. Wegener, *J. Appl. Phys.* **32**, 125 (1961).

[174a] V. Gerold and F. Meier, *in* "Direct Observation of Imperfections in Crystals" (J. B. Newkirk and J. H. Wernick, eds.), p. 509. Wiley (Interscience), New York, 1962.

Since the characteristic X-radiation generally used in these experiments does not penetrate very deeply into the specimen, only a relatively thin surface region is explored. The penetration depth evidently depends on the atomic weight of the specimens; it is about 50 microns in lithium fluoride, which is especially favorable, but for elements heavier than aluminium it is reduced to less than 1 micron.

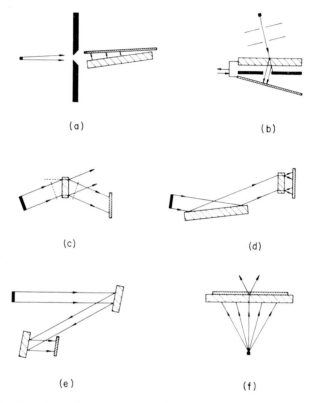

(a)

(b)

(c)

(d)

(e)

(f)

FIG. 56. Experimental arrangements used in different X-ray methods for revealing dislocations. (a) Newkirk's technique [J. Newkirk, *Trans. A.I.M.E.* **215**, 483 (1959)] (improved Berg-Barett technique). (b) Lang's technique [A. R. Lang, *Acta Cryst.* **12**, 249 (1959)]. (c) Barth and Hosemann's technique [H. Barth and R. Hosemann, *Z. Naturforsch.* **13a**, 792 (1958)]. (d) Authier's technique [A. Authier, *J. Phys. Radium* **21**, 655 (1960); Thesis, University of Paris, 1961]. (e) Double crystal goniometer [U. Bonse, *Z. Physik* **153**, 278 (1958); U. Bonse and E. Kappler, *Z. Naturforsch.* **13a**, 348 (1958)]. (f) Borrmann's technique [G. Borrmann, W. Hartwig, and H. Irmler, *Z. Naturforsch.* **13a**, 423 (1958); G. Borrmann and G. Hildebrandt, *Z. Naturforsch.* **11a**, 585 (1956)].

The contrast arises as a consequence of enhanced diffraction from the region around the dislocations which therefore appear as dark lines (the effect is largest for strong low-order reflection). This is in fact due to a decrease in primary extinction in the vicinity of the dislocation. The contrast formation can be understood intuitively as follows. For a perfect crystal the amplitude of the Bragg reflected beam results from interference between the simply diffracted and multiply diffracted beams. The twice reflected beam, for instance, has the same direction as the incident beam but since on each reflection a phase change of $\pi/2$ occurs, it will be in opposite phase with the incident beam and destructive interference will occur. This phenomenon causes a continuous interplay of waves in the diffracted and incident beams, resulting in a rapid attenuation of the diffracted beam, superimposed on the normal absorption.

If an imperfection is present which destroys the strict periodicity, the once diffracted beam is no longer incident under the exact Bragg angle in the vicinity of the imperfection and it emerges with a larger amplitude since it is only partly diffracted again into the crystal. This mechanism leads therefore to enhancement of the diffracted intensity for regions in the vicinity of the dislocation.

In lithium fluoride, Newkirk was able to reveal dislocations in sub-boundaries and in glide bands (Fig. 57). Comparison with etch patterns for the same crystal showed a one to one correspondence between etch pits and dislocation images. Also in silicon which is a specially suitable material for this type of experiments, similar results

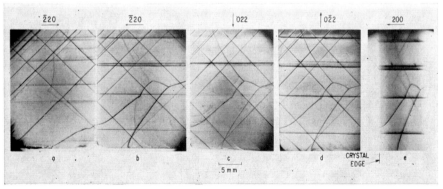

FIG. 57. Sub-boundaries and slip bands in lithium fluoride photographed by the improved Berg-Barett technique. The diffraction vectors are indicated for each picture. All pictures refer to the same crystal face. (Courtesy of Newkirk.)

were obtained. Using different reflections it is possible to determine the direction of Burgers vectors. The criterion for the absence of contrast is to a first approximation the same as in electron diffraction $\bar{n} \cdot \bar{b} = 0$ ($\bar{n}$: diffraction vector) [see Section 9, d, (2)].

c. Transmission Method

Greatest detail is presumably obtained with the technique developed by Lang.[157] A well-collimated narrow beam of penetrating characteristic radiation is sent through a relatively thin, nonabsorbing crystal, set under the Bragg angle for a set of lattice planes approximately perpendicular to the crystal slab. The directly transmitted beam is intercepted by a shield, while the diffracted beam is allowed to pass through a slit in the shield. A photographic plate mounted normal to the diffracted beam records the distribution of intensity in the diffracted beam. Figure 58 shows an example of the type of images that is obtained.

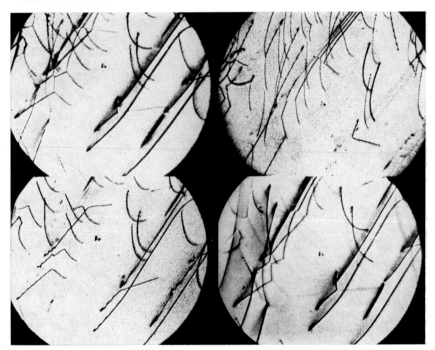

FIG. 58.   Diffraction topographs of silicon single crystal; the 4 photographs represent the same area but seen with different diffraction vectors operating. Notice the absence of contrast for certain dislocation segments. (Courtesy of A. R. Lang.)

Only a relatively narrow region is explored for a given setting. In order to scan a large region the specimen and the plate are moved, simultaneously, as shown by the arrows in Fig. 56b.

Many of the remarks made above are valid to the same extent here. However, it is clear that this method reveals the dislocations throughout the whole specimen.

Contrast arises again as a consequence of enhanced Bragg scattering around the imperfection due to a local decrease in primary extinction; the dislocation lines usually show up as dark lines on the negative. Occasionally if $\mu t \sim 10$[174b] dislocations may show up in inverted contrast due to the anomalous transmission discussed below. In order to obtain high resolution the horizontal divergence[174c] has to be kept small by means of slits and the plate has to be as close as possible to the crystal. Reconstruction in 3 dimensions of the dislocation pattern can be performed by using a stereo technique. Diffraction topographs are made using reflections from both sides of the same set of lattice planes, i.e., using hkl and $\bar{h}\bar{k}\bar{l}$ reflections.

A variant of Lang's method developed by Carlson and Wegener[173] allows a larger area to be mapped without the necessity for simultaneous movement of specimen and film. More efficient use is made of the line focus of an ordinary X-ray tube. The horizontal divergency of the wide beam is limited by a Soler slit and ordinary slits. Drawbacks of the method, however, are that the images are double as a consequence of the doublet structure of the $K\alpha$ line. In order to avoid the direct beam the distance crystal-plate has to be relatively large, resulting in some loss of definition.

A similar method has been used by Amelinckx et al.[70] in their study of small crystals of silicon carbide and by Webb[175] for mapping dislocations in whiskers. For instance, Fig. 59 shows dislocations in a sodium chloride whisker.[175]

d. Double Crystal Goniometer

A particularly accurate and sensitive method has been developed by Bonse and Kappler.[168] A nearly perfect germanium crystal is set

---

[174b] μ: absorption coefficient; t: thickness.

[174c] The horizontal plane contains the incident beam and the normal to the reflecting planes.

[175] W. W. Webb, in "Growth and Perfection of Crystals" (R. H. Doremus et al., eds.), p. 230. Wiley, New York, 1958.

under the Bragg angle in order to produce a highly parallel and strictly monochromatic X-ray beam as shown in Fig. 56e. A second crystal is set for the same Bragg reflection. A very slight misorientation or a very slight change in lattice parameter in some regions of the second crystal brings these regions in or out of the reflecting position: Misorientations of the order of 1 sec of arc can be revealed. The inten-

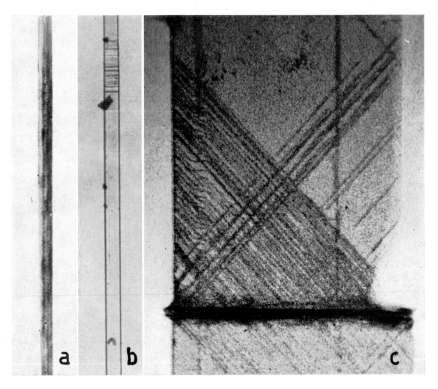

Fig. 59.   Dislocations in sodium chloride whiskers, revealed by X-ray diffraction. (a) and (b) The same whisker observed optically and in X-ray diffraction contrast. (Courtesy of W. W. Webb.) Axial dislocations are visible. (c) Sodium chloride platelet observed in diffraction contrast. Notice the glide traces. (Courtesy of W. W. Webb.)

sity distribution reflected from the second crystal is again recorded photographically on a plate as close and as parallel as possible to the specimen. In this way it is possible to record the strain field around the dislocation in a quantitative manner (Fig. 60). In germanium some contrast is still observed up to distances as large as 15 microns from the core of the dislocation. As in the reflection method discussed in Section 7b only dislocations near the surface can be revealed.

The quantitative interpretation of the image is based on the rocking-

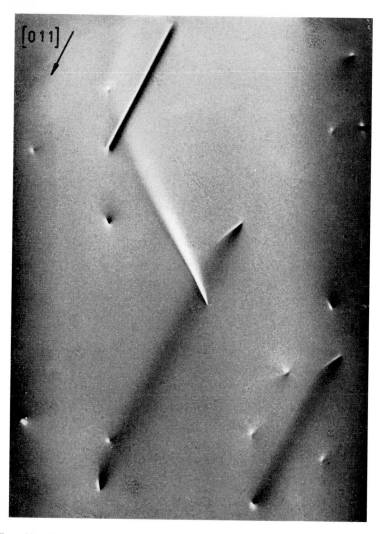

FIG. 60. Photographic recording of strain field around dislocations in germanium, as observed by means of the double crystal goniometer. The surface of the specimen is a (111) plane; CuKα radiation was used. Notice dislocations seen end-on; other dislocations are parallel to the surface but at different depths. (Courtesy of Bonse.)

curve. In the particular case of the 2 germanium crystals used by the authors, the width at one-half maximum of the rocking curve is between 5 and 10 sec. of arc and the smallest detectable intensity difference (10 %) corresponds to an orientation difference of about 1 sec. The slope $\partial I/\partial \delta$ of the rocking curve determines the sensitivity to orientation differences. The corresponding sensitivity to lattice parameter changes, under the assumption that the crystal lattices are perfectly parallel, is deduced by differentiating the Bragg condition

$$\Delta d/d = \operatorname{cotg} \vartheta \Delta \vartheta$$

and hence

$$\partial I/\partial \left(\frac{\Delta d}{d}\right) \simeq \operatorname{tg}\vartheta \partial I/\partial(\Delta \vartheta). \tag{7.1}$$

For the reflections used [(711) and (551)] tg $\vartheta \simeq 4$ and the smallest detectable relative parameter change is 8×10^{-7}. It is clear that around a dislocation both parameter and orientation changes occur. It is therefore convenient to introduce a quantity

$$\delta = \frac{\Delta d}{d} + \operatorname{cotg} \vartheta \Delta \vartheta \tag{7.2}$$

which describes the total "deformation" and which is "translated" by the X-rays into a change in reflected intensity. Using the displacement function for dislocations [see Section 9e (4) (a)] it is possible to calculate δ as a function of the space coordinates and of the operating Bragg reflections. Lines of equal diffracted intensity can then be drawn. Of greatest interest is the locus $\delta = \delta_{min}$ where δ_{min} is the smallest deformation detectable; for $d = 4 \times 10^{-8}$ cm δ_{min} is about 8×10^{-7}. Some loci $\delta = \delta_{min}$ lying in the plane perpendicular to the dislocation are shown in Fig. 61 for different reflections. The essential features of the observed images agree well with the theoretical picture.

e. Anomalous Transmission

The method requiring the minimum of special equipment is due to Borrmann et al.[165] Use is made of widely divergent (30°) beam X-rays, emerging preferentially from a point focus, for instance, from a micro focus tube. The crystal is chosen sufficiently thick so that the normal absorption is large; a silicon crystal of about 1 mm thick is adequate, μt is then about 15 (μ = normal absorption coefficient, t = thickness).

The crystal is placed at a small distance ($\pm$ 3 cm) from the focal spot of the X-ray tube and the photographic plate is mounted in contact with the crystal. A schematic view of the set up is shown in

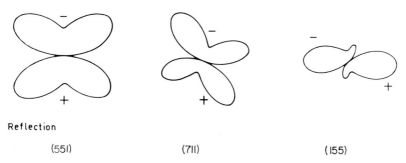

Reflection

(551)   (711)   (155)

Fig. 61. Curves of minimum observable intensity as deduced from the displacement function of the dislocation. These curves are to be compared with the images of the dislocations seen end-on in Fig. 60. [After U. Bonse, *Z. Physik* **153**, 278 (1958).]

Fig. 56f. Due to the divergence of the beam the Bragg condition is exactly fulfilled for a number of sets of planes. Along these planes there is a so-called "anomalous" transmission of the stationary waves. The wave field consists in fact of 2 waves. For one of them the nodes correspond with the nuclear position, while for the other the nodes are exactly in between nuclear positions. The second wave suffers a much larger absorption than the first. The situation is similar to that encountered in the anomalous absorption of electrons, to be discussed below.

The stationary crystal waves split into 2 beams on emerging from the exit face, the diffracted and the transmitted beam, as shown schematically in Fig. 56f. The slightest deviations from the exact periodicity destroy the condition for anomalous transmission. This was demonstrated convincingly by Borrmann and Hildebrandt.[166] A temperature difference of 0.6°C between the 2 end faces of a calcite crystal is sufficient to reduce the anomalously transmitted intensity by 2 orders of magnitude.

The presence of a dislocation is sufficient to destroy locally the transmission condition; as a consequence, dislocations will "cast shadows," i.e., white lines on the negative. An example is shown in Fig. 62a. Since the plate is in contact with the specimen, the images of diffracted and transmitted beam are not appreciably separated, and sharp images can be obtained.

A combination of the Borrmann and Lang technique has been used by Barth and Hosemann[170] and by Authier.[171,172] Authier uses a monochromatic and parallel beam of X-rays, obtained by reflection against a plane monochromator (a cleaved rhomb of calcite) as the incident beam instead of the divergent beam (Fig. 56d). The crystal to be investigated is then set under the Bragg angle; the plate is at some distance behind the crystal. Under these circumstances the images of the dislocations consist of a double black-white streak; they are observed in both the diffracted and transmitted beam, the contrast being complementary. Silicon crystals up to 4 mm thick were examined in this way.

Barth and Hosemann[170] on the other hand use no monochromator (Fig. 56c), and they record the diffracted image only.

Gerold and Meier[174a] have used the same technique to study the dislocation distribution in a (111) slab of germanium. They demonstrated the use of the anomalous transmission technique for Burgers vector determination. Making exposures along 2 different lattice planes, they obtain a spatial representation of the dislocations.

f. Discussion of the Method

A severe limitation of the method is at present the resolution, which is about 5 microns for Lang's method and about 2 microns for the reflection methods. The resolution is determined by the width of the dislocation images on the one hand, and by the resolving power of the photographic emulsion on the other hand. At small dislocation densities the images may be as wide as 10 microns. Fortunately as the dislocation density increases, the strain fields of the dislocations extend less far and the images become correspondingly sharper. The practical resolution limit seems to be 5 microns. As a consequence only relatively perfect crystals can be examined. The fact that continuous observation is not yet possible eliminates the possibility of studying dynamic phenomena. The development of a high-resolution electronic observation method would improve the situation. The main specific advantages of the method over the others (electron diffraction) is the possibility of examining relatively thick specimens of nearly perfect crystals non destructively.

(a)

(b)

FIG. 62. (a) Dislocation "shadows" as observed by anomalous transmission through a thick silicon crystal (20x). (Courtesy of Borrmann.) (b) Dislocations and stacking faults in silicon revealed by the Lang method. (Courtesy of Kohra.)

g. Examples of Applications

Papers published so far have been concerned mainly with demonstrating the possibilities of the technique. Most of the information obtained so far was already available from other observations. We summarize the results here.

(1) *Dislocations in whiskers* [see also Section 6c (3)]. Since the lattice twist measurement, discussed in Section 7h only gives the twist due to the excess screw components of 1 sign, the complete dislocation content of the whiskers is not known. Should the whisker contain, for instance, 2 screws of opposite sign, no lattice twist should be detected. Webb used a modified Lang method to map dislocations in whiskers and small platelets. The specimen is placed at large distance ($\sim$60 cm) from the focal spot of the X-ray tube and the plate is placed very close to the specimen ($\sim$0.1 to 0.5 cm). The results for NaCl agree well with these obtained by decoration methods.[175,176]

(2) *Stacking fault loops in zinc.* Fourdeux et al.[177] observed stacking fault loops in zinc, both by electron microscopy and X-ray diffraction. The X-ray diffraction work was performed on small zinc platelets grown by sublimation.

With ($1\bar{1}00$) type reflections contrast is observed within circular areas; the contrast is absent for ($1\bar{2}10$)-type reflections. It was concluded that this is due to the presence of stacking faults within the loop. They also conclude that the direction of the Burgers vector should be $\langle 0001 \rangle$. This conclusion is possibly not justified as can be inferred from later work. Much smaller loops, resulting from vacancy condensation, have been observed by electron microscopy and the Burgers vector for these loops is inclined with respect to the basal plane [Section 10d, (5)]. Of course it is possible that both kinds of loops have a different Burgers vector.

(3) *Dislocations in silicon carbide.* Dislocations in thin silicon carbide platelets were studied by a combination of etching and X-ray diffraction by Amelinckx et al.[70] The two methods provide complementary information: etch pits reveal dislocations emerging in the basal plane, whereas the X-ray method reveals mainly dislocations in the basal plane. When dislocations are revealed by both methods,

---

[176] W. W. Webb, R. D. Dragsdorf, and W. D. Forgeng, *Phys. Rev.* **108**, 498 (1957).

[177] A. Fourdeux, A. Berghezan, and W. W. Webb, *J. Appl. Phys.* **31**, 918 (1960).

[178] F. C. Frank and A. R. Lang, *Phil. Mag.* [8] **4**, 383 (1959).

corresponding results are obtained. The Burgers vectors of glide dislocations were found to be [11$\bar{2}$0].

(4) *Dislocations in diamond.* A diffraction topograph of a diamond crystal was obtained by Frank and Lang.[178] The crystals appears to contain mainly dislocations radiating out from a center of strain somewhere in the middle of the crystal.

(5) *Substructure in lithium fluoride.* Yoshimatsu and Kohra[179] have made a detailed study using the Newkirk technique, of the substructure of lithium fluoride and its changes after annealing. In particular Yoshimatsu[180a] has studied the polygonization of bent lithium fluoride single crystals; by examining the surface perpendicular to the bending axis it is possible to follow the growth of domains on heat treatment.

(6) *Dislocations in silicon.* Since silicon is a choice material for transmission studies, a large amount of work has been done on this substance.

The most extensive study is probably due to Jenkinson and Lang[180b] (Fig. 58) who studied the dislocation distribution in floating zone grown silicon. From stereo pictures the spatial configuration of dislocations could be determined. From this it was concluded that dislocations are mainly generated by plastic flow, either from internal or from surface sources. The average dislocation density is found to lie between 3×10^3 and 3×10^4/cm². Burgers vectors were determined by the use of different diffraction vectors, and dislocation reactions were studied. Local high densities of dislocations (10^5) appear to have nucleated at Lomer-Cottrell barriers.

Schwuttke[180c] made a study of oxygen containing crystals. It turns out that the segregation of oxygen in "bands" can also be revealed by the diffraction contrast.

Using the Lang technique, Kohra and Yoshimatsu[180d] have in-

[179] M. Yoshimatsu and K. Kohra, *J. Phys. Soc. Japan* 15, 1760 (1960).

[180a] M. Yoshimatsu, *J. Phys. Soc. Japan* 16, 2246 (1961).

[180b] A. E. Jenkinson and A. R. Lang, *in* "Direct Observations of Imperfections in Crystals" (J. B. Newkirk and J. H. Wernick, eds.), p. 471. Wiley (Interscience), New York, 1962.

[180c] G. H. Schwuttke, *in* "Direct Observations of Imperfections in Crystals" (J. B. Newkirk and J. H. Wernick, eds.), p. 497. Wiley (Interscience), New York, 1962.

[180d] K. Kohra, M. Yoshimatsu, and I. Shimuzu, *in* "Direct Observations of Imperfections in Crystals" (J. B. Newkirk and J. H. Wernick, eds.), p. 461. Wiley (Interscience), New York, 1962; K. Kohra and M. Yoshimatsu, *J. Phys. Soc. Japan* 17, 1041, (1962).

vestigated stacking faults in pulled silicon crystals. They have shown by contrast experiments, using different reflections, that the extended regions producing fringe contrast are due to stacking faults rather than to twin boundaries.

However, it was not possible to distinguish between faults associated with Frank sessiles and faults connected with Shockley partials. An example of images due to stacking faults in silicon is shown in Fig. 62b.

The authors also studied the effect of heat treatment on the dislocation configuration; the number of dislocations is found to decrease with increasing heat treatment.

h. Lattice Twist in Whiskers

Eshelby[181] has shown that a whisker with cross-sectional area A containing a screw dislocation with Burgers vector b along its axis should exhibit a lattice twist of α radians per unit length given by the relation

$$\alpha = kb/A$$

where k is a numerical factor depending on the shape of the cross section (for a circle $k = 1$). This results from the image forces due to the free surface. The screw dislocation turns out to be in stable equilibrium along the axis of the rod. For very thin whiskers the lattice twist may become quite appreciable (several degrees per millimeter) and the phenomenon offers therefore the attractive possibility of determining the magnitude of b either by X-ray diffraction[182,183] or by purely optical means.[184] Webb[182] showed that lattice twist can easily be detected by means of the Laue method. For whiskers with well-developed flat side faces, the lattice twist can be detected by means of light reflection. This was done, for instance, for silicon carbide by Hamilton[184] and for sodium chloride whiskers by Jeszenszky and Hartmann.[185] Values for b equal to several elementary vectors were found in some cases, suggesting that several parallel dislocations are present parallel to the axis. This is in agreement with results obtained by decoration methods.[141,143]

[181] J. Eshelby, *J. Appl. Phys.* **24**, 176 (1953).
[182] R. D. Dragsdorf and W. W. Webb, *J. Appl. Phys.* **29**, 817 (1958).
[183] R. G. Trueting, *Acta Met.* **5**, 173 (1957).
[184] D. R. Hamilton, Westinghouse Res. Lab. Report 406-FF-377-P2 (1959).
[185] B. Jeszenszky and E. Hartmann, *Nature* **189**, 213 (1961).

8. BIREFRINGENCE

This method is based on the fact that an isotropic crystal becomes birefringent when stressed. The birefringence can be observed in the polarizing microscope between crossed nicols, eventually supplemented with an additional "sensitive tint" plate. For instance, it is well-known that deformed alkali halide crystals, e.g., NaCl, can be made strongly birefringent by deformation. However, the resolution of the method is poor and the observation of individual dislocations has been difficult because the concentration is too high in most crystals.

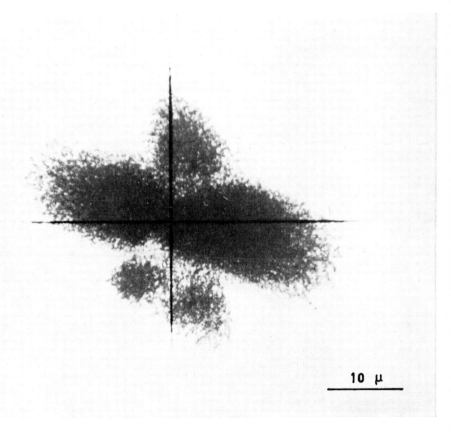

FIG. 63.   Stress field around a dislocation in silicon as observed in infrared radiation between crossed polarizers. (Courtesy of W. L. Bond)

Kear and Pratt[186] succeeded in correlating the birefringence bands in lithium fluoride slighly deformed by quenching with the corresponding etch pattern. Also they were able to identify the compression and extension side of the edge dislocations, which show up in different colors. This was done by applying a small compressive external stress and observing the resulting color changes; a decrease in birefringence in a given region then points to dilatation and vice versa.

The most striking observations in this field were made by Bond and Andrus.[187] Exploiting the small dislocation density of carefully prepared silicon single crystals on the one hand, and its large strain-optic coefficients on the other hand, they were able to reveal the stress field associated with a single edge dislocation seen end-on. Figure 63 reproduces one of their observations which can be compared (after reflection with respect to the slip plane) with Fig. 64 representing the calculated contours.

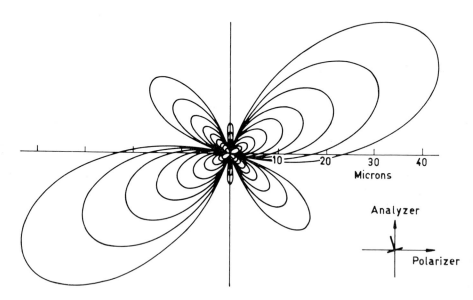

Fig. 64. Contours of equal transmitted intensity for an edge dislocation. This image has to be compared with Fig. 63. [After W. L. Bond and J. Andrus, *Phys. Rev.* **101**, 1211 (1956).]

[186] B. H. Kear, A. Taylor, and P. L. Pratt, *Phil. Mag.* [8] **4**, 665 (1959).
[187] W. L. Bond and J. Andrus, *Phys. Rev.* **101**, 1211 (1956).

Bullough[188] has discussed these results in detail and points out some discrepancies which may mean that the edge dislocation seen in Fig. 63 was some kind of macrodislocation. The same author also

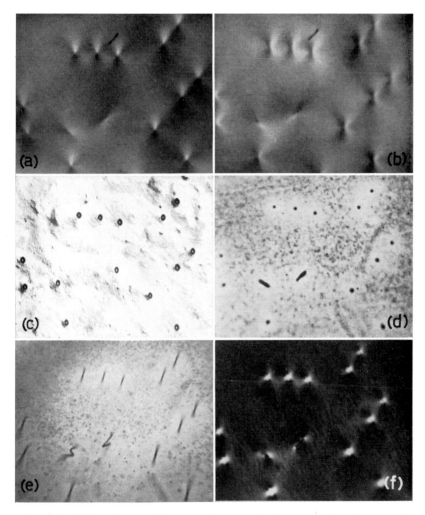

Fig. 65. Dislocations in silicon seen by stress birefringence and other methods. All pictures refer to the same area. (a) Stress pattern around dislocations ($\alpha = 0°$). (b) Stress pattern around dislocations ($\alpha = 45°$). (c) Etch pits. (d) Decorated dislocations seen end-on. (e) Decorated dislocation seen at small inclination. (f) Stress pattern around decorated dislocations. (Courtesy of Indenbohm.)

calculated the birefringence pattern to be expected for a tilt boundary; no observations are available so far.

Observations of the same kind have been performed by Indenbohm and Chernycheva[189] on Rochelle salt. The authors find that to account for their results the extra half-plane should be about 100 A wide. Evidently this means that the defect which they observe is more macroscopic than a single dislocation.

Recently Indenbohm et al.[190] have extended Bond's observations on Silicon. They demonstrated the one to one correspondence of etch figures and dislocation images as seen between crossed polarizers. The curves of equal transmitted light intensity around an edge dislocation seen end-on in an isotropic medium are in polar coordinates given by

$$r = C \cos \vartheta \cos 2(\vartheta - \alpha)$$

where the origin of the angle ϑ is along the glide plane, α is the angle between the glide plane and the plane of the polarizer, C is a constant proportional to the edge component of the Burgers vector and it involves elastic and photoelastic constants. The neutral line corresponds to $\vartheta = (\pi/2)$, i.e., it is perpendicular to the glide plane. The curves of equal intensity have a "rosette" shape. The pattern is especially simple if $\alpha = 45°$; Fig. 65b was made under these conditions. The 4 "petals" of the rosette are clearly visible; 2 are light and 2 are dark.

The same dislocations were then decorated with copper following the Dash procedure. Figure 65c shows the result; it is apparent that a region depleted in copper is formed around the dislocations. This is concluded from the increased transmission in the vicinity of the dislocations. When now viewed between crossed polarizers it appears that the stress pattern has considerably changed. The cross of the isoclines has now been rotated over 45° with respect to its previous position; it is parallel to the plane of polarization of analyzer and polarizer; furthermore the 4 "petals" of the rosette appear light. The pattern suggests that stress has been relieved.

[188] R. Bullough, Phys. Rev. 110, 620 (1958).

[189] V. L. Indenbohm and M. A. Chernycheva, Dokl. Akad. Nauk S.S.S.R. 111, 596 (1956); V. L. Indenbohm and G. E. Tomilovskii, Kristallografiya 2, 190 (1957).

[190] V. L. Indenbohm, Fis. Tverd. Tela 4, 231 (1962); Dokl. Akad. Nauk S.S.S.R. 141, 1360 (1961).

III. Thin Film Methods

9. ELECTRON MICROSCOPY IN TRANSMISSION: THEORY

a. Introduction

We now discuss the most recent and probably the most versatile method for studying dislocations: transmission electron microscopy. The method was developed mainly at the Cavendish Laboratory in Cambridge by Hirsch, Whelan, Howie, and co-workers[191-193] and independently by Bollmann[194] The basis for the approach had been developed before by Heidenreich[195] and Castaing.[196]

The contrast observed in thin crystalline specimens is diffraction contrast and not absorption contrast as is usually observed in replica work.

We will summarize the theory necessary to understand the origin of the contrast so as to allow its use in obtaining the maximum of information from it. We proceed in three steps; first an intuitive picture will be given, which is sufficient for many purposes. In the second place the kinematical approach will be discussed, and finally the dynamical theory will be summarized.

b. Discussion of the Method

(1) *Observation technique.* The observation technique is in fact very simple. With many microscopes it is possible to view the specimen first on the intermediate screen at low magnification and select

[191] P. B. Hirsch, R. W. Horne, and M. J. Whelan, *Phil. Mag.* [8] 1, 677 (1956).

[192] P. B. Hirsch, A. Howie, and M. J. Whelan, *Phil. Trans. Roy. Soc.* **A252**, 499 (1960).

[193] P. B. Hirsch, A. Howie, and M. J Whelan, *Proc. 4th Intern. Conf. on Electron Microscopy, Berlin, 1958* 1, 527 (1960).

[194] W. Bollmann, *Phys. Rev.* **103**, 1588 (1956).

[195] R. D. Heidenreich, *J. Appl. Phys.* **20**, 993 (1949).

[196] R. Castaing, *Proc. 3rd Intern. Conf. on Electron Microscopy, London, 1954* p. 379 (1956).

suitable transparent areas. The diffraction pattern is then made and then an object aperture is put around the selected beam, usually the direct beam. The image is then magnified at the desired level.

It is necessary to change the tilt angle continuously in order to obtain optimum contrast and to make sure that all details are revealed. Selected area diffraction patterns are usually taken with each image of interest since they allow a determination of the crystallographic orientation of the sample, and of the diffraction conditions. The most intense spots are responsible for the contrast. Dark field images are often useful in determining Burgers vectors. They are obtained by selecting a diffracted beam by means of the object aperture. Whereas in the direct beam the contrast is "subtractive," it is "additive" in the dark field image. Especially in thin samples "not much intensity" is diffracted away and the dark field image may have better contrast (Fig. 66).

Dark field images are due to nonaxial beams and therefore may be

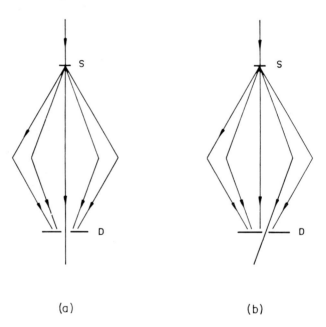

(a)   (b)

FIG. 66. Schematic representation of the beam path in the electron microscope; S is the specimen: D is the selector diaphragm. (a) The direct beam is selected and a bright field image is obtained. (b) A diffracted beam is selected and a dark field image is obtained.

slightly deformed; this can be avoided by tilting the electron gun slightly. In dark field the contrast producing beam is usually known unambiguously; this is not always the case in bright field where several beams may contribute.

In specimens which are poor heat conductors, care must be taken to use optimum illumination conditions, otherwise the specimen will heat up considerably. One can even melt uranium oxide specimens in the microscope (melting point $\sim$2700°C). Under the proper conditions a metal specimen will heat up only 10–20°C.

A number of attachments are available for most microscopes: cooling stages, heating stages, straining devices, double tilting, automatic tilting,[197] etc. Most commercial microscopes work under a maximum accelerating voltage of 100 kv. Recently an electron microscope working under a maximum voltage of 1.5 Mv has been constructed.[198] It permits observation of foils of greater thickness; for instance, aluminum foils of up to several microns can be studied. A 300-kv microscope was constructed by a Japanese group.[198a]

(2) *Thinning techniques.*   The main practical problem that arises in the application of this technique is the preparation of suitable specimens ($\pm$1000 − 2000 A thick). An excellent review of thinning techniques applicable to metals has been presented by Kelly and Nutting.[199] We will only summarize some of the essential features here.

(a) Electropolishing.[200] Metals are usually thinned in successive stages by rolling, eventually followed or alternated with annealing treatments, and finally by electropolishing. A typical specimen used for electrothinning is about 1 cm wide, 5 cm long, and 1/10 mm thick. The specimen is usually coated partly with a protecting varnish,

[197] A. Berghezan and A. Fourdeux, *Proc. 4th Intern. Conf. on Electron Microscopy, Berlin, 1958* p. 567 (1960); M. J. Whelan, *ibid.* p. 96; M. Watanabe, I. Okazaki, G. Honjo, K. Mihama, *ibid.* p. 90; I. Okazaki, M. Watanabe, and K. Mihama, *ibid.* p. 93; W. Loebe, O. Schott, and F. Wilke, *Proc. European Reg. Conf. on Electron Microscopy, Delft, 1960*, Vol. 1, p. 134 (1961); P. J. Jackson and J. W. Matthews, *ibid.* p. 137; A. Berghezan and A. Fourdeux, *Compt. Rend.* **247**, 1194 (1958).

[198] G. Dupouy, F. Perrier, and R. Fabre, *Compt. Rend.* **252**, 627 (1961); G. Dupouy and F. Perrier, *ibid.* **253**, 2435 (1961); G. Dupouy, F. Perrier, and C. Durrieur, *ibid.* **251**, 2836 (1960).

[198a] K. Kobayashi, H. Hashimoto, E. Suito, S. Shimadzu, and M. Iwanaga, *Japan. J. Appl. Phys.* **2**, 47 (1963).

[199] P. M. Kelly and J. Nutting, *J. Inst. Metals* **87**, 385 (1958-1959).

[200] H. M. Tomlinson, *Phil. Mag.* [8] **3**, 867 (1958).

leaving a window as shown in Fig. 67. It is mounted as the anode in an electrolytic cell with a suitable electrolyte. The cathode, as used by Bollmann, consists of 2 stainless steel points, 1 on each side of the anode specimen, and at a variable distance from it. The polishing conditions are determined by trial and error variation of: (1) the voltage across the cell; (2) the position of the electrodes; (3) the temperature of the bath; etc.

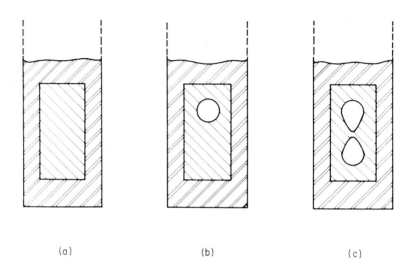

(a)   (b)   (c)

Fig. 67. Preparation of samples by the "window" method. (a) The edge of the specimen is varnished. (b) A hole is polished. (c) A second hole is electropolished; specimens are taken from the bridge.

Usually the procedure is as follows. The pointed cathodes are first put close to the specimen and a hole is formed in the foil. A second hole is made not far from the first. The points are then mounted somewhat further and the bridge between the two holes is polished untill it starts disappearing. Transparent specimens can then be cut from the thin part around the edges of the bridge.

The preparation of thin foils from bulk specimens requires more elaborate techniques, if one wants to avoid changing the structure. The bulk material is first sliced by acid sawing or by spark cutting, and then further thinned by electropolishing. During this last operation it is sometimes useful to move the specimen between the electrodes in order to have uniform removal over the whole area. The final thinning can be done in the way described above.

In special circumstances it may be required to preserve one face of a specimen unattacked, e.g., if one wishes to compare the internal arrangement of dislocations with slip lines. This can be done by depositing a lacquer on one side.[201]

(b) Cleavage. Simple cleavage is useful for a number of layer structures: graphite, mica, bismuth telluride, indium selenide.[202] In order to avoid heavy deformation the following procedure is convenient. A relatively thick undeformed specimen is backed on one side by a rather thick layer of plastic (e.g. *Palavit*). From the other side layers are pulled off by means of adhesive tape. The flakes which adhere to the plastic are recovered after dissolving it. In the process of pulling off layers flakes are sometimes formed which stick neither to the adhesive tape, nor to the palavit. These are usually the best specimens.

(c) Crystal growth. A number of substances can be grown directly in platelets thin enough to transmit electrons, especially by sublimation. This is particularly easy for some hexagonal metals, zinc, cadmium, magnesium, etc.,[203-205] and also for lamellar crystals: tin disulfide, chromium chloride, chromium bromide, tin diselenide, zinc sulfide,[206] nickel bromide.[207] The specimens so obtained are usually very perfect, often dislocation-free. In many cases they can be selected according to thickness by looking for interference colors in reflected light. Many oxides can also be obtained by sublimation; in particular, molybdenum and tungsten trioxide[207a] are easily grown.

Solution-grown lead iodide[208] and cadmium iodide[202] are often thin enough. Silver platelets transparent for electrons can be grown by chemical reduction of silver perchlorate.[208a]

[201] P. B. Hirsch, P. Partridge, and R. Segall, *Phil. Mag.* [8] **4**, 721 (1959).

[202] S. Amelinckx and P. Delavignette, *in* "Direct observation of imperfections in Crystals" (J. B. Newkirk and J. H. Wernick, eds.), p. 295. Wiley (Interscience), New York, 1962.

[203] P. B. Price, *Phil. Mag.* [8] **5**, 473 (1960).

[204] P. B. Price, *Phil. Mag.* [8] **5**, 417 (1960).

[205] P. B. Price, *Phil. Mag.* [8] **5**, 873 (1960).

[206] H. Blank, P. Delavignette, and S. Amelinckx, *Phys. Stat. Solidi* **2**, 1660 (1962).

[207] P. B. Price and J. Nadeau, *J. Appl. Phys.* **33**, 1543 (1962).

[207a] V. Marinkovic, private communication (1962).

[208] A. J. Forty, *Phil. Mag.* [8] **5**, 787 (1960).

[208a] T. H. Alden, *Phil. Mag.* [8] **7**, 1435 (1962).

(d) Chemical thinning.   Chemical polishing of cleaved or sawn slices is used for many nonmetallic crystals, for instance, germanium,[209] silicon,[210] magnesium oxide,[145] calcium fluoride,[211] and uranium dioxide.[212]

(e) Evaporated films.   Thin metal foils can be prepared conveniently by evaporation under high vacuum on a substrate that gives rise to epitaxy. In this way thin films of gold and silver have been prepared. Such films appear to contain a high concentration of stacking faults.[213]

(f) Beaten foils.   Some of the very first observations were made on beaten foils of aluminum and gold which were further thinned chemically.[214] Such foils are usually rather deformed, but they are quite suitable for study after an anneal. Annealed foils of beaten platinum were extensively used for radiation damage studies.[215]

(g) Chemical etching.   Chemical etching of metal foils often produces a surface structure which may interfere with the observations. Therefore it is of limited application but it has been used for pure metals like aluminum and gold.[214]

(h) Ion bombardment.   Castaing[216] has developed a method to thin bulk material by ion bombardment. Critical adjustment of the ion energy is required to produce a structureless undamaged surface (3000 v for aluminum). The method is slow; the removal rate is about $10\,\mu$ in 24 hours.

(i) Preparation from the melt.   Takahashi et al.[217a] have developed a method for preparing thin metallic films directly from the melt. Under vacuum or in a protective atmosphere an elongated loop of

[209] G. Geach, B. A. Irving, and R. Philips, *Research (London)* 10, No. 10 (1957).

[210] E. Aerts, P. Delavignette, R. Siems, and S. Amelinckx, *J. Appl. Phys.* 33, 3078 (1962).

[211] E. Schüller, and S. Amelinckx, *Naturwissenschaften* 21, 591 (1960).

[212] H. Blank, Private communication, 1962.

[213] V. A. Phillips, *Phil. Mag.* [8] 5, 571 (1960).

[214] P. B. Hirsch, A. Kelly, and J. W. Menter, *Proc. 3rd Intern. Conf. on Electron Microscopy, London, 1954* p. 231 (1956).

[215] E. Ruedl, P. Delavignette, and S. Amelinckx, *J. Nucl. Mater.*, 6, 46 (1962).

[216] R. D. Castaing, *Rev. Met. (Paris)* 52, 669 (1955); *Proc. 3rd Intern. Conf. on Electron Microscopy, London, 1954* p. 379 (1956).

[217a] N. Takahashi and K. Ashinuma, *J. Inst. Metals* 87, 19 (1958); *Proc. 3rd Intern. Conf. on Electron Microscopy, London, 1954* p. 388 (1956); *J. Electromicroscopy (Tokyo)* 7, 37 (1959); N. Takahashi, *J. Appl. Phys.* 31, 1287 (1960).

wire is immersed in the liquid metal and then withdrawn at a rate of about 2 cm/sec. By capillarity a thin film of metal is formed and solidifies immediately. No further treatment is necessary. The method becomes increasingly difficult to apply for metals with a high melting point.

It appears that the behavior of foils of a few hundred Angstroms is representative for the bulk, at least for the alloys studied so far.

(3) *Specimen mounting.* The specimen is mounted on a grid with large mesh size; often no grid at all is used. Some substances have a tendency to fall off the grid. A useful trick is to deposit some adhesive on the grids by immersing them in a very dilute solution of tape adhesive in amyl acetate; after drying it is sufficient to press the grid very lightly on the crystal flake to make it adhere firmly.

c. Introduction to Diffraction

(1) *Interference conditions.* We start by giving a short summary of the necessary elements of diffraction theory.

The amplitude of the spherical wavelet scattered by an isolated atom can be represented by

$$a(r) = \left(\frac{1}{r}\right) f(\vartheta)\, e^{2\pi i k r} \tag{9.1}$$

where $f(\vartheta)$ is the atomic scattering factor for the radiation used, e.g., X-rays or electrons. It is a function of the scattering angle 2ϑ and of $\lambda = 1/k$. For X-rays, tables of scattering factors are available [217b]; in the case of electrons one has

$$f_e(\vartheta) = \frac{me^2\lambda}{2h^2}\left(\frac{z - f_x(\vartheta)}{\sin^2 \vartheta}\right) \tag{9.2}$$

where $f_x(\vartheta)$ is the atomic scattering factor for X-rays, z is the atomic number, and the other symbols have their usual meaning. Whereas X-rays are only scattered by the electron cloud, electrons are scattered by the Coulomb field of the nucleus as well. The scattering factor for electrons is much larger (almost 10^4 times) than for X-rays, but it falls off more rapidly with increasing scattering angle.[217c]

We now consider an assembly of atoms, as in the crystal structure,

[217b] "International Tables for X-Ray Crystallography" (Intern. Union Cryst.), Vol. 3. Kynoch Press, Birmingham, England, 1962.
[217c] *Ibid.* Vol. 3, p. 216.

based on a perfect lattice with node points at $\bar{r} = n_1\bar{a}_1 + n_2\bar{a}_2 + n_3\bar{a}_3$ where the n_i are integers or zero. Let the incident radiation be described by a wave vector $\bar{k}_0$ and the diffracted wave by $\bar{k}$ (Fig. 68). For elastic scattering $\bar{k}$ and $\bar{k}_0$ only differ in direction. The path difference of the wave scattered by an atom A_n at $\bar{r}_n$ and having a

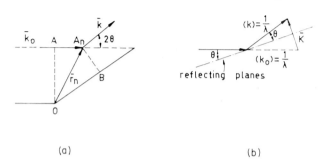

(a)   (b)

FIG. 68.   Illustrating the relation between the wave vector of the incident beam, $\bar{k}_0$ of the diffracted beam, $\bar{k}$, and $\bar{K}$, the normal to the set of reflecting planes. (a) $OB - AA_n$ is the path difference between the 2 waves shown. (b) $\bar{K}$ is perpendicular to the set of reflecting planes.

scattering factor f_n as compared to an atom at the origin 0 is $OB - AA_n$ or $\lambda(\bar{k} - \bar{k}_0) \cdot \bar{r}_n$. The resultant amplitude scattered by the assembly in a point far from the crystal as compared to the crystal dimensions, is then found by summing the wavelets scattered by all atoms taking into account the phase differences:

$$A = \sum_n f_n \exp [2\pi i(\bar{k} - \bar{k}_0) \cdot \bar{r}_n] \tag{9.3}$$

or

$$A = \sum_n f_n \exp (2\pi i\bar{K} \cdot \bar{r}_n) \tag{9.4}$$

with $\bar{K} = \bar{k} - \bar{k}_0$. From Fig. 68b it is clear that $\bar{K}$ is perpendicular to the reflecting lattice planes, its length is $2 \sin \vartheta/\lambda$.

It is now convenient to introduce the concept of the reciprocal lattice defined by the vectors $\bar{A}_1$, $\bar{A}_2$, $\bar{A}_3$ satisfying the following relations

$$\bar{a}_i \cdot \bar{A}_j = \delta_{ij} \tag{9.5}$$

where δ_{ij} is the well-known Kronecker symbol. The vector $\bar{K}$, like any other vector, can be considered as a vector in reciprocal space

$$\bar{K} = l_1 \bar{A}_1 + l_2 \bar{A}_2 + l_3 \bar{A}_3$$

and
$$A = \sum_n f_n \exp \left(2\pi \sum_{i=1}^{3} n_i l_i \right). \tag{9.6}$$

From this expression it is clear that the amplitude will be large, i.e., all wavelets will be in phase if all the sums in the exponents of Eq. (9.6) are integers whatever the values of the integers n_i. This is only possible if the l_i are also integers, i.e., if $\bar{K} = \bar{n}$ where $\bar{n}$ is a "reciprocal lattice" vector corresponding to the lattice planes (hkl). One then has $|\bar{n}| = 1/d_{hkl}$. The statement

$$\bar{K} = \bar{n} \tag{9.7}$$

corresponds to the Laue conditions, which are also equivalent to the Bragg conditions since

$$K = 1/d_{hkl} = 2 \sin \vartheta / \lambda. \tag{9.8}$$

Using the reciprocal lattice concept and Ewald's sphere there is a simple geometrical interpretation of Eq. (9.7). Let 0 be the origin of the reciprocal lattice (Fig. 69a). We draw a vector $-k_0$ from 0, this gives the point C. With C as a center we construct a sphere of

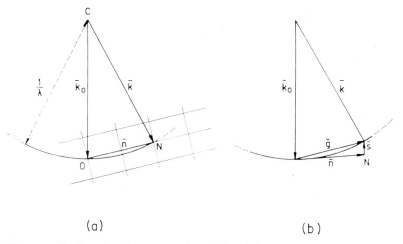

(a)   (b)

FIG. 69.   Reciprocal lattice construction of the diffracted beam. (a) The reflection condition is exactly satisfied: N is on Ewald's sphere. (b) The interference error is s (s is negative in the case shown).

radius $1/\lambda$. If the sphere passes through the point N of the reciprocal lattice, the diffracted beam will have direction CN.

The wavelength of the 100 kv electron beam is of the order of 0.04 A; the radius of the reflecting sphere is quite large as compared to the mesh size of the reciprocal lattice, which is about 1 reciprocal angstrom. Therefore the reflecting sphere can be considered as almost planar for many practical purposes.

(2) *Diffraction by a perfect crystal.* We now inquire about the magnitude of the amplitude of the wave diffracted in directions which differ slightly from the exact Bragg directions. This deviation can be described by a vector $\bar{s}$ in reciprocal space, which gives the distance of the considered reciprocal lattice point to a point on Ewald's sphere. We call $\bar{s}$: the interference error (Fig. 69b). We now have $\bar{K} = \bar{k} - \bar{k}_0 = \bar{n} + \bar{s}$ and formula (9.4) becomes

$$A = \sum_n f_n \exp\left[2\pi i(\bar{n} + \bar{s}) \cdot \bar{r}_n\right]. \tag{9.9}$$

The $\bar{r}_n$ can be expressed as the sum of 2 vectors

$$\bar{r}_n = \bar{r}_j + \bar{\rho}_k \tag{9.10}$$

where the $\bar{r}_j$ join the origin of the lattice to the origin of the jth unit cell; the $\bar{\rho}_k$ join the origin of the unit cell to the kth atom in the unit cell. We thus have

$$A(\bar{s}) = \sum_{j,k} f_k \exp\left[2\pi i(\bar{n} + \bar{s}) \cdot (\bar{r}_j + \bar{\rho}_k)\right] \tag{9.11}$$

or

$$A(\bar{s}) = \sum_j F \exp\left(2\pi i\left[(\bar{n} + \bar{s}) \cdot \bar{r}_j\right]\right) \tag{9.12}$$

where the F is the structure amplitude for 1 unit cell defined by[217d]

$$F = \sum_k f_k \exp\left[2\pi i(\bar{n} + \bar{s}) \cdot \bar{\rho}_k\right] = \sum_k f_k \exp\left(2\pi i \bar{g} \cdot \bar{\rho}_k\right) \tag{9.13}$$

where $\bar{g} = \bar{n} + \bar{s}$.
The amplitude is then

$$A = F \sum_j \exp\left(2\pi i \bar{s} \cdot \bar{r}\right). \tag{9.14}$$

[217d] In x-ray diffraction the structure amplitude is usually defined as:

$$F = \sum_k f_k \exp\left(2\pi i\bar{n} \cdot \bar{\ }_k\right)$$

In order to perform this triple summation we write explicitly

$$\bar{s} \cdot \bar{r}_j = s_x x_j + s_y y_j + s_z z_j = s_x ka + s_y lb + s_z mc$$

where a, b, and c are the edges of the unit cell, s_x, s_y, and s_z are the components of the vector $\bar{s}$; k, l, and m are integers. The summation has now to be performed over k, l, and m. We assume the crystal to be a parallelepiped with edges $N_1 a = (2r_1 + 1)a$, $N_2 b = (2r_2 + 1)b$, and $N_3 c = (2r_3 + 1)c$ where the r_i are integers.

Formula (9.14) then reduces to

$$A = F \sum_{k=-r_1}^{r_1} \sum_{l=-r_2}^{r_2} \sum_{m=-r_3}^{r_3} \exp\left[2\pi i (s_x ka + s_y lb + s_z mc)\right].$$

Performing the summation yields

$$A = F \frac{\sin \pi N_1 s_x a}{\sin \pi s_x a} \cdot \frac{\sin \pi N_2 s_y b}{\sin \pi s_y b} \cdot \frac{\sin \pi N_3 s_z c}{\sin \pi s_z c}.$$

Since $\pi s_x a$, $\pi s_y b$, and $\pi s_z c$ are small with respect to unity, one can write

$$A = F \frac{\sin \pi N_1 s_x a}{\pi s_x a} \cdot \frac{\sin \pi N_2 s_y b}{\pi s_y b} \cdot \frac{\sin \pi N_3 s_z c}{\pi s_z c}.$$

For a thin foil with the z-axis perpendicular to the foil plane, N_1 and N_2 are large numbers, whereas N_3 is relatively small. From the definition of Dirac's delta function,

$$\delta(x) = \lim_{g \to \infty} \frac{\sin gx}{\pi x},$$

one can see that the first two factors become approximately $\delta(s_x a) = (1/a)\delta(s_x)$ and $\delta(s_y b) = (1/b)\delta(s_y)$. From this it is clear that

$$A \sim \frac{F}{abc} \frac{\sin \pi N_3 s_z c}{\pi s} = \frac{F}{V} \frac{\sin \pi ts}{\pi s}$$

where $s_z = s$, $N_3 c = t$ (t: thickness of the foil), and $V = abc$ is the volume of the unit cell. For the intensity one obtains

$$I_s \sim \frac{F^2}{V^2} \frac{\sin^2 \pi ts}{(\pi s)^2}. \tag{9.15}$$

The same formula can also be obtained by approximating the summation by means of an integration. The origin is chosen in the center of the foil. We will, for simplicity, only consider the z direction:

$$A \sim F \sum_j e^{2\pi i s z_j} = F \int_{-t/2}^{+t/2} e^{2\pi i s z} \, dz \simeq \frac{\sin \pi t s}{\pi s} . \qquad (9.16)$$

If the origin is chosen at the entrance face this evidently becomes:

$$A \sim F \frac{\sin \pi t_s}{\pi \rho} e^{\pi i s t}. \qquad (9.16a)$$

Omitting the constant factor we find for I_s (Fig. 70)

$$I_s \sim \frac{\sin^2 \pi t s}{(\pi s)^2} . \qquad (9.17)$$

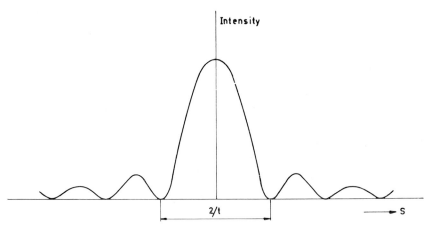

FIG. 70. Diffracted intensity versus the interference error according to the kinematical approximation.

In the other directions the integration again leads to delta functions. The crystal can now be "dissected" into small prisms of material parallel to the direction of the diffracted beam and the amplitude of the diffracted wave is calculated at the back surface of the foil for one such column. The columns are assumed to diffract independently of each other. This follows from the fact that the electrons are only diffracted over a small angle. Under the usual conditions the diffracted and the incident beam form only small angles with the foil normal. The column can therefore be taken as perpendicular to the foil (see Fig. 71). It is further assumed that there is only one strongly diffracted beam, i.e., that s is small only for one reciprocal lattice point. The amplitude of the wave diffracted by a column of crystal is then given by formula (9.12).

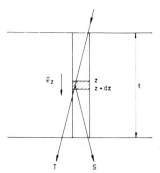

FIG. 71.   The column approximation; the crystal is dissected into small columns perpendicular to the foil plane.

In the reciprocal lattice picture it is convenient to imagine that the points have become rods with an intensity distribution along the z direction, i.e., in the direction perpendicular to the foil plane, given by (9.17). Also it is now clear that it is sufficient to consider vectors s along this rod, since the intensity will be vanishingly small outside the rod, at least for a plate-like crystal. The intensity of the diffracted beam is now a function of s, i.e., of the exact point where the Ewald sphere intersects the rod. The zero's of the expression (9.17) are given by

$$st = k \qquad (9.18)$$

where k is an integer; the s difference between two successive zeros is constant: $\Delta s = 1/t$. In thin foils the zeros are sufficiently far apart to be resolved as a fine structure of the diffraction spots especially if a slightly nonparallel incident beam is used.[218] This will be discussed in detail further below since the phenomenon provides a means for measuring t. If we take as an order of magnitude for $t \simeq 200$ A and for $d_{hkl} = 2$ A, the angular range of reflection is $\delta\vartheta \sim d_{hkl}/t \simeq 10^{-2}$ rad, i.e., about half a degree, which is of the same order of magnitude as the Bragg angles themselves.

[218] W. Kossel and G. Möllenstedt, *Naturwissenschaften* **26**, 660 (1936); *Ann. Physik* [5] **36**, 113 (1939).

For thick crystals the beam will be diffracted away completely at a depth determined by the extinction distance t_e to be defined later; t_e is also of the order of magnitude of a few hundred angstroms, and therefore the angular width of the reflections remains the same as if the crystal had only a thickness t_e. This will be shown in detail later.

(3) *Amplitude-phase diagram.* A convenient graphical way of representing the amplitude of the diffracted wave has been used by Hirsch, Howie, and Whelan.[192] It makes use of the Fresnel construction or the amplitude-phase diagram for representing the summation (9.16). The amplitude scattered by a crystal block between z and $z + dz$ (Fig. 71) in a given column can be represented by a vector of length dz and having a phase angle $2\pi sz$ with respect to the wave diffracted by the block at the origin, which is chosen, for instance, in the middle of the column. For a perfect foil there results a diagram such as the one in Fig. 72.[192] The magnitude of the resulting amplitude is given by PP'. The angle between successive vectors is constant for a perfect foil and since the length of the vectors is also a constant these vectors are chords of a circle with radius $1/2\pi s$. The limiting form of the amplitude phase diagram is therefore a circle with radius $1/2\pi s$ as in Fig. 72. The length of the arc is equal to the thickness of the crystal, this evidently means that in this approximation the

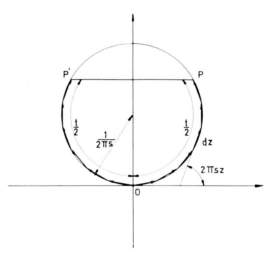

FIG. 72. Phase amplitude diagram for the amplitude of the wave diffracted by a perfect crystal. The resulting amplitude is given by PP'.

amplitude of the diffracted wave is zero for $t = 2k\pi \times (1/2\pi s) = k/s$ in agreement with formula (9.18).

The intensity of the diffracted beam becomes a periodic function of the thickness, the period being $t'_e = 1/s$. This relation would imply that t'_e becomes infinite for $s \to 0$, i.e., for the exact Bragg position. However the kinematical theory is only applicable for s sufficiently large. We will see further that the dynamical theory is required to describe the behavior in the neighborhood of $s = 0$. In this theory t'_e has to be replaced by some quantity t_e which is the limiting value of t'_e for $s \to 0$: t_e is called the extinction distance.

(4) *Extinction contours.*[195]   Practically all specimens used in transmission work are slightly bent and nonuniform in thickness. As a consequence of this, so-called extinction contours, visible in the image as dark bands, result. They are the geometrical loci for $t = $ constant or for $s = $ constant (inclination contours).

(a) Thickness contours.   Along lines where the crystal is $(r + \frac{1}{2})t'_e$ thick ($r = $ integer) the diffracted intensity is a maximum and black lines will be observed. They result from the periodic nature of the transmitted intensity as given by formula (9.17). The contours of this type are not very mobile when the specimen tilt is being changed. However, their appearance and their periodicity is sensitive to which reflection is operating. The latter effect can only be explained on the basis of the dynamical theory. The difference in foil thickness between the points where two successive fringes occur is equal to an extinction distance, which is of the order of a few hundred angstroms, in most metals, but may be over 1000 A in more complicated structures [e.g., 1100 A for the $(11\bar{2}0)$ reflection in graphite]. Knowing the extinction distance the number of fringes gives an easy measure of the foil thickness (Fig. 73b, c).

(b) Inclination contours.   The geometrical loci of the points of the foil for which the Bragg condition for a given set of planes is satisfied form continuous curves. Along these curves much intensity is diffracted away and they will therefore appear as dark lines. Subsidiary maxima in the diffracted intensity may cause visible subsidiary contours. In nearly perfect foils these contours are broad and hardly visible, for metal specimens they are usually much sharper and may possibly be mistaken for dislocation images. In thick metal specimens the contours also become broad, as a consequence of anomalous absorption (Fig. 141).

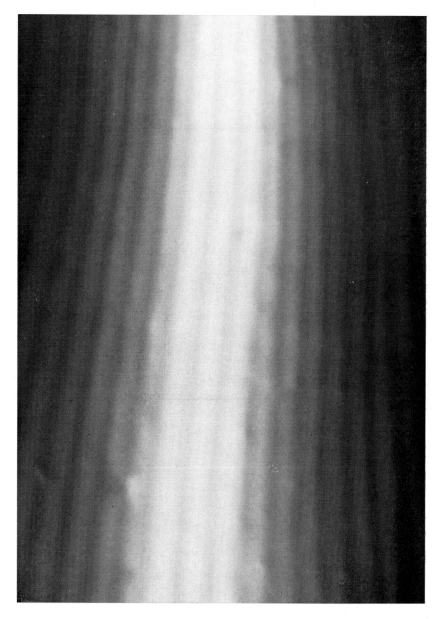

FIG. 73. (a) Bent contour exhibiting subsidiary extrema corresponding to Fig. 71. The crystal foil is graphite.

The inclination contours are very sensitive to changes in orientation; on tilting the specimen they move rapidly over the surface. In the reciprocal lattice language one can say that along the main extinction contour $s \simeq 0$ for the considered reflection. On passing the contour s changes sign, i.e., the deviation from the exact Bragg condition changes sense. This is the reason why the contrast side for a given dislocation changes on crossing over a contour.

In favorable circumstances the extinction contour may exhibit the fine structure predicted by Eq. (9.17), i.e., the subsidiary maxima may also be visible. An example of this is shown in Fig. 73a. From the spacing of the fine structure the foil thickness can be computed.

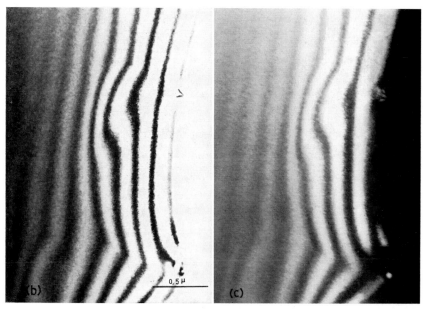

Fig. 73.   (b) Thickness contours in silicon crystal thinned chemically: bright field image. (c) The same area as in (b) but in dark field.

(5) *Experimental determination of the parameter s.*   For a number of problems it is of interest to know the magnitude and more specifically the *sign of s*. (Fig. 74). There is a simple method of determining both It is based on the use of the diffraction pattern. If specimens are not too thin, so-called Kikuchi lines will appear in the diffraction pattern. These lines are due to Bragg reflection of inelastically scattered

electrons. The interpretation given by von Laue[219] seems to be well-established. The origin of the inelastically scattered electrons is less well understood, but this need not bother us here. Suppose that the crystal foil in Fig. 74a is in the exact Bragg position for the set of planes (hko). The Kikuchi line due to this reflection then passes exactly through the diffraction spot due to the same reflection; this can be deduced from Fig. 74a. If now the foil is rotated over an angle $\Delta\vartheta$ about an axis perpendicular to the reciprocal lattice row the position of the diffraction spot will change slightly. The beam of electrons producing the Kikuchi line will also rotate over $\Delta\vartheta$. Therefore the Kikuchi line itself is displaced with respect to the diffraction spot over a distance Δx given by the expression $\Delta x = L\Delta\vartheta$, where L

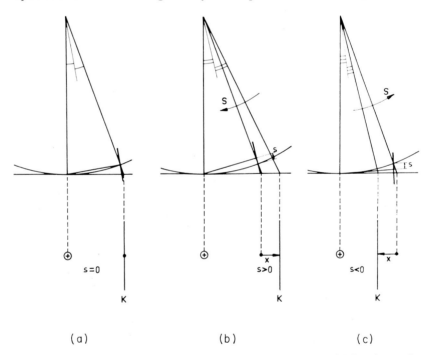

(a)   (b)   (c)

FIG. 74. Relative position of the diffraction spot and the Kikuchi line due to the same reflection. (a) $s = 0$; the reciprocal lattice point is on Ewald's sphere and the Kikuchi line passes through the spot. (b) $s > 0$: the reciprocal lattice point is inside Ewald's sphere and the Kikuchi line is further from the center than the spot. (c) $s < 0$: If the reciprocal lattice point is outside the reflecting sphere the Kikuchi line is closer to the center than the spot. The arrow S indicates the sense into which the lattice has to be rotated in order to bring the crystal into the exact reflecting position.

is the effective camera length, i.e., the distance specimen plate, if no lenses are used. On the other hand the change in s is given by $\Delta s = |\bar{n}| \, \Delta\vartheta$. Elimination of $\Delta\vartheta$ gives $\Delta s = |\bar{n}| \, \Delta x/L$. The effective camera length can be eliminated by expressing it in quantities which can be measured directly on the plate: $L = x/2\vartheta$ where x is the distance of the spot from the center of the pattern. Since 2ϑ is further given by Bragg's law $2\, d_{hkl}\vartheta = \lambda$ we obtain finally

$$\Delta s = \frac{\Delta x}{x} \cdot \frac{\lambda}{d^2_{hkl}}. \tag{9.19}$$

This formula becomes a little more complicated if the axis of rotation is not perpendicular to the reciprocal lattice row.

The displacement is inward if the reciprocal lattice point is rotated towards the interior of the Ewald sphere; on the contrary it is outward if the reciprocal lattice point moves out of Ewald's sphere. If we call s positive for the reciprocal lattice point inside the sphere, we can summarize the situation by means of Figs. 74a,b,c.

d. Origin of Contrast at Imperfections

(1) *Intuitive picture.* Suppose that the thin foil F of Fig. 75 contains an edge dislocation in E. A set of lattice planes has been drawn; we suppose that the foil is oriented in such a way that the Bragg condition $2\, d_{hkl} \sin \vartheta = k\lambda$ for reflection against that particular

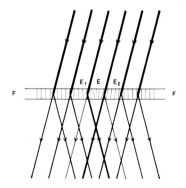

FIG. 75. Diffraction of electrons by a foil containing an edge dislocation in E. In E_2 less intensity is diffracted away than in the perfect part, while in E_1 more intensity is diffracted away. In E_1 a black line will be observed in the bright field image.

[219] M. von Laue, *Ann. Physik* [5] **25**, 569 (1936); *Physik. Z.* **37**, 544 (1936).

set of lattice planes is almost satisfied. Since we are considering a thin foil this condition is considerably relaxed because points in reciprocal space have become rods perpendicular to the foil plane. It is therefore very easy to have some diffraction although Ewald's sphere does not pass exactly through a reciprocal lattice point. Suppose, for instance, that the situation is as shown in Fig. 74c, i.e., that the reciprocal lattice point under consideration is outside the Ewald sphere. A rotation in the sense indicated by the arrow would bring it on the sphere, i.e., would bring the lattice planes indicated in Fig. 75 exactly into the Bragg position.

Under these circumstances a considerable intensity is diffracted away in the perfect part of the foil; the remaining part is transmitted.

Let us now focus attention on the region in the immediate vicinity of the edge dislocation. As a consequence of the geometry of the dislocation the set of lattice planes considered will be deformed. This deformation is such that the Bragg condition will be satisfied better on the left than on the right side of the dislocation. In other words, the reciprocal lattice point will be brought nearer to the Ewald sphere at the left and it will go away from it at the right of the dislocation. As a result more intensity will be diffracted say at E_1 and less at E_2. In the electron microscope under normal conditions of observation a diaphragm intercepts the diffracted beams and only the directly transmitted beam is allowed to fall on the fluorescent screen. A lack of intensity will now be found, at the image of E_1, i.e., a black line will be observed slightly *to the side* of the dislocation.

A similar reasoning can be applied to the case of a pure screw dislocation as shown in Fig. 76. When attention is focused on the set of lattice planes perpendicular to the Burgers vector, i.e., to the axis of the helical surface, it is clear that to the right and left of it the lattice planes are rotated in the opposite sense. More intensity will again be diffracted away on one side, and less on the other side, as compared to the intensity diffracted by the perfect material. Therefore the dislocation will again be imaged as a black line on one side of the real position of the dislocation.

(2) *Determination of the direction of the Burgers vector.* The intuitive picture given in the previous paragraph is sufficient for understanding how the directions of Burgers vectors can be determined by the use of contrast effects. It follows immediately from the picture given above that contrast effects are to be expected only for those

lattice planes which are deformed by the presence of the dislocation. To a first approximation one can assume that the lattice planes parallel to the Burgers vector of a dislocation remain flat since all displacements are parallel to such planes. Hence reflection against such planes will not give rise to contrast. This is most easily visualized for an edge dislocation. Such a dislocation can be considered as a superposition of *plane configurations* as drawn in Fig. 77a. The glide plane of an edge dislocation, although passing through $\bar{b}$, is nevertheless slightly curved in the immediate vicinity of the dislocation and hence a slight contrast effect is to be expected.

For a screw dislocation all displacements are parallel to the axis: the strain pattern has axial symmetry. Therefore it is again clear that planes parallel to the axis will remain flat as in Fig. 78b. In summarizing we can say that no contrast will be observed for planes for which the diffraction vector $\bar{n}$ satisfies the relation $\bar{n} \cdot \bar{b} = 0$. The direction of $\bar{b}$ can be deduced by determining 2 sets of planes having diffraction vectors $\bar{n}_1$ and $\bar{n}_2$ for which this condition is satisfied; $\bar{b}$ is parallel to $\bar{n}_1 \times \bar{n}_2$. An example of a Burgers vector determination is shown in Fig. 79.

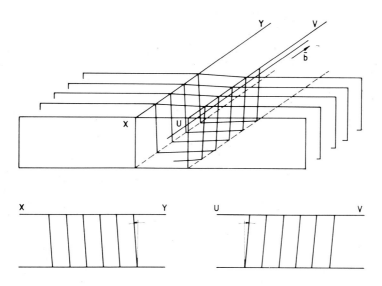

FIG. 76. Deformation of lattice planes in the vicinity of a screw dislocation. Left XY and right UV of the dislocation axis the lattice planes are inclined in opposite sense.

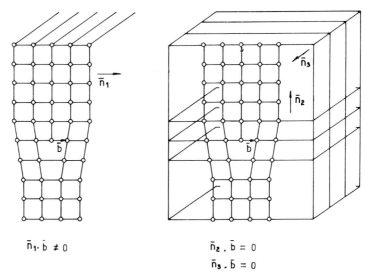

$$\bar{n}_1 \cdot \bar{b} \neq 0$$

$$\bar{n}_2 \cdot \bar{b} = 0$$
$$\bar{n}_3 \cdot \bar{b} = 0$$

FIG. 77. Schematic view of edge dislocations with the 3 main diffraction vectors indicated. Planes normal to $\bar{n}_3$ and $\bar{n}_2$ remain flat, while the planes normal to $\bar{n}_1$ are deformed by the presence of the dislocation; $\bar{n}_2 \bar{b} = \bar{n}_3 \bar{b} = 0$ while $\bar{n}_1 \bar{b} \neq 0$. In the compressed region the lattice parameter is somewhat smaller than in the perfect part; in the expanded region it is somewhat larger.

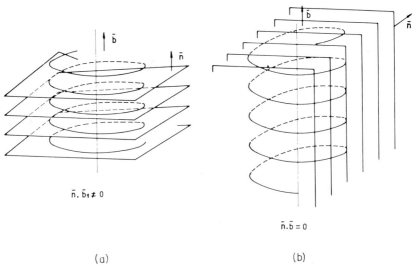

$$\bar{n} \cdot \bar{b}_1 \neq 0$$

$$\bar{n} \cdot \bar{b} = 0$$

(a)   (b)

FIG. 78. Deformation of lattice planes around a screw. (a) The planes perpendicular to $\bar{b}$ are deformed into a helicoidal surface $\bar{n} \cdot \bar{b} \neq 0$. (b) The planes parallel to $\bar{b}$ remain flat; notice that $\bar{n} \cdot \bar{b} = 0$.

(3) *Contrast at dislocation seen end on.* An adaptation of the intuitive picture given in the previous paragraph allows us to see qualitatively how contrast can arise from dislocations perpendicular to the foil plane.

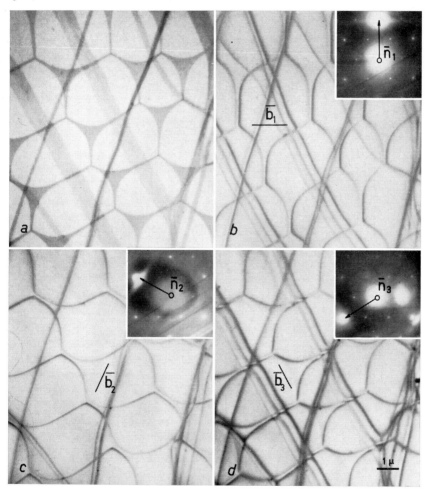

FIG. 79. Example of Burgers vector determination for a network of extended and contracted nodes in graphite. The insets show the relevant parts of the diffraction patterns. (a) Stacking fault contrast. (b) First family of dislocations absent. (c) Second family of dislocations absent. (d) Third family of dislocations absent. The directions of the diffraction vector and of the Burgers vectors are indicated in each case. The situation is especially simple since we know that b is in the foil plane.

Let us first consider the edge dislocation of Fig. 77. Suppose that contrast comes from reflections against planes of the type {100}. In the compressed region of the dislocation the spacing d_c is smaller than the normal spacing $d_c < d_{100}$ while in the expanded region $d_e > d_{100}$. Suppose now that the Bragg condition is nearly satisfied for the set of planes {100} in the perfect region, the deviation being such that the angle of incidence is too small. The Bragg condition for this reflection will then be better satisfied in the expanded region and not as good in the compressed region. Therefore, more intensity will be diffracted away from the expanded region than from the compressed region and the result will be that the former will appear darker, while the latter region will appear lighter than the background.

For reflections against {010}, on the other hand, no contrast would arise since the spacing of these planes to a first approximation is not affected by the presence of the dislocation. The condition for lack of contrast can be formulated as $\bar{n} \cdot \bar{b} = 0$.

It is clear that for a dislocation of the opposite sign the black and white regions will be inverted. Inversion of the contrast would also take place when the extinction contour corresponding to the contrast producing reflection passes at the other side of the dislocation, since this corresponds to a deviation in the opposite sense of the angle of incidence from the Bragg condition.

We now consider a screw dislocation seen end on. To a first approximation all displacements around a screw dislocation are parallel to the line, at least in an infinite solid, and as a consequence no contrast is to be expected if the screw dislocation is parallel to the incident beam. Ruedl et al.[215] considered the possibility that contrast could be produced by the lattice twist around a dislocation in a thin plate. This lattice twist, as shown by Eshelby[181,219a] is essentially a consequence of the finite dimensions of the crystal, i.e., of surface relaxations. For a right-handed screw the lattice twist transforms the lattice planes passing through the screw axis into left-handed helices; the reverse is true for a left-handed screw as represented schematically in Fig. 79[(1)].

Consider in particular a right-handed screw. Along the line AB, parallel to the diffraction vector $\bar{n}$, all lattice displacements caused by the screw are perpendicular to $\bar{n}$, and hence the diffracting planes remain flat. As a consequence the background intensity will be less

[219a] J. D. Eshelby and A. N. Stroh, *Phil. Mag.* [7] **42**, 1401 (1951).

observed along the line AB. Along CD on the other hand, the planes under consideration are sloping in C upwards to the right and in D upwards to the left. For a positive s-value in the perfect material, the lattice planes in the vicinity of C are rotated towards the Bragg orientation, i.e., s decreases and hence more intensity is diffracted

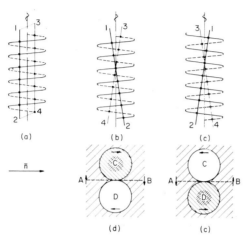

(a)   (b)   (c)

(d)   (c)

FIG. 79[1].   Contrast produced at screw dislocations seen end on. (a) Displacements around a screw dislocation if surface relaxation is neglected. All planes through the axis remain flat. (b) After relaxation the presence of a right-handed screw causes the lattice planes passing through the axis to become left-handed helicoidal surfaces. (c) The presence of a left-handed screw causes the lattice planes passing through the axis to become right-handed helicoidal surfaces. (d) Schematic view of contrast to be expected at right-handed screw ($s > 0$). (e) Schematic view of contrast to be expected at left-handed screw ($s > 0$).

away: this region will therefore appear black in the bright field image. On the other hand the same reasoning shows that in D, s increases and hence this region will appear brighter in the bright field image.

This conclusion can be formulated in a general way, as follows: for positive s and for a right-handed screw the region at the left of the line of no contrast, when looking in the sense indicated by the diffraction vector, will be dark, while the region at the right will be bright in the bright field image. The contrast side changes with the sign of s and with the sign of b (i.e., with the sense of winding of the screw).

The considerations of this paragraph clearly allow a determination of the sign of dislocations seen end on, as well for edges as for screws (see also Section 9, e, (5), 3. The distinction between edges and screws can be made by using the following remarks. The line of no contrast for screws is always parallel to the diffraction vector, and the image never goes out of contrast. For edge dislocations, extinction (at least a pronounced decrease in contrast) should occur for $\bar{n}.\bar{b} = 0$.

The reasoning given here would be valid only for a thin crystal or for large deviations from the Bragg position, where the kinematical theory can be applied. In practice however, "thick" foils are used and anomalous absorption has to be taken into account. This modifies somewhat the reasoning, but the result is essentially unaltered (see Section 9, i, (5), (c)).

e. Diffraction by an Imperfect Crystal: Kinematical Theory[192]

(1) *General formulation.* We will consider only deviations of the atoms or better, of the unit cells, from their ideal position such that the composition of the crystal does not change. Such deviations can be described by a vector $\bar{R}(\bar{r}_j)$ for the unit cell at $\bar{r}_j$. The amplitude of the beam diffracted by the imperfect crystal now becomes [from Eq. (9.12)]

$$A = F \sum_j \exp\left[2\pi i(\bar{n} + \bar{s}) \cdot (\bar{r}_j + \bar{R})\right]. \tag{9.20}$$

Taking into account that $\bar{s} \cdot \bar{R}$ is small with respect to the other terms in the exponent, and that $\bar{n} \cdot \bar{r}_j$ is an integer, this simplifies to

$$A = F \sum_j \exp\left[2\pi i(\bar{n} \cdot \bar{R} + \bar{s} \cdot \bar{r}_j)\right]. \tag{9.21}$$

The product $\bar{n} \cdot \bar{r}_j$ is an integer m because $\bar{r}_j$ is a lattice vector and $\bar{n}$ is a reciprocal lattice vector, the factor $e^{2\pi i m}$ is therefore unity. We can now use the same approximation as above and replace the summation by an integration over the column, taking into account that $\bar{s} \cdot \bar{r}_j = sz$

$$A \sim \int_{-t/2}^{+t/2} e^{i\alpha(z)} e^{2\pi i s z} dz \tag{9.22}$$

where

$$\alpha(z) = 2\pi \bar{n} \cdot \bar{R}(\bar{r}_j).$$

(2) *Contrast due to a stacking fault.* For a stacking fault the factor α is constant; it is zero on one side of the fault plane and α at the other side.

For the face-centered cubic lattice $\bar{R} = (a/6)$ [112] and for a diffraction vector $\bar{n} = [hkl]$, α becomes $\pi/3(h + k + 2l)$, since h, k, and l (integers) have to be all even or all odd. For integer values of h, k, and l this reduces to $\pm 2\pi/3$ and zero. Let us consider a stacking fault parallel to the foil plane (Fig. 80). It is now easy to evaluate the

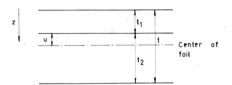

FIG. 80.   Stacking fault parallel to foil plane: illustrating the notations used

integral for this simple case. Let us assume that t_1 and t_2 are the distances of the foil surfaces to the fault plane and let $t = t_1 + t_2$ be the total thickness. The amplitude is then given by:

$$A = \int_0^{t_1} e^{2\pi i s z}\, dz + e^{i\alpha} \int_{t_1}^{t_1+t_2} e^{2\pi i s z}\, dz. \qquad (9.23)$$

Performing the integration and multiplying with the complex conjugate immediately leads to the value for the intensity

$$I(s, t) = AA^* = \frac{1}{\pi^2 s^2} \{1 - \cos(\alpha + \pi st)\cos \pi st$$
$$+ \cos 2\pi su\,[\cos(\alpha + \pi st) - \cos \pi st]\} \qquad (9.24)$$

with

$$2u = t_1 - t_2.$$

From this expression it is clear that the observed contrast will be a function of the thickness and of the depth of the fault plane. Let us assume for simplicity that the fault is in the middle of the foil, i.e., $t_1 = t_2$ or $u = 0$; it is then obvious that the intensity is a periodic function of t, the period being $2/s$. In a similar way by keeping t a constant and displacing the fault with respect to the surface it becomes clear that the contrast depends again in a periodic way on the position of the fault, the periodicity in u is now $1/s = t_e'$.

For a fault which is inclined with respect to the foil plane, the depth periodicity will become visible as fringes parallel to the line of intersection of fault plane and surface, and having the depth periodicity

t'_e . Figure 81 shows, e.g., such fringes due to a stacking fault in stainless steel. A more refined theory discussed further below is required
to understand the details of the fringe pattern.

For a fault plane parallel to the foil plane a uniform shade difference
will be visible as, for instance, in Fig. 82. The intensity of the diffracted

FIG. 81. Stacking fault fringes in stainless steel foil. The fringes are parallel to
the surfaces of the foil. The fault plane is inclined with respect to the surface. (Courtesy
of P. B. Hirsch.)

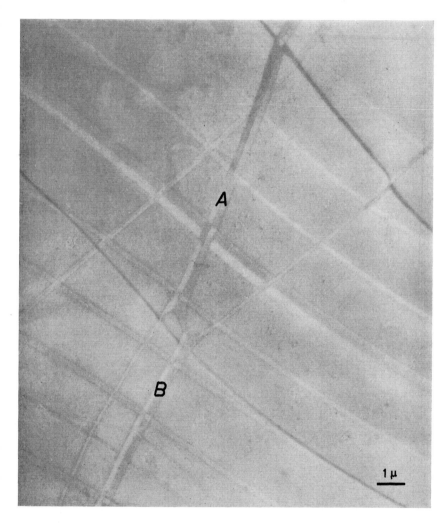

FIG. 82. Uniform shade difference at stacking fault ribbons in graphite. The ribbons are parallel to the foil plane. Notice the contrast at overlapping stacking fault ribbons. Contrast differs in the 2 halves of the triple ribbons; in the overlapping region the contrast is visibly different. [After S. Amelinckx and P. Delavignette, *J. Appl. Phys.* **31**, 2126 (1960).]

beam is given by the relation (9.24). We can define the contrast of the stacking fault as the difference ΔI in intensity diffracted by the perfect [Eq. (9.17)] and by the faulted region. We obtain for the bright field picture:

$$\Delta I = \frac{1}{\pi^2 s^2} \left[\cos \pi st - \cos 2\pi su \right] \left[\cos \pi st - \cos (\alpha + \pi st) \right] \qquad (9.25)$$

or

$$\Delta I = \frac{4}{\pi^2 s^2} \sin \pi st_1 \sin \pi st_2 \sin \left(\pi st + \frac{\alpha}{2} \right) \sin \frac{\alpha}{2}. \qquad (9.26)$$

The fault contrast will be zero if either, $t_1 = kt'_e$ or $t_2 = kt'_e$; (k: integer) one of these conditions is *sufficient* to make the stacking fault invisible. Since $t'_e = 1/s$, i.e., a function of the specimen orientation, these conditions can easily be satisfied by tilting the specimen. The contrast will be maximum if all the sine factors are unity, i.e., if the following conditions are satisfied simultaneously

$$t_1 = (2k \pm \tfrac{1}{2}) t'_e; \; t_2 = (2k' \pm \tfrac{1}{2}) t'_e; \; t = \left(k'' - \frac{\alpha}{2\pi} \pm \tfrac{1}{2} \right) t'_e, \qquad (9.27)$$

The upper sign makes the sine factor $+1$, the lower sign makes them -1. We call the contrast "normal" when the stacking fault appears black on a light background: ΔI is then positive. The following sign combinations in relation (9.27) are the conditions for maximum normal contrast: $+ + +$; $- - +$; $- + -$; $+ - -$. For "inverted" contrast the fault shows up lighter than the background. ΔI is then negative; optimum inverted contrast occurs for the following sign combinations in relation (9.27): $- - -$, $- + +$; $+ - +$; $+ + -$.

There is a simple way, based on the use of the amplitude phase diagram introduced above, to deduce the amplitude of the diffracted wave in a given situation. The sudden phase change over α at the fault plane is represented as a rotation of the two circles, representing the perfect crystal above and below the fault. Such a diagram is shown in Fig. 83. The arc OP_1 of length t_1 as measured along the circle O_1 represents part I of the crystal, while the arc OP_2 of length t_2 measured along circle O_2, represents part II. The two circles have a radius of $1/2\pi s$ and they intersect at the origin under an angle α (120° in the particular case shown). The diffracted amplitude is proportional to the length of $P_1 P_2$. The diffracted amplitude for the same crystal without faults would be $P_1 P'_2$ where $OP'_2 = t_2$ measured along circle O_1. A measure for the contrast is the difference in length $\Delta A = P_1P_2 - P_1P_2'$. It is easy to check that the conditions for the disappearance of contrast are the same as deduced above. The condition $t_2 = k't_e'$ means, for instance, that P_2 as well as P_2' coincides with O; the contrast is then evidently zero.

The diagram also demonstrates that if either t_1 or t_2 increases by t_e' the contrast remains unchanged since the points P_1 and P_2 remain unchanged. Evidently this means that there is a depth periodicity with period t_e'. Hirsch et al.[192] have a simple picture to visualize the origin of the fringes in the case of an inclined fault. They suppose that the periodic function (9.24) describes the scattered intensity in the foil as in Fig. 84. A black fringe will appear in the bright field image

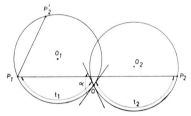

FIG. 83. Phase amplitude diagram for a foil containing a stacking fault. The circles intersect under an angle of 120°. The resultant amplitude is given by P_1P_2. The thicknesses t_1 and t_2 are measured along the circles. Without stacking fault the amplitude would be P_1P_2.

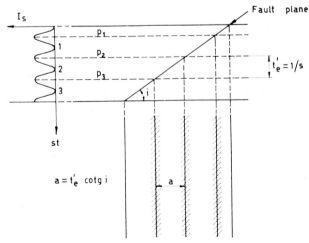

Fig. 84. Origin of stacking fault fringes for an inclined fault plane. Where I_s has a maximum a dark fringe will be observed in the bright field image.

each time the fault plane intersects a plane of the family p_i of Fig. 84. The distance a between fringes is $a = t'_e \cot g\, i$. The same picture also describes adequately the periodic nature of the intensity, which leaves the back surface. It is, however, inadequate in its simple form to account for the details deduced from formulas (9.24) and (9.26).

(3) *Overlapping stacking faults.* We will now sketch briefly the problem of overlapping stacking faults. The notation defined by Fig. 85a will be used; further we call α_1 and α_2 the phase shifts intro-

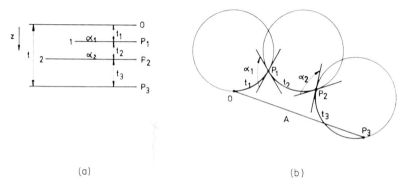

(a)   (b)

Fig. 85.   Overlapping stacking faults. (a) Illustrating the notations used. (b) Phase-amplitude diagram for a foil containing 2 overlapping stacking faults. The final amplitude is given by OP_3.

duced by the 2 faults separately. The amplitude of the wave diffracted can then be calculated in the region of overlap by means of the expression

$$A = \int_0^{t_1} e^{2\pi i s z}\, dz + e^{i\alpha_1} \int_{t_1}^{t_1+t_2} e^{2\pi i s z}\, dz + e^{i(\alpha_1+\alpha_2)} \int_{t_1+t_2}^{t=t_1+t_2+t_3} e^{2\pi i s z}\, dz. \quad (9.28)$$

The intensity is found as above by taking AA^*; the result is given below as formula (9.29). In the region where 1 stacking fault is present formula (9.28) is still valid, *mutatis mutandis*. From Eq. (9.28) it is clear that in general the regions of overlap will exhibit another shade than the region where only 1 stacking fault is present. It is also possible that the region of overlap has the same shade as the perfect region. Examples of contrast at overlapping stacking faults parallel to the foil plane are visible in Fig. 82 which refers to a graphite sample. The two faults in the threefold ribbon correspond to $\alpha_1 = +\,120°$

and $\alpha_2 = -120°$, respectively. The contrast at overlapping stacking faults evidently can be analyzed using the phase-amplitude diagram. An example of this is shown in Fig. 85b, which refers to the situation of Fig. 85a.

The amplitude scattered by a crystal containing 2 overlapping stacking faults is given by [from Eq. (9.28)]

$$\pi s A = e^{\pi i s t_1} \sin \pi s t_1 + e^{\pi i s (2t_1 + t_2)} e^{i\alpha_1} \sin \pi s t_2$$

$$+ e^{\pi i s (2t_1 + 2t_2 + t_3)} e^{i(\alpha_1 + \alpha_2)} \sin \pi s t_3. \tag{9.29}$$

Consider the special case where the 2 faults have displacement vectors of the same magnitude but opposite in direction. One might expect that the net result would be to eliminate all contrast effects due to the stacking fault. This is not true in general as can immediately be noticed by putting $\alpha_1 = -\alpha_2$. The resulting amplitude is not the same as for a crystal without faults and having the same total thickness $t = t_1 + t_2 + t_3$.

However, if the distance between the fault planes is equal to an integral number of extinction distances, i.e., $t_2 = k/s = k t_e'$ (k: integer) Eq. (9.29) reduces to

$$\pi s A = e^{\pi i s t_1} (\sin \pi s t_1 + e^{\pi i s (t_1 + t_3)} \sin \pi s t_3) \tag{9.30}$$

or

$$\pi s A = e^{\pi i s (t_1 + t_3)} \sin \pi s (t_1 + t_3) \tag{9.31}$$

which is equivalent to the expression for the amplitude scattered by a perfect crystal of thickness $t_1 + t_3$. Since t_2 is equal to an integral number of extinction distances the scattered intensity does not change if a layer of thickness t_2 is added, i.e., it is the same as for a crystal of thickness t. It is concluded that stacking faults separated by an integral number of extinction distances cannot be detected.

If the 2 stacking faults are at distances from the foil surface equal to an integral number of extinction distances, e.g., $t_1 = k t_e'$ and $t_3 = k' t_e'$, it is clear from Eq. (9.29) that the scattered intensity is the same as for a perfect foil of thickness t_2 or t since this only adds an integral number of t_e'. Hence there will be no contrast. If one stacking fault is at an integral number of extinction distances from the surface, e.g., $t_1 = k t_e'$, the crystal scatters as if it contained only the other stacking fault and had a thickness $t_2 + t_3$ or t. Hence there is again no contrast due to such a fault.

Formula (9.29) can easily be generalized to

$$A = \frac{1}{\pi s} \sum_{j=1}^{n} \sin \pi s t_j \; e^{\pi i s(t_j + 2t_{j-1} + 2t_{j-2} + \ldots)} \; e^{i(\alpha_{j-1} + \alpha_{j-2} + \ldots)} \qquad (9.32)$$

where the faults are in planes t_j and produce phase shifts α_j.[219d]

We discuss here the fringe pattern for a particularly simple but nevertheless interesting example of overlapping stacking faults observed in a face centered cubic alloy of low stacking fault energy: copper-18% gallium.[219b] The stacking faults in this alloy have been shown to be of the intrinsic type (see appendix A). Figure 85[(1)] shows two intersecting faults symmetrically inclined with respect to the foil plane, which is approximately a (110) plane.

The reference system is chosen in the following way. The y-axis is parallel to the foil, in the middle of it, along the projection of the line of intersection of the two fault planes. The origin is put at the point of intersection of the y-axis with the intersection line of the two fault planes. The z-axis is perpendicular to the foil plane and the x-axis is perpendicular to the y and to the z-axis. In this reference system the two fault planes have as equations:

$$z = \lambda y + \mu x \qquad \text{and} \qquad z = \lambda y - \mu x$$

The two faults being both intrinsic and being inclined in opposite sense with respect to the foil give a different sign to $\alpha(= \pm 2\pi/3)$, because for one of the faults the diffraction vector is left whilst it is right for the second one.[219c] The second fault met by the electron beam thus annihilates the phase shift introduced by the first one. In this special case the formula (9.28) reduces to

$$A(x, y) = \int_{-z_0}^{+z_0} e^{2\pi i s z} \, dz + (e^{i\alpha} - 1) \int_{\lambda y - \mu x}^{\lambda y + \mu x} e^{2\pi i s z} \, dz \qquad (9.32.1)$$

where $2z_0$ is the thickness of the foil. After integration this becomes:

$$A(x, y) = \frac{\sin 2\pi s z_0}{\pi s} + (e^{i\alpha} - 1) \, e^{2\pi i s \lambda} \frac{\sin 2\pi s \mu x}{\pi s} \qquad (9.32.2)$$

[219b] A. Art, R. Gevers, and S. Amelinckx, *Phys. Stat. Solid.* 3, 697 (1963); *ibid.* (to be published).

[219c] For a definition of left and right see Appendix A. Further evidence that the geometry is correct is obtained by observing that the nature of the first fringes for both faults is different. In Appendix A it is shown that this means that α has a different sign for both.

[219d] $\alpha_0 = 0$; $\alpha_{-k} = 0$; $t_0 = 0$; $t_{-k} = 0$.

FIG. 85[1]. Intersecting intrinsic stacking faults in a copper-18 % gallium alloy. (Courtesy of A. Art.)

which leads to an expression for the intensity of the diffracted beam

$$(\pi s)^2 \, I(x, y) = \sin^2 2\pi s z_0 + 4 \sin^2 2\pi s \mu x \sin^2 \frac{\alpha}{2}$$

$$- 4 \sin 2\pi s \mu x \sin 2\pi s z_0 \sin \left(2\pi s \lambda y + \frac{\alpha}{2}\right) \sin \frac{\alpha}{2}. \quad (9.32.3)$$

The first term represents the background intensity in the perfect part of the foil. For $x = 0$ i.e., along the intersection line of the two faults and in general for $\sin 2\pi s \mu x = 0$, I reduces to the background intensity. The locus of points for which this is true is therefore a system of fringes parallel to the y-axis. The periodicity in the x-direction is $1/\mu s$. Along the lines $x = $ constant, except along those for which $x = k/2 \, \mu s$, the intensity is further modulated by the factor $\sin [2\pi s \lambda y + (\alpha/2)]$ with period $1/s\lambda$. In foil thicknesses such that $\sin 2\pi s z_0 = 0$ the modulation along the fringes $x = $ constant is absent.

A convenient method to discuss the fringe profiles makes use of a graphical representation for the amplitude $A(x, y)$ in the complex plane. One can rewrite (9.32.2) as:

$$A(x, y) = A_0(2z_0) + \exp \left[i \left(\frac{\pi}{2} + \frac{\alpha}{2} + 2\pi s \lambda y\right)\right] A_1(x) \quad (9.32.4)$$

where

$$A_0(2z_0) = \frac{\sin 2\pi s z_0}{\pi s} \quad \text{and} \quad A_1(x) = 2 \sin \frac{\alpha}{2} \frac{\sin 2\pi s \mu x}{\pi s} \quad (9.32.5)$$

The vector $\bar{A}(x, y)$ in the complex plane is the sum of two vectors (1) OB $= A_0(2z_0)$ measured along the real axis and (2) $A_1(x) = $ BC measured along the direction which forms an angle $\vartheta = (\pi/2) + (\alpha/2) + 2\pi s \lambda y$ with the real axis. The final amplitude is given by OC (Fig. 85[2], a). It is clear that as y varies the point C describes a circle with radius $|A_1(x)|$ and the amplitude OC $\equiv |A(x, y)|$ oscillates between two extreme values OD and OF.

For values of x for which $\sin 2\pi s \mu x = 0$, i.e., for $x = k/2\mu s$, the radius of the circle reduces to zero and OC $\equiv$ OB $= A_0(z_0)$ which is the amplitude of the wave scattered by the perfect part of the crystal. One expects therefore white fringes at distances $1/2\mu s$. The real periodicity in the x-direction is, however, $1/\mu s$. This follows from the fact that the black fringes, for which $\sin 2\pi s \mu x = \pm 1$ are still modulated. If the situation is as shown in Fig. 85[2], a, i.e., if

$$A_1(x) < A_0(2z_0) \quad \text{or} \quad 2 \sin \frac{\alpha}{2} \sin 2\pi s \mu x < \sin 2\pi s z_0$$

the amplitude fluctuates between two rather different extreme values OB and OF with a period equal to $1/s\lambda$. For successive black fringes, i.e., for sin $2\pi s\mu n = +1$ and -1 the modulation has a phase difference of π. The modulation will have optimum contrast if the following conditions are satisfied:

$$\begin{cases} 2 \sin \dfrac{\alpha}{2} = \pm \sin 2\pi s z_0 \\ |\sin 2\pi s \mu x| = 1 \end{cases}$$

The circle now passes through 0. In this case the minima will be zero and the maxima equal to $2\bar{A}_0(2z_0)$. On the other hand if the thickness of the foil is such that sin $2\pi s z_0 = 0$ the center of the circle coincides with 0 and whatever the values of x there will be no modulation of the amplitude. Figure 85[(2)] (b) shows the image predicted by the kinematical theory for a general case, it can be compared with the observed pattern of Fig. 85[(1)]. The qualitative behavior is well explained, but the dynamical theory taking into account absorption is required to explain the details.[219b]

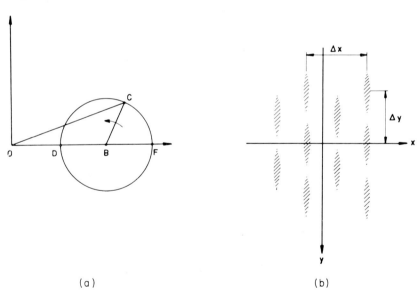

(a)   (b)

FIG. 85[(2)].   (a) Complex plane construction for the fringe system observed at the intersecting faults of Fig. 85[(1)], according to the kinematical theory. (b) Image predicted according to the kinematical theory for the intersecting faults of Fig. 85[(1)]. The periods are $\Delta x = 1/\mu s$; $\Delta y = 1/s\lambda$.

(4) *Contrast at dislocations.* (a) perfect screw dislocation. The discussion will be based on the reference system shown in Fig. 86. The dislocation is parallel to the foil plane, and situated along the y axis. The z axis is perpendicular to the foil, which is bounded by the planes $z = -t_1$ and $z = +t_2$. The displacement function $\bar{R}$ now becomes a continuous function of $\bar{r}$; it describes the displacement

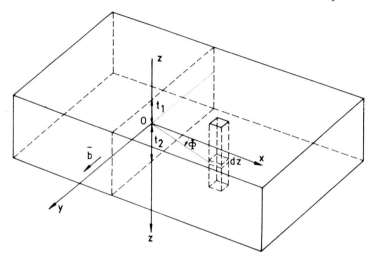

FIG. 86. Illustrating the notation used in calculating the contrast due to a screw dislocation parallel to the foil plane. The dislocation is along the y axis. The z axis is perpendicular to the foil.

caused by the presence of the dislocation. The screw dislocation is characterized by the simplest displacement function since all displacements are parallel to y; for an infinite isotropic medium this function is:

$$\bar{R} = \bar{b}\phi/2\pi \qquad (9.33)$$

where $\bar{b}$ is the Burgers vector and ϕ is the angle shown in Fig. 86. With cartesian coordinates one obtains:

$$\alpha = 2\pi\bar{n} \cdot \bar{R} = \bar{n} \cdot \bar{b} \operatorname{arctg} \frac{z}{x}. \qquad (9.34)$$

The quantity $\bar{n} \cdot \bar{b}$ has a special significance and it is noted n[219e]; for a perfect dislocation n is an integer whose magnitude depends on $\bar{n}$, i.e., on the reflection which is operating.

[219e] We propose to call this quantity the "image order".

For values of z large in absolute value as compared to x, which we assume to be positive, α is $+ (\pi/2)n$ or $- (\pi/2)n$ depending on whether z is positive or negative. From this the physical meaning of n becomes clear; $n\pi$ is the phase difference between waves diffracted above and below the dislocation. In general n is a small integer because in view of their intensity the beams corresponding to small $\bar{n}$ vectors (i.e., small Bragg angles) are the most important ones in producing contrast. This is a consequence of the fact that the scattering factor for electrons is strongly peaked forward.

The amplitude scattered by a column of crystal at the position x (the amplitude does not depend on y) is given by Eq. (9.22) where now a has the value given by Eq. (9.34); it becomes

$$A = \int_{-t_1}^{+t_2} e^{in \, \text{arctg}(z/x)} \, e^{2\pi i s z} \, dz. \qquad (9.35)$$

The integral has been evaluated for different values of the parameters $\beta = 2\pi s x$ (to be introduced later) and n, using an approximation, which consists in taking the integration limits as infinite after having performed one partial integration. However the discussion of these results is more profitable after having demonstrated the application of the amplitude phase diagram to the case of a dislocation.

(b) Amplitude-phase diagram for a perfect screw dislocation. In the case of a perfect crystal the phase difference between 2 crystal blocks at z and $z + dz$ was a constant and equal to $2\pi s dz$. This is no longer true: a supplementary phase difference results from the displacements. This additional phase shift will either be added or subtracted depending on the sign of x, s, and z; further its magnitude depends on x and z, since it is equal to

$$[nx/(x^2 + z^2)] \, dx = nd \left[\text{arctg} \left(\frac{z}{x} \right) \right]. \qquad (9.36)$$

For $z \gg x$, i.e., for z large enough, this quantity becomes zero and the final shape of the diagram will again be a circle with radius $1/2 \pi s$ as for the perfect crystal. This is of course physically evident: far enough from the dislocation the lattice is again perfect.

For small values of z, i.e., near the dislocation, we have to consider two distinct cases for the shape of the diagram:

(i) The supplementary phase shift is additive: this will be the case if s and nx have the same sign or if $n\beta > 0$; n arctg (z/x) then has the same sign as $2\pi s z$ for positive as well as for negative z. The angle between 2 successive vectors is now *larger* than $2\pi s\, dz$ near the origin of the curve; as z becomes larger it approaches the value $2\pi s\, dz$. Therefore the curve will be a woundup spiral which gradually tends to a circle as shown in Fig. 87. The magnitude of the vector that joins the end points of the curve is proportional to the diffracted amplitude at the point x (Fig. 87).

Fig. 87.   Phase amplitude diagram for a foil containing a screw dislocation: $n\beta > 0$ ($n = 1$, $\beta = +1$). The diagram is a woundup spiral.

(ii) If on the other hand s and nx have opposite sign, i.e., for $n\beta < 0$, n arctg (z/x) has the opposite sign of $2\pi s z$ and the resulting phase difference between successive vectors will be smaller than $2\pi s\, dz$ for the perfect crystal by the quantity $n[x/(x^2 + z^2)]dz$. For large z the latter expression again reduces to zero and the curve approaches a circle with radius $(2\pi s)^{-1}$. The diagram is now an unwound spiral of the type represented in Fig. 88. The diffracted amplitude is found again by taking an arc proportional to t_1 in the negative sense and an arc proportional to t_2 in the positive sense along the curve, and joining the two end points.

When going from one side of the dislocation to the other side x changes sign and the spirals corresponding to different sides of the dislocation will therefore be of different type: one wound up and one unwound. It is now clear that the vector representing the diffracted

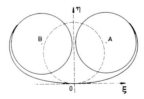

Fig. 88.   Phase amplitude diagram for a foil containing a screw dislocation $n\beta < 0$ ($n = 1$, $\beta = -1$). The diagram is an unwound spiral.

amplitude will be largest for an unwound spiral; therefore this is the side on which a dark line will be observed in the dark field image. Since the shape of the diagram depends only on the value of $\beta = 2\pi s x$ and on n, and not on the value of s and x separately, we have only to construct curves for different values of these two parameters.

The graphical construction of a sufficient number of curves allows us in principle to obtain image profiles (this is the diffracted intensity as a function of β) with a moderate accuracy. Some typical curves are shown in Fig. 89. For large values of z or for large thicknesses the

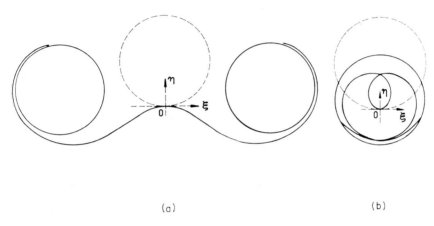

(a)   (b)

FIG. 89.   Phase amplitude diagrams for the case $n = 2$. (a) $\beta = -1$. (b) $\beta = +1$.

amplitude will oscillate between two extremes as the image point A goes around the final circle. We will discuss this point further below.

(c) Analytical method.   We will now discuss an analytical method which permits a numerical computation of the phase-amplitude diagram.

The integral (9.35) is in fact an alternative formulation of the amplitude-phase diagram. The integration is difficult to perform with finite integration limits, but it becomes feasible when extended, after one partial integration, between minus and plus infinity. The justification for this approximation lies in the fact that the contribution for $z \gg x$ is small anyway. It is sufficient to consider the integral

$$\bar{A} = \xi + i\eta = \int_0^{t_2} e^{\pm i n \phi} e^{2\pi i s z} \, dz \qquad (9.37)$$

where $\phi = \alpha/n = \text{arctg } z/|x|$. The plus and minus signs correspond respectively to the positive and negative values of x. The significance of Eq. (9.37) is as follows: $\bar{A}$ is the vector joining the origin of the complex plane to the end point of the amplitude-phase diagram. With the help of Eq. (9.37) some geometrical properties of the phase-amplitude diagram can easily be derived.[220] We have

$$\xi = \int_0^z \cos (2\pi sz \pm n\phi)\, dz; \quad d\xi = \cos (2\pi sz \pm n\phi)\, dz$$

$$\eta = \int_0^z \sin (2\pi sz \pm n\phi)\, dz; \quad d\eta = \sin (2\pi sz \pm n\phi)\, dz \tag{9.38}$$

and hence

$$dS^2 = (d\xi)^2 + (d\eta)^2 = dz^2 \tag{9.39}$$

or

$$S = z \tag{9.40}$$

confirming that the length of the spiral arc is equal to z.

Further one can calculate the radius of curvature

$$\rho = \frac{(1 + (d\eta/d\xi)^2)^{3/2}}{d^2\eta/d\xi^2} = \frac{1}{2\pi s}\frac{1}{1 \pm n/\beta(1 + u^2)} \tag{9.41}$$

where $u = z/|x|$; $\beta = 2\pi s |x|$. From this formula it is evident that for large z, i.e., for large u, the radius of curvature will be

$$\rho_\infty = 1/2\pi s$$

i.e., the same as for the final circle which confirms that the spiral approaches asymptotically to the final circle. Furthermore it is clear that depending on the sign of $n\beta$, ρ will be smaller or larger than ρ_∞, i.e., that the spiral will approach to the asymptotic circle from the inside or from the outside; in other words: the spiral will be wound up or unwound.

We can integrate by parts and find

$$\bar{A} = \frac{1}{2\pi i s} \left[e^{2\pi i sz \pm in\, \text{arctg}(z/|x|)} \right]_0^{t_2} \pm \frac{n}{2\pi s} \int_0^{t_2 \to \infty} e^{2\pi i sz \pm in\, \text{arctg}(z/|x|)}\, d\, \text{arctg}\frac{z}{|x|}. \tag{9.42}$$

[220] R. Gevers, *Phil. Mag.* [8] **8**, 769, (1963).

The first term oscillates at the upper limit, while at the lower limit it reduces to $i/2\pi s$. The integral forming the second part can be expressed more conveniently by taking ϕ as the integration variable. Leaving out the oscillating part we obtain the following integral

$$\bar{A}_c = \frac{i}{2\pi s} \mp \frac{n}{2\pi s} \int_0^{\pi/2} e^{2\pi i s z \pm i n \phi} \, d\phi. \tag{9.43}$$

The vector $\bar{A}$ can then be considered as the sum of two vectors: $\bar{A}_c$ going to the center C of the final circle and $\bar{A}_1$ going from C to the end point and therefore rotating around C as z increases. For sufficiently large z, arctg $z/|x|$ is $\pi/2$ and the vector $\bar{A}_1$ has a constant length and its end point describes the final circle. The components of vector $A_c = \xi_{n_c} + i\eta_{n_c}$ are the coordinates of the center of the final circle; $\xi_{n,c}$ and $\eta_{n,c}$ can be calculated by evaluating the real and imaginary parts of the integral (9.43). The results are

$$\left. \begin{array}{l} \xi_{n,c} = \mp \dfrac{n}{2\pi s} \left(1 \pm \dfrac{d}{d\,|\,\beta\,|} \right)^n \displaystyle\int_0^\infty \dfrac{\cos |\,\beta\,|\, u}{(1 + u^2)^{1/2(n+2)}} \, du \\[4mm] \eta_{n,c} = \dfrac{1}{2\pi s} \mp \dfrac{n}{2\pi s} \left(1 \pm \dfrac{d}{d\,|\,\beta\,|} \right)^n \displaystyle\int_0^\infty \dfrac{\sin |\,\beta\,|\, u}{(1 + u^2)^{1/2(n+2)}} \, du. \end{array} \right\} \tag{9.44}$$

The value of $\xi_{n,c}$ and $\eta_{n,c}$ can be expressed in terms of modified Bessel functions and modified Struve functions. For the detailed calculation of the integrals we refer to the original paper.[192]

The loci of the centers of the limiting circles have been calculated for different values of n. Some examples are shown in Fig. 90. The values of β are indicated along the curves. It is to be noted that for n even the loci corresponding to β positive reduce to a vertical line since $\xi_{n,c} \equiv 0$. For n odd the loci for positive and negative values of β are mirror images with respect to the η-axis.

At this point a new approximation has to be accepted before image profiles can be constructed. We are now interested in the variation of the amplitude PP_1, in Fig. 87, for instance, as x varies. From formula (9.41) it is clear that the distance of the point P_1 to the circle becomes negligible as soon as $[n/\beta(1 + u^2)] \ll 1$. Not too far from the dislocation, which is the region of interest for us, $|x| \ll t_2$ and arctg $t_2/|x|$ is practically constant and equal to $\pi/2$. The variation in amplitude, due to the fact that with changing x the position of P_1 along the final circle changes also, is therefore small. The main variation in amplitude with x will be due to the change in the position of the center of the final circle. Therefore it is considered a reasonable approximation to accept the variation of $2\ \xi_{n,c}$ as a measure for the variation of the diffracted amplitude. It is equivalent to assuming that the diffracted amplitude varies as the difference of vector $OA(\!=\!A_c)$ and its mirror image $OB(\!=\!A_c')$ with respect to the η-axis (see Fig. 88).

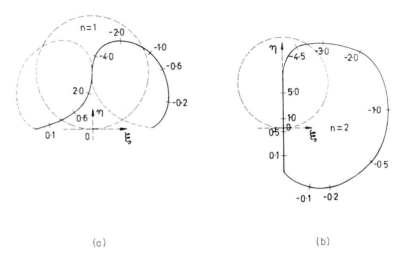

(a)   (b)

Fig. 90. Loci of the centers of the limiting circles for the phase amplitude diagrams corresponding to: (a) $n = 1$; (b) $n = 2$.

It is further clear that the diffracted amplitude will be proportional to the scale of the amplitude diagram. This scale is determined by the radius of the final circle which is $1/2\pi s$.

As a convenient measure of the diffracted intensity Hirsch et al.[192,221] propose $(4\pi s\xi_{n,c})^2$. This quantity has been deduced as a function of β from the loci of Fig. 90 for different values of n.

The results are shown in Fig. 91.

(d) Discussion of the image profiles. The profiles confirm what was already derived on intuitive grounds, that the image of the dislocation is asymmetrical: It occurs for β negative since more intensity is scattered away at that side. The peak displacement with respect to the position of the dislocation is of the order of the image

[221] M. J. Whelan, J. Inst. Metals, 87, 392 (1958-1959).

width and therefore appreciable. It increases with n; for $n = 3$ and $n = 4$ the curve presents 2 minima. Multiple images due to a different cause will be discussed below.

In a sense the profiles of Fig. 91 are not to scale since the decrease of the atomic scattering factor for electrons with increasing scattering angle has been neglected. As a consequence of this, the peaks should be reduced in height, for higher values of n.

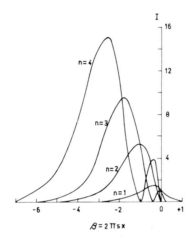

FIG. 91. Intensity profiles of screw dislocation images for different values of $n = \bar{n} \cdot \bar{b} = 1, 2, 3, 4$. The intensity variation is taken proportional to the variation of the square of the distance between the centers of the final circles. The abcissa is $2\pi s x$; for constant s this is proportional to the distance from the dislocation. [After P. B. Hirsch, A. Howie, and M. J. Whelan, *Phil. Trans. Roy. Soc.* **A252**, 499 (1960).]

Another effect also affects the height. It was mentioned above that the scattered intensity is proportional to $1/s^2$ or to $t_e'^2$. It will be shown further that the value of t_e' is limited to some maximum t_e the extinction depth, the value of which depends on the reflection. The extinction depth increases with the order of the reflection. Smaller effective values of s are therefore possible with higher order reflections. The general result will be that the intensity profiles become about of equal height for all values of n; the difference in width remains, however.

(e) Contrast due to edge dislocations.[222] The displacement function for an edge dislocation along the y axis with slip plane parallel to the

[222] R. Gevers, *Phil. Mag.* [8] **7**, 59 (1962).

foil plane is given by

$$\bar{R} = \frac{b}{2\pi} \left[\phi + \frac{\sin 2\phi}{4(1-\nu)} \right] \bar{e}_x - \frac{b}{2\pi} \left[\frac{1-2\nu}{2(1-\nu)} \ln r + \frac{\cos 2\phi}{4(1-\nu)} \right] \bar{e}_z \tag{9.45}$$

where $\phi = \text{arctg}(z/x)$, $\bar{e}_x$ and $\bar{e}_z$ are unit vectors respectively along x and z axes. The second term gives a displacement vector perpendicular to the foil; since the diffraction vector $\bar{n}$ is in the foil plane the dot product with $\bar{e}_z$ will be zero and it is therefore sufficient to consider the first term

$$\bar{n} \cdot \bar{R} = \frac{n}{2\pi} \left[\phi + \frac{\sin 2\phi}{4(1-\nu)} \right] \tag{9.46}$$

with $n = \bar{n} \cdot \bar{b}$.

One has now to evaluate the integral (9.22) with (9.46) as an expression for $\bar{n} \cdot \bar{R}$ in $\alpha = 2\pi \bar{n} \bar{R}$. Gevers [222] has shown that the calculation can be reduced to the summation of a series involving integrals already calculated for the screw dislocation.

Use is made of the following expansion

$$e^{it\sin\varphi} = \sum_{-\infty}^{+\infty} J_r(t) e^{ir\varphi}. \tag{9.47}$$

One obtains at once for the integral corresponding to n for the edge dislocation

$$A_n^{(e)} = \sum_{r=-\infty}^{+\infty} J_r \left[\frac{n}{4\pi(1-\nu)} \right] \int_{\text{column}} e^{i(n+2r)\phi} e^{2\pi i sz} dz \tag{9.48}$$

where the J_r are Bessel functions of order r. The remaining integrals are of the same type as Eq. (9.37) since r is an integer. They can be noted conveniently as[222a]

$$A_n^{(s)} = \int_{\text{column}} e^{in\phi} e^{2\pi i sz} dz. \tag{9.49}$$

They were calculated for $n = 1, 2, 3, 4$ by Hirsch et al.[192] and for larger values by Gevers.[222]

With this notation one obtains the following series expansion

$$A_n^{(e)} = \sum_{r=-\infty}^{+\infty} J_r \left[\frac{n}{4(1-\nu)} \right] A_{n+2r}^{(s)} \tag{9.50}$$

which in principle solves the problem.

[222a] The upper index (s) refers to "screw."

Gevers[222] has shown that this series can be cut off after a reasonable number of terms with errors smaller than 1%.

Image profiles can now be calculated by the same procedure outlined in the previous paragraph for the screw dislocation, and therefore subject to the same approximations. The results are shown in Fig. 92.

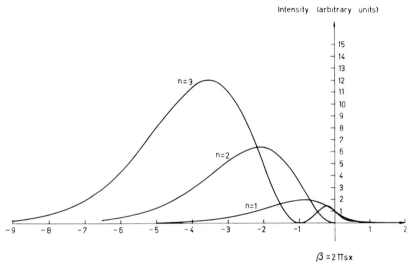

FIG. 92. Image profiles for edge dislocations according to the kinematical approximation. When compared with the image profile of Fig. 91 it becomes evident that they have twice the width. [After R. Gevers, *Phil. Mag.* [8] **7**, 59 (1962).]

As already pointed out by Hirsch *et al.*[192] the image turns out to be about twice as wide as that for the screw dislocations with the same n value; the peak displacement is also about twice as large. The height of the maximum is somewhat larger for the edge than for the screw; the ratio of $I_{max}^{(e)}/I_{max}^{(s)}$ depends on the value of n; one has, respectively, 1.11, 1.20, 1.22, and 1.28 for $n = 1, 2, 3, 4$. The peak position and I_{max}^e also depend on v (Poisson's ratio) which was taken as $1/3$ in computing the curves.

(f) *Mixed dislocations with slip plane parallel to the foil plane.*[223a] The kinematical theory for a mixed dislocation with slip plane parallel to the foil plane has been developed by Gevers. It is assumed that the strain field of the mixed dislocation with Burgers vector b is in fact the superposition of the strain field of a screw with Burgers

223a R. Gevers, *Phil. Mag.* [8] **7**, 651 (1962).

vector $\bar{b} \cos \vartheta$ and an edge with a vector $\bar{b} \sin \vartheta$ where ϑ is the angle between line and Burgers vector. The displacements for the screw and for the edge part are then, respectively,

$$\bar{R}_s = b \cos \vartheta \cdot \frac{\phi}{2\pi} \bar{e}_y \tag{9.51}$$

$$\bar{R}_e = \frac{b \sin \vartheta}{2\pi} \left[\phi + \frac{\sin 2\phi}{4(1-\nu)} \right] \bar{e}_x + \bar{R}_{e,z} \bar{e}_z \tag{9.52}$$

where $\bar{R}_{e,z}$ is the z component of the displacement; its explicit form is of no importance since it does not enter into the calculation, because $\bar{n}$ is perpendicular to $\bar{e}_z$.

We now have

$$\alpha = 2\pi \bar{n} \cdot \bar{R} = 2\pi \bar{n} \cdot (\bar{R}_s + \bar{R}_e) = n\phi + \delta \sin 2\phi \tag{9.53}$$

where

$$n = \bar{n} \cdot \bar{b} \quad \text{and} \quad \delta = \frac{n}{4(1-\nu)} \sin \vartheta \sin \gamma \tag{9.54}$$

where γ is the angle between the dislocation line and the diffraction vector. Using the procedure outlined in the previous paragraph one obtains for the amplitude $A_n^{(\vartheta, \gamma)}$ diffracted by a column

$$A_n^{(\vartheta, \gamma)} = \sum_{r=-\infty}^{+\infty} J_r(\delta) A_{n+2r}^{(s)} \tag{9.55}$$

where the meaning of the $A_{n+2r}^{(s)}$ is defined by Eq. (9.49). This expression allows us to calculate the image profile by the use of known integrals. Some results of the computation are shown in Figs. 93 and 94. An important feature of the calculation is that the profile no longer depends in a unique way on n as was the case for the pure screw and for the pure edge, but on the particular diffraction vector $\bar{n}$. This comes about because δ, the argument of the Bessel functions, depends on the diffraction vector as well as on n. It is convenient to write

$$\delta = \delta_e \cdot \eta \tag{9.56}$$

where

$$\delta_e = \frac{n}{4(1-\nu)} \quad \text{and} \quad \eta = \sin \vartheta \sin \gamma = (\tfrac{1}{2}) \left[1 - \frac{\cos (2\vartheta + \epsilon)}{\cos \epsilon} \right] \tag{9.57}$$

with $\epsilon = \gamma - \vartheta$; δ_e is the value of δ for the pure edge orientation.

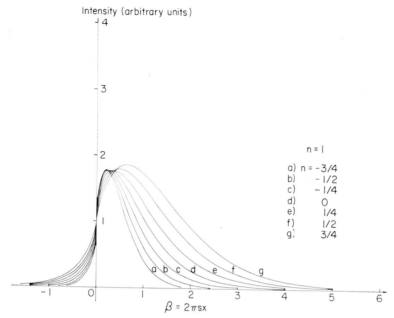

FIG. 93. Image profiles for mixed dislocations parallel to the foil plane according to the kinematical approximation: $n = 1$; $\eta = -3/4, -1/2, -1/4, 0, 1/4, 1/2, 3/4$. [After R. Gevers, *Phil. Mag.* [8] **7**, 651 (1962).]

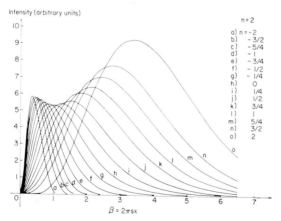

FIG. 94. Image profiles for mixed dislocations parallel to the foil plane, according to the kinematical approximation. $n = 2$; $\eta = -2, -3/2 - 5/4, -1, -3/4, -1/2, -1/4$ and the corresponding positive values. [After R. Gevers, *Phil. Mag.* [8] **7**, 651 (1962).]

Profiles are shown for $n = 1$ and $n = 2$ and for the different η values. Three important conclusions emerge from these calculations.

(1) The image is narrowest if the dislocation bissects the acute angle formed by the Burgers vector and the diffraction vector. This profile is narrower than for a pure screw with the same diffraction vector.

(2) The image is broadest for a dislocation line bissecting the obtuse angle formed by the Burgers vector and the diffraction vector; this profile is broader than for the pure edge.

(3) Even if $n = 0$ there will be some contrast for high-order reflections and for a 45° dislocation; the image should be a weak asymmetrical double line.

(g) Dislocation inclined with respect to the foil plane — oscillating contrast.[192] The use of the phase-amplitude diagram allows us to deduce qualitatively the contrast to be expected for a dislocation which is inclined with respect to the foil surface. We assume that the phase-amplitude d agram for the contrast side of the dislocation segment in the middle of the foil is PP' as shown in Fig. 95a. We further suppose that the foil is thick enough so that the image points P and P' are on the final circles for both surfaces. The amplitude of the scattered wave is then given by the segment PP'.

The dislocation will now be dissected into small segments parallel to the foil, but at varying depths. The phase-amplitude diagram for a segment at distance t_1 below the foil center is derived from that in the center by moving the points P to P_1 and P' to P_1', in such a way that $PP_1 = P'P_1'$, the distance PP_1, being proportional to t_1. On changing the level continuously the point P describes one final circle and the point P' the other one in the opposite sense, in such a way that $PP_1 = P'P_1'$. It is clear that in this process the resultant amplitude oscillates with a depth periodicity, equal to the length of the circle, i.e., $2\pi \times (1/2\pi s) = (1/s) = t_e'$ between two extreme values P_2P_2' and PP'. Oscillations with the same period occur for the phase-amplitude diagram at the other side of the dislocation.

Such oscillatory contrast frequently has been observed at dislocations not too far below the surface and inclined with respect to it. Examples are reproduced in Fig. 96. The oscillating character is found to depend on the operating reflection. This can be understood in the following way. For another operating reflection t_e' may be different because of differences in s and the diagram for a segment in the middle of the foil may now be as shown in Fig. 95,b.[223b] If now the points P and P' move on the final circles no oscillations will occur.

The two cases shown are the extremes; the first gives maximum amplitude of the oscillations, the second gives none. For intermediate cases the oscillations will be less pronounced. It is clear that there will always be oscillations if one of the end points is near to the origin, or, what is equivalent, if the considered dislocation segment is near

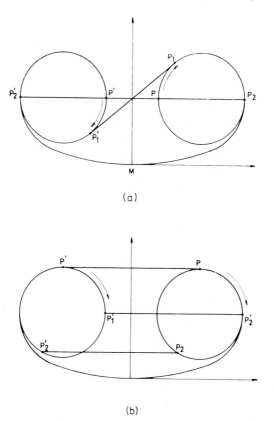

(a)

(b)

FIG. 95.   Phase-amplitude diagram for a screw dislocation inclined with respect to the foil surface. (a) Diagram for a situation where large oscillations occur. The amplitude oscillates between a minimum PP' and a maximum P_2P_2'. (b) Diagram for a situation which does not give rise to oscillations: the amplitude PP' remains constant.

[223b] We will show further that on the basis of the dynamical theory t_e' has to be replaced by a quantity t_e which does explicitly depend on the operating reflection.

Fig. 96. Oscillatory contrast observed at dislocations which approach gradually the surface of a tin disulfide crystal.

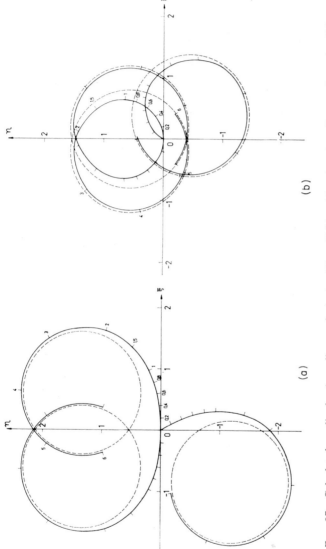

(a)

(b)

FIG. 97. Calculated amplitude–phase diagram for $|n| = 2/3$ and $|\beta| = |2\pi sx| = 1$. x distance from the dislocation, the stacking fault corresponds to $x > 0$; $1/2\pi |s|$ is chosen as a unit. The values of $\gamma = |2\pi sz|$ are indicated on the diagram. (a) $n\beta < 0$, i.e., $n = -2/3$, $\beta = +1$ or $n = 2/3$, $\beta = -1$. (b) $n\beta > 0$, i.e., $n = -2/3$, $\beta = -1$ or $n = 2/3$, $\beta = +1$. The left-handed spiral is rotated counterclockwise over an angle $\pm 4\pi/3$ ($\pm$ if $n\beta \gtrless 0$) in the region $x > 0$; it is not rotated in the region $x < 0$.

to the surface. For a detailed treatment of the effect the dynamical theory is required.

It may be worth while pointing out that this effect is very useful for estimating the distance of a dislocation from the foil surface.

(h) Partial dislocations.[220] The Burgers vector is no longer a lattice vector and the product is therefore not necessarily an integer. In the case of e.g., the face-centered cubic metals $b = (a/6)$ [112] and $\bar{n} \cdot \bar{b}$ can adopt values which are multiples of 1/3. The same is true for graphite. Furthermore one side of the dislocation is a stacking fault; for columns intersecting the stacking fault a sudden phase change occurs.

The effect of this on the phase-amplitude diagram is to cause a

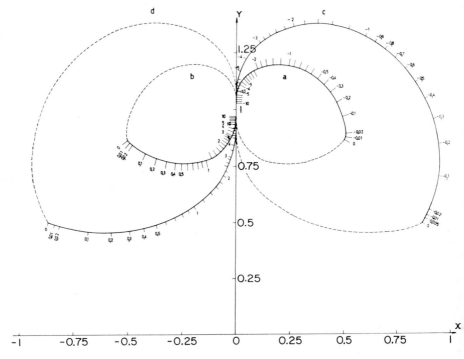

FIG. 98.   Loci of the centers of the final circle (solid line) and the initial circle (broken line) of the amplitude-phase diagrams. The values of $\beta(=2\pi sx)$ are marked (on the locus). The stacking fault corresponds to $x > 0$ (x: distance from the disloca-tion). (a) and (b) $n = 1/3$, or $n = -1/3$ if $\beta \rightarrow -\beta$. (c) and (d) $n = 2/3$, or $n = -2/3$ if $\beta \rightarrow -\beta$.

relative rotation over 120° of the 2 branches of the spiral as shown in Fig. 97.

Since n is no longer an integer the analytical calculations made for perfect dislocations cannot be applied. Gevers[220] has extended these calculations to include also fractional values of n. Figure 98 represents, for instance, the geometrical loci of the centers of the final circles for $n = 1/3$ and $n = 2/3$. He also derived the amplitude profiles for screw dislocations with n a multiple of $1/3$, using the same assumption as Hirsch *et al.*[192] for perfect dislocations (Fig. 99). An important conclusion is that dislocations for which $n = \pm 1/3$ would presumably not be visible, or in any case would produce no line contrast. This

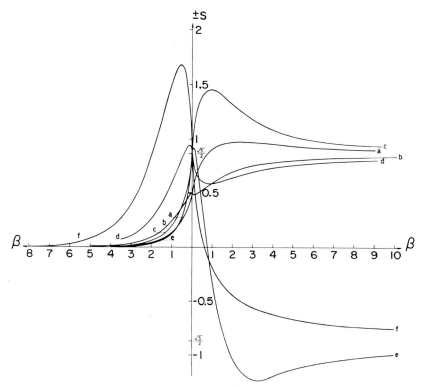

FIG. 99(*a*). Dark field image profiles of partial dislocations according to the kinematical approximation. Only the steady part of the amplitude is shown ($\beta = 2\pi s \,|\, x \,|$). The stacking fault is at the right. Only curves for $s > 0$ are shown; those for $s < 0$ obtained by changing the sign of n. (a) $n = -1/3$. (b) $n = 1/3$. (c) $n = -2/3$. (d) $n = 2/3$. (e) $n = -4/3$. (f) $n = 4/3$.

explains the observation that extended nodes in graphite may exhibit a double extinction, i.e., the simultaneous absence of contrast at 2 partials with a different Burgers vector, violating apparently the $\bar{n} \cdot \bar{b} = 0$ criterion for the absence of contrast as shown in Fig. 100. The quantities $\bar{n} \cdot \bar{b}$ for the 3 partials of the node can be derived from the diffraction pattern: they are 1/3, 1/3, and $-$ 2/3.

(5) *Use of the image profiles.* (a) Determination of the exact position of the dislocation line. As shown above the image is situated at one side of the actual position of the dislocation line, the displacement being of the order of the image width. In order to determine this side it is necessary to know the actual position. This can be done by looking for a foil orientation where two reflections at opposite sides of the central spot but, having s values with the same sign, produce contrast. The corresponding images are then one at each

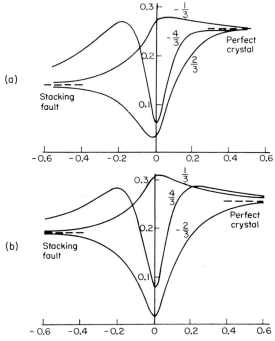

FIG. 99(*b*). Dynamical theory. Computed bright field image profiles for partial dislocations. The depth is chosen in such a way that on the left the intensity is asymptotic to that in a bright fringe of the stacking fault image. (a) $t = 6t_e$; $s = 0$ and $\tau_e = 10t_e$ and $\tau_e = \tau_0$; $y = 1.7t_e$. (b) The same values are chosen, except that $y = 1.2t_e$. [After A. Howie and M. J. Whelan, *Proc. Roy. Soc.* **A267**, 206 (1962).]

side of the dislocation; they form a so-called double image; the actual position is then roughly in the middle. One can also make photographs of the 2 images *successively* with respect to some marks on the surface of the film; e.g., dust particles. At the passage of an extinction contour the image often changes side as a consequence of a change in the sign of s. This phenomenon can clearly be used also to determine the contrast side. Photographs illustrating the described features are reproduced in Fig. 101.

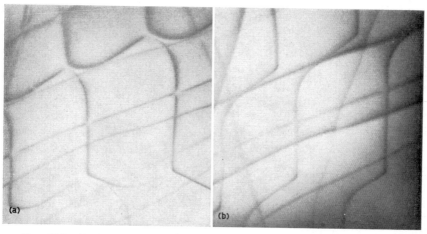

FIG. 100.   Double extinction in extended nodes in graphite. For 2 of the dislocations the value of $\bar{n} \cdot b = 1/3$. (a) Two partials are visible. (b) Only one family of partials is visible.

(b) Determination of the sign of the Burgers vector. A careful study of contrast effects allows us to determine the sign of the Burgers vector, i.e., to decide whether the supplementary half-plane of an edge dislocation is above or below the glide plane with respect to a given reference. In the case of screw dislocations it is possible to determine whether it is left- or right-handed. The method implies knowledge of the sign of s [Section 9c (5)]. Once the sign of s is determined, we know in fact in which sense the foil (or the lattice) should be rotated in order to bring the contrast producing planes into the exact Bragg condition. Let us assume that this sense is as indicated by S in Fig. 102 which also represents 2 edge dislocations of opposite sign. The lattice rotations caused by the presence of these dislocations are also indicated. It is now clear that the contrast side,

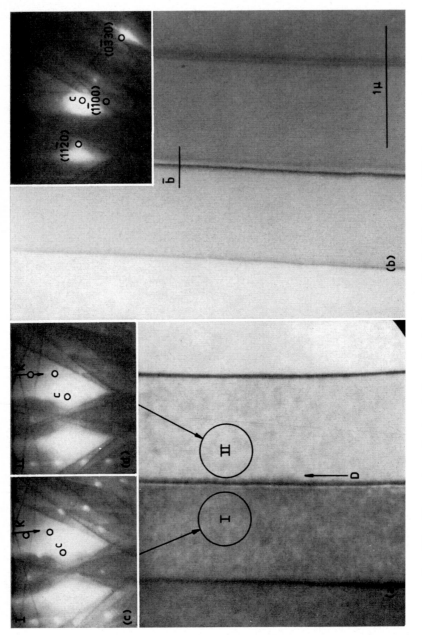

FIG. 101. Observations allowing to determine the image side of a dislocation. (a) Single contrast due to the $(11\bar{2}0)$ reflection. (b) Double contrast due to the reflections $(11\bar{2}0)$ and $(0\bar{3}30)$. The dislocation is somewhere between the 2 images. [After R. Siems, P. Delavignette, and S. Amelinckx, *Phys. Stat. Solidi* **2**, 421 (1962).]

which is the side where the lattice rotation is in the sense S, is different for the 2 dislocations. The relation image-dislocation line-diffraction pattern is shown schematically in Fig. 102a,b and this allows us to decide immediately the sign of the Burgers vector.

For screw dislocations the procedure is analogous. The reasoning can be summarized by Fig. 102c,d. In the case of a right-handed screw (Fig. 102c) the lattice rotations are as indicated by the arrows; they are in the opposite sense for the left-handed screw of Fig. 102d.

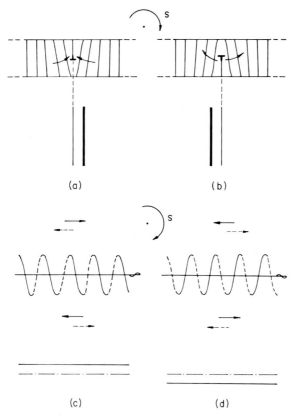

(a)   (b)

(c)   (d)

FIG. 102.   Illustrating the sign determination of the Burgers vector. The arrow S indicates in which sense the lattice has to be rotated in order to bring it into the Bragg condition. The image is represented as a heavy line. The drawing is made assuming that S is as indicated when looking along the same direction and sense as in Figs. 101 and 74. (a) and (b) for the positive and the negative edge dislocation, (c) and (d) for the right- and the left-handed screw.

The relation between the image and the position of the dislocation and with respect to the diffraction pattern is then as shown in Fig. 102c,d.

The contrast side of course has to be determined by means of the methods outlined in Section 9,e, (5), (a). The method outlined here was employed to determine the sign of edge dislocations giving rise to bending of a thin foil,[223c] and it was also applied to dislocations in magnesium oxide[223d].

The question whether loops are due to the condensation of vacancies or of interstitials can be answered by the use of the procedure just outlined, provided the loop plane is known. This can in many cases be determined by mounting the specimen in two different inclinations with respect to the incident beam.[224a] The change in loop width then allows us to deduce the loop plane if the indices are known or can be assumed.[224b]

The procedure was applied to loops in deformed and annealed magnesium oxide,[224a] to loops in fission fragment irradiated platinum[224c] and to loops in α-bombarded aluminium.[224d] We refer to these papers for a detailed description of the different steps involved.[224e]

(c) Sign determination of edge dislocations seen end on. The methods outlined before for the sign determination of the Burgers vectors do not apply to the case of a dislocation seen end-on, since the contrast is due to changes in lattice parameter. However the method can be adapted. We assume again that the sign of the interference error s has been determined for the contrast producing reflection by means of the method outlined in Section 9c (5). If s is negative an increase in lattice spacing will bring the crystal towards

[223c] R. Siems, P. Delavignette, and S. Amelinckx, *Phys. Stat. Solidi* 2, 421 (1962).

[223d] G. W. Groves and M. J. Whelan, *Phil. Mag.* [8] 7, 1603 (1962).

[224a] G. Groves and A. Kelly, *Phil. Mag.* [8] 6, 1527 (1961).

[224b] When determining the sign of Burgers vectors the relative orientation of image and diffraction pattern has to be known accurately. It is therefore necessary to study carefully the ray path in the microscope used in order to take into account the image inversions and rotations. A useful trick is to tilt the specimen in a known sense and watch the sense of movement of the Kikuchi lines.

[224c] E. Ruedl, P. Delavignette, and S. Amelinckx, *Proc. I.A.E.A. Symp. on Radiation Damage in Solids and Reactor Materials, Venice, 1962*, Vol. 1, p. 363. I.A.E.A., Vienna, 1962.

[224d] D. J. Mazey, R. S. Barnes, and A. Howie, *Phil. Mag.* [8] 7, 1861 (1962).

[224e] See also Section 10a(9) for other methods.

the exact Bragg condition. If s is positive a local decrease in lattice parameter will increase locally the scattered intensity and a dark region will result in the bright field image. For reflections against the set of planes drawn in Fig. 77, the upper part will appear lighter and the lower part darker than the background, if s is negative. The reverse will of course be true if the sign of s is changed, or if the sign of the dislocation changes.

An alternative procedure is based on the use of bright and dark field images in absorbing crystals (see Section 9i (4) (6)).

f. Dynamical Theory

It would be out of place to give here a detailed account of the dynamical theory of electron diffraction. We refer to Bethe,[225] McGillavry[226] and Howie and Whelan[227] for such a treatment. We shall only briefly derive some results which are of direct application to the observation of dislocations in the electron microscope. We shall more specially draw attention to the results which differ qualitatively from those obtained by the kinematical approach, and which may be of importance for a correct interpretation of the observations.

(1) *Diffraction by a perfect crystal.* Two main approaches have been used: one can obtain the scattered and transmitted waves as solutions of Schrödinger's equation,[225,226] or one can use a system of coupled first-order differential equations for transmitted and scattered amplitude.[227] We shall briefly sketch the two approaches which, as will be shown, are in fact equivalent.

(a) Solution of Schrödinger's equation by the method of Bethe-Dispersion surface. Schrödinger's equation is

$$\Delta\Psi(\bar{r}) + \frac{8\pi^2 me}{h^2}[E + V(\bar{r})]\Psi(\bar{r}) = 0 \qquad (9.58)$$

where eE is the energy of the incident electrons.

The crystal potential $V(\bar{r})$ can be expanded in a triple Fourier series of the form

$$\frac{2me}{h^2}V(\bar{r}) = \frac{2me}{h^2}V_0 + \sum_{n} V_n \exp(2\pi i\bar{n}\cdot\bar{r}) \qquad (9.59)$$

[225] H. A. Bethe, *Ann. Physik* [4] **87**, 55 (1928).
[226] C. H. McGillavry, *Physica* **7**, 329 (1940).
[227] A. Howie and M. J. Whelan, *Proc. Roy. Soc.* **A263**, 217 (1961).

where $\bar{n}$ is a reciprocal lattice vector; V_0 is the mean inner potential. We further put

$$\frac{2me}{h^2}(E + V_0) = K^2. \tag{9.60}$$

We will also use the vector K which is the incident electron wave vector corrected for refraction, by a medium with mean inner potential V_0.

We shall now make the simplifying assumption that there are only 2 crystal waves to be considered: the incident wave with wave vector $\bar{k}_0$ and one strongly scattered wave with wave vector $\bar{k}_n = \bar{k}_0 + \bar{n}$ where N is the reciprocal lattice node point which is near to the reflecting sphere.

The wave function for the electrons is of the Bloch form; it can be written as

$$\Psi(\bar{r}) = \psi_0 \exp(2\pi i \bar{k}_0 \cdot \bar{r}) + \psi_n \exp(2\pi i \bar{k}_n \cdot \bar{r}). \tag{9.61}$$

Substituting the expressions (9.59), (9.60), and (9.61) into (9.58) and putting the coefficients of the exponential functions $\exp(2\pi i \bar{k}_0 \bar{r})$ and $\exp(2\pi i \bar{k}_n \cdot \bar{r})$ equal to zero separately, leads to the set of linear homogeneous equations, determining the ratio of ψ_0 and ψ_n

$$(K^2 - k_0^2)\psi_0 + V_{-n}\psi_n = 0 \tag{9.62}$$

$$V_n\psi_0 + (K^2 - k_n^2)\psi_n = 0. \tag{9.63}$$

There will be a nonzero solution only for the amplitudes ψ_0 and ψ_n if the following determinant is zero

$$\begin{vmatrix} K^2 - k_0^2 & V_{-n} \\ V_n & K^2 - k_n^2 \end{vmatrix} = 0. \tag{9.64}$$

If Eq. (9.64) is satisfied the ratio R_n which is called the "reflection coefficient" can be found.

$$R_n = \psi_n/\psi_0 = (K^2 - k_0^2)/V_{-n} = V_n/(K^2 - k_n^2). \tag{9.65}$$

The relation (9.64) can be written as

$$(K^2 - k_0^2)(K^2 - k_n^2) = |V_n|^2. \tag{9.66}$$

In k space this equation represents a surface called the "dispersion surface." Its approximate shape can be deduced in the following way. All the vectors $\bar{K}$, $\bar{k}_0$, $\bar{k}_n$ differ in length only by small quantities. Therefore we can write approximately $K + k_0 \simeq 2K$ and $K + k_n \simeq 2K$; Eq. (9.66) then reduces to

$$(K - k_0)(K - k_n) = |V_n|^2/4K^2. \tag{9.67}$$

It is now evident that a cross section of this surface with the plane determined by $\bar{k}_0$ and $\bar{n}$ will have a hyperbolic shape in the vicinity of the Brillouin zone boundary. The asymptotic curves, i.e., for large k, are given by $K = k_0$ and $K = k_n$. They are large circles with radius K and with centers in 0 (the origin of the reciprocal space) and in N (the reciprocal lattice point). In order to satisfy relation (9.67) the vectors $\bar{K}$, $\bar{k}_0$, and $\bar{k}_n$ have to be related geometrically in the way shown in Fig. 103a; an infinite number of combinations satisfy relation (9.67). However we still have to take into account the boundary conditions. These are continuity : (1) of the wave functions and, (2) of the derivatives in the direction of the normal to the interface. It is easy to satisfy the first requirement by expressing that the tangential components of the wave vectors have to be equal on either side of the interface. In electron diffraction the perpendicular components are much larger than the differences between these perpendicular components. Bethe[228,229] has shown that under these conditions the tangential continuity of vectors implies continuity of the normal derivatives. The boundary conditions therefore reduce to $\bar{K}_t = \bar{k}_{0t}$. This condition amounts to giving k_{0t} the tangential component; we then have to calculate the possible $\bar{k}_{0p}$, the perpendicular components. Graphically it is clear that there are 2 points on the dispersion surface corresponding to a given k_{0t} (D and D'), i.e., there will be two values for k_{0p}, we will call them $k_0^{(+)}$ and $k_0^{(-)}$.

Now we shall calculate these 2 values. The deviation from the exact Bragg position can be described by the parameter x which is the distance from the Brillouin zone boundary to the point C on the dispersion surface corresponding to the reciprocal lattice point N. With reference to Fig. 103 $\bar{x}$ can be related to $\bar{s}$, which is the vector NG giving the distance of the reciprocal lattice point N to Ewald's

[228] H. A. Bethe, *Ann. Physik* [4] **87**, 55 (1928).
[229] M. J. Whelan and P. B. Hirsch, *Phil. Mag.* [8] **2**, 1121, 1303 (1957).

sphere. It is clear that, if $\delta\vartheta$ represents a small rotation of the reciprocal lattice around O, $\delta\vartheta = xk_0$ and $s = n\delta\vartheta$ and hence

$$s \simeq nx/k_0 = nx/K. \qquad (9.68)$$

One now has from Fig. 103

$$K^2 - k_0^2 = \overline{OC^2} - \overline{OD^2} = \overline{BC^2} - \overline{BD^2} = (\overline{BC} + \overline{BD})\,\overline{CD} \simeq 2BC \cdot CD. \qquad (9.69)$$

The vector $\overline{CD} = \Delta\bar{e}$ is now introduced; $\bar{e}$ is the unit vector along CB which is perpendicular to the foil plane.

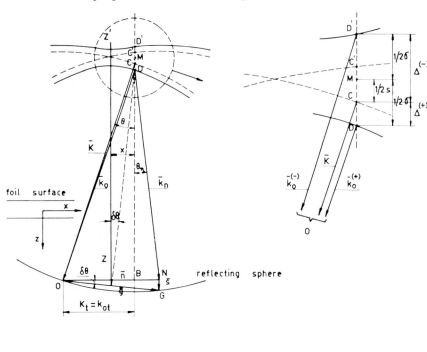

(a)   (b)

Fig. 103. Part of the dispersion surface in the vicinity of the Brillouin zone boundary illustrating the relation between the different wave vectors. The dotted arcs are the intersection of the asymptotic spheres with the plane of the drawing. These spheres have radii equal to K and they have O and N as centers. In (a) part of the drawing is surrounded by a dotted circle; this part is shown in detail in (b).

One finds

$$K^2 - k_0^2 = 2K\Delta \cos \vartheta$$

$$K^2 - k_n^2 = (K^2 - k_0^2) - (k_n^2 - k_0^2)$$ (9.70)

$$= 2K\Delta \cos \vartheta - [\overline{BN^2} + \overline{BD^2} - (\overline{OB^2} + \overline{BD^2})]$$

$$= 2K\Delta \cos \vartheta - [\overline{BN^2} - \overline{OB^2}]$$

$$= 2K\Delta \cos \vartheta + \left[\left(\frac{n}{2} + x \right)^2 - \left(\frac{n}{2} - x \right)^2 \right]$$ (9.71)

or finally

$$K^2 - k_n^2 = 2K\Delta \cos \vartheta + 2nx.$$ (9.72)

Substitution of these expressions into the dispersion equation gives a relation which allows us to determine the possible values of Δ.

$$\Delta \left(\Delta + \frac{nx}{K \cos \vartheta} \right) = \frac{|V_n|^2}{4K^2 \cos^2 \vartheta}.$$ (9.73)

In view of Eq. (9.68) one has

$$nx/K \cos \vartheta \simeq s.$$ (9.74)

Further we introduce the concept of "extinction distance" t_e defined by

$$t_e = K \cos \vartheta / |V_n|$$ (9.75)

Equation (9.73) can then be written as

$$\Delta^2 + s\Delta - \left(\frac{1}{4} \right) t_e^{-2} = 0.$$ (9.76)

The roots of this quadratic equation are evidently

$$\Delta^{(\pm)} = -\left(\frac{1}{2} \right) s \pm \left(\frac{1}{2} \right) t_e^{-1} \sqrt{1 + (st_e)^2} = \left(\frac{1}{2} \right)(-s \pm \sigma)$$ (9.77)

where σ, defined by

$$\sigma t_e = \sqrt{1 + (st_e)^2},$$ (9.78)

was introduced.

The 2 values of Δ determine the allowed values of k_0; the geometry of the situation is represented in Fig. 103a,b.

$$\bar{k}_0^{(+)} = \bar{K} - \Delta^{(+)}\bar{e}; \quad \bar{k}_0^{(-)} = \bar{K} - \Delta^{(-)}\bar{e}.$$ (9.79)

Substituting these values for $\bar{k}_0$ into Eq. (9.65) gives the 2 values of the reflection coefficient

$$R_n^{(\pm)} = (K^2 - k_0^2)/|\,V_{-n}\,| = (2K\cos\vartheta\varDelta^{(\pm)})/|\,V_{-n}\,| = 2t_e\varDelta^{(\pm)}\,e^{i\phi_{-n}}.$$

$$(9.80)$$

By a proper choice of the origin of the reference system the phase factor $e^{i\phi_n}$ of the Fourier coefficient can be reduced to unity, and one obtains

$$R_n^{(\pm)} = t_e(-s \pm \sigma).$$

$$(9.81)$$

The wave functions corresponding with the 2 values of R_n are now

$$\psi^{(\pm)} = \exp(2\pi i \bar{k}_0^{(\pm)} \cdot \bar{r})\,[1 + R_n^{(\pm)}\exp(2\pi i \bar{n} \cdot \bar{r})].$$

$$(9.82)$$

The first terms represent waves in the incident direction; the second terms represent diffracted waves. The total wave function is a linear combination of these two solutions

$$\Psi = A^{(+)}\psi^{(+)} + A^{(-)}\psi^{(-)}.$$

$$(9.83)$$

The arbitrary constants $A^{(\pm)}$ have to be determined by the initial values. The scattered wave has to have zero amplitude at the entrance face $\bar{r} \cdot \bar{e} = z = 0$ while the amplitude of the transmitted wave is unity since back reflection at the entrance face is neglected. This leads to the system of linear equations

$$A^{(+)} + A^{(-)} = 1$$
$$A^{(+)}R_n^{(+)} + A^{(-)}R_n^{(-)} = 0$$

$$(9.84)$$

with solutions

$$A^{(+)} = R_n^{(-)}/(R_n^{(-)} - R_n^{(+)}) = (\tfrac{1}{2})\,[1 + (s/\sigma)]$$
$$A^{(-)} = 1 - A^{(+)} = (\tfrac{1}{2})\,[1 - (s/\sigma)].$$

$$(9.85)$$

In the kinematical approximation $\sigma \to |\,s\,|$, and one of the coefficients becomes zero, which evidently means that only 1 of the 2 waves making up the diffracted and transmitted beams remains. We now derive explicit expressions for the amplitudes of the transmitted and the scattered waves.

(i) Transmitted beam. The amplitude of the transmitted beam is given by

$$\psi_T = A^{(+)} \exp\left(2\pi i \bar{k}_0^{(+)} \cdot \bar{r}\right) + A^{(-)} \exp\left(2\pi i \bar{k}_0^{(-)} \cdot \bar{r}\right)$$

or

$$\psi_T = \exp\left[\pi i(\bar{k}_0^{(+)} + \bar{k}_0^{(-)}) \cdot \bar{r}\right] \{A^{(+)} \exp\left[\pi i(\bar{k}_0^{(+)} - \bar{k}_0^{(-)}) \cdot \bar{r}\right]$$
$$+ A^{(-)} \exp\left[-\pi i(\bar{k}_0^{(+)} - \bar{k}_0^{(-)}) \cdot \bar{r}\right]\}. \quad (9.86)$$

We now calculate $k_0^{(+)} \pm k_0^{(-)}$ separately

$$(\tfrac{1}{2})(\bar{k}_0^{(+)} + \bar{k}_0^{(-)}) = K - (\tfrac{1}{2})(\varDelta^{(+)} + \varDelta^{(-)})\,\bar{e} = \bar{K} + (\tfrac{1}{2})s\bar{e} \quad (9.87)$$

$$(\tfrac{1}{2})(\bar{k}_0^{(+)} - \bar{k}_0^{(-)}) = (\tfrac{1}{2})(\varDelta^{(-)} - \varDelta^{(+)})\,\bar{e} = -(\tfrac{1}{2})t_e^{-1}\sqrt{1 + (st_e)^2} = -(\tfrac{1}{2})\sigma\bar{e}.$$

From this follows

$$\exp\left[\pi i(\bar{k}_0^{(+)} + \bar{k}_0^{(-)}) \cdot \bar{r}\right] = \exp\left(2\pi i\bar{K} \cdot \bar{r}\right) \exp\left(\pi i s\bar{r} \cdot \bar{e}\right)$$
$$= \exp\left(2\pi i\bar{K} \cdot \bar{r}\right) \exp\left(\pi i s z\right)$$

since $\bar{r} \cdot \bar{e} = z$.

Finally the amplitude of the transmitted wave is given by

$$\psi_T = \exp\left(2\pi i\bar{K} \cdot \bar{r}\right) \exp\left(\pi i s z\right)$$
$$\times \left[\left(\tfrac{1}{2}\right)\left(1 + \frac{s}{\sigma}\right) \exp\left(-\pi i\sigma z\right) + \left(\tfrac{1}{2}\right)\left(1 - \frac{s}{\sigma}\right) \exp\left(\pi i\sigma z\right)\right] \quad (9.88)$$

or

$$\psi_T = \exp\left(2\pi i\bar{K} \cdot \bar{r}\right) \exp\left(\pi i s z\right) \left[\cos \pi\sigma z - \frac{is}{\sigma} \sin \pi\sigma z\right]. \quad (9.89)$$

(ii) Scattered beam. The amplitude of the scattered wave is calculated in the same way. It is given by:

$$\psi_S = A^{(+)}R^{(+)} \exp\left(2\pi i\bar{k}_n^{(+)} \cdot \bar{r}\right) + A^{(-)}R^{(-)} \exp\left(2\pi i\bar{k}_n^{(-)} \cdot \bar{r}\right). \quad (9.90)$$

Since $\bar{k}_n = \bar{k}_0 + \bar{n}$ one obtains

$$\psi_S = \exp\left[\pi i(\bar{k}_n^{(+)} + \bar{k}_n^{(-)}) \cdot \bar{r}\right] \{A^{(+)}R^{(+)} \exp\left[\pi i(\bar{k}_n^{(+)} - \bar{k}_n^{(-)}) \cdot \bar{r}\right]$$
$$+ A^{(-)}R^{(-)} \exp\left[-\pi i(\bar{k}_n^{(+)} - \bar{k}_n^{(-)}) \cdot \bar{r}\right]\}. \quad (9.91)$$

Some of the coefficients are now calculated separately

$$A^{(+)}R^{(+)} = \left(\frac{1}{2}\right)\left(1 + \frac{s}{\sigma}\right) t_e(-s + \sigma) = \left(\frac{1}{2}\right)(\sigma t_e)^{-1}$$

$$A^{(-)}R^{(-)} = -\left(\tfrac{1}{2}\right)(\sigma t_e)^{-1} \quad (9.92)$$

$$\left.\begin{array}{l} \bar{k}_n^{(+)} + \bar{k}_n^{(-)} = \bar{k}_0^{(+)} + \bar{k}_0^{(-)} + 2\bar{n} = 2\bar{K} + 2\bar{n} + s\bar{e} \\[4pt] \bar{k}_n^{(+)} - \bar{k}_n^{(-)} = \bar{k}_0^{(+)} - \bar{k}_0^{(-)} = -\sigma\bar{e}. \end{array}\right\} \quad (9.93)$$

Putting these into Eq. (9.91) yields:

$$\psi_S = \exp\left[2\pi i(\bar{K} + \bar{n})\cdot\bar{r}\right] \exp\left(\pi isz\right) \left\{\left(\tfrac{1}{2}\right)(\sigma t_e)^{-1}\left[\exp\left(-\pi i\sigma z\right) - \exp\left(\pi i\sigma z\right)\right]\right\}$$

or after some transformation

$$\psi_S = \exp\left(2\pi i\bar{K}_n\cdot\bar{r}\right)\exp\left(\pi isz\right)i(\sigma t_e)^{-1}\sin\pi\sigma z \quad (9.94)$$

where $\bar{K}_n = \bar{K} + \bar{n}$.

(iii) Complete wave function. The complete wave function is now

$$\Psi = \exp\left(2\pi i\bar{K}\cdot\bar{r}\right)\left[T' + \exp\left(2\pi i\bar{n}\cdot\bar{r}\right)S'\right] \quad (9.95)$$

where

$$T' = \left(\cos\pi\sigma z - \frac{is}{\sigma}\sin\pi\sigma z\right)\exp\left(\pi isz\right) \quad (9.96)$$

$$S' = \frac{i}{\sigma t_e}\sin\pi\sigma z\exp\left(\pi isz\right). \quad (9.97)$$

A discussion of these equations is given below.

(b) System of linear differential equations: Darwin-Howie-Whelan method.[227] We now discuss an alternative way of solving Schrödinger's equation which leads to a simple system of ordinary linear first-order differential equations for the amplitudes of transmitted and scattered waves. This second method is particularly useful for discussing the diffraction effects due to imperfections.

One starts again with Schrödinger's equation in the form (9.58). Instead of Eq. (9.61) we now propose as a solution of the Bloch form:[230a]

$$\Psi(\bar{r}) = \sum_n \psi_n(\bar{r})\, e^{2\pi i k_n\cdot r} \quad (9.98)$$

[230a] Notice that the ψ_n are functions of $\bar{r}$; this was not the case in the previous paragraph. The solutions are strictly speaking *not* of the Bloch form. This is of course connected with the fact that a *finite* crystal or a deformed crystal is *not* perfectly periodic and hence a Bloch solution is not expected.

All diffracted beams are now taken into account.

The wave transmitted in the incident direction corresponds in particular to the term in $n = 0$:

$$\Psi_0(\bar{r}) = \psi_0(\bar{r})\, e^{2\pi i \bar{k}_0 \cdot \bar{r}} \qquad (9.99)$$

while the other waves represent diffracted waves.

$$\Psi_n(\bar{r}) = \psi_n(\bar{r})\, e^{2\pi i \bar{k}_n \cdot \bar{r}}. \qquad (9.100)$$

Substitution of Eqs. (9.98) and (9.59) into (9.58) and identifying the terms containing the same exponential separately leads to the following set of equations.[230b]

$$(4\pi^2)^{-1}\Delta[\psi_n(\bar{r})\, e^{2\pi i \bar{k}_n \cdot \bar{r}}] + K^2 \psi_n(\bar{r})\, e^{2\pi i \bar{k}_n \cdot \bar{r}} = -\sum_m V_{n-m}\psi_m\, e^{2\pi i \bar{k}_n \cdot \bar{r}}. \qquad (9.101)$$

We now make use of the following identity to transform the left-hand side of Eq. (9.101)

$$\Delta uv = u\Delta v + v\Delta u + 2\nabla u \cdot \nabla v.$$

One finds

$$\Delta \Psi_n = e^{2\pi i \bar{k}_n \cdot \bar{r}}\Delta \psi_n + 4\pi i e^{2\pi i \bar{k}_n \cdot \bar{r}}\,\bar{k}_n \cdot \nabla \psi_n - 4\pi^2 k_n^2 \psi_n\, e^{2\pi i \bar{k}_n \cdot \bar{r}}. \qquad (9.102)$$

Substituting this expression in Eq. (9.101) and leaving out common factors, we obtain the following infinite set of first-order differential equations

$$\frac{i}{\pi}\,\bar{k}_n \cdot \nabla \psi_n + (K^2 - k_n^2)\,\psi_n(\bar{r}) = -\sum V_{n-m}\psi_m. \qquad (9.103)$$

We have hereby neglected $\Delta \psi_n$; this is justified since $\Delta \psi_n$ is much smaller than $\bar{k}_n \cdot \nabla \psi_n$ and even smaller as compared to the other terms. This follows from the fact that k_n is very large for electrons.

The geometrical significance of the left-hand side of Eq. (9.103) will be derived with reference to Fig. 103a. The line ZZ' is the Brillouin zone boundary of N. The deviation from the exact reflecting position is described either by the parameter x_n, the distance from the zone boundary to C, or by s_n as before. We shall count s_n positive

---

[230b] Mathematically speaking this procedure is subject to criticism. In the context used here it can be justified on physical arguments.

if it has the same sense as the incident wave vector $\bar{k}_0$, in other words, if the reciprocal lattice point is inside Ewald's sphere. With this convention and the reference system of Fig. 103, x_n and s_n have the same sign.

The z axis is chosen parallel to the zone boundary and in the same sense as $\bar{k}_0$. We then have

$$\bar{k}_n \cdot \nabla \psi_n = \bar{k}_{n,z} \frac{\partial \psi_n}{\partial z} + \bar{k}_{n,x} \frac{\partial \psi_n}{\partial x} \simeq k_n \left(\frac{\partial \psi_n}{\partial z} + \vartheta_n \frac{\partial \psi_n}{\partial x} \right) \qquad (9.104)$$

where we have made the approximation $k_n \simeq k_{n,z}$ and

$$\vartheta_n \simeq \mathrm{tg}\, \vartheta_n = \frac{k_{n,x}}{k_{n,z}} . \quad \text{230c}$$

Further we have $\vartheta_0 = - \vartheta_n (n = 0)$. This approximation is justified because in electron diffraction the angles ϑ_n are very small ($\vartheta_n \simeq 10^{-2}$ rad).

The second term of the left-hand side can be transformed into

$$K^2 - k_n^2 \simeq k_0^2 - k_n^2 = (k_0 + k_n)(k_0 - k_n) \simeq 2k_0 s_n \simeq 2K s_n \qquad (9.105)$$

since $k_n \simeq k_0 = K$. Further one has $s_0 = 0 \cdot (K^2 - k_0^2 = 0!)$.

Equation (9.103) then finally becomes

$$\frac{\partial \psi_n}{\partial z} + \vartheta_n \frac{\partial \psi_n}{\partial x} = 2\pi i \left(s_n \psi_n + \sum_m \frac{V_{n-m}}{2k_0} \psi_m \right). \qquad (9.106)$$

[230c] The relation (9.104) can still be generalized to the case where the deviation from the exact Bragg condition is described, by means of 2 parameters instead of the single one x (or s). A complete description of the deviation from the exact reflecting position is given by the vector $\bar{\rho} = \overline{CO}$ where C is the projection of the center of Ewald's sphere on the reciprocal lattice plane parallel to the foil plane $\overline{CN} = \bar{n} - \bar{\rho}$. The expression (9.104) is then

$$\bar{k}_n \cdot \nabla \psi_n = k_n \left(\frac{\partial \psi_n}{\partial z} - \frac{\rho}{k_0} \nabla_2 \psi_n \right)$$

where ∇_2 means

$$\bar{e}_x \frac{\partial}{\partial x} + \bar{e}_y \frac{\partial}{\partial y} .$$

The y direction is perpendicular to the x direction. If C is situated on the line ON, or if the y dependence can be neglected, this expression is equivalent to (9.104).

The quantities $V_{n-m} \mid 2k_0 = V_{n-m} \mid 2K$ are directly related to the extinction distance:[230d]

$$\frac{V_{n-m}}{2k_0} = \left(\frac{1}{2}\right) t_{n-m}^{-1}. \tag{9.107}$$

We now adopt a further simplification. The term containing the factor ϑ will be neglected. This is equivalent to the "column" approximation introduced earlier in the kinematical approach. It is now clear that this approximation is justified, because not only are the angles ϑ_n small, but moreover the amplitudes do not change very much with x, except near to a dislocation line. In perfect crystals the procedure is exact since there is no x or y dependence of ψ.

For this column approximation it is possible to formulate the system in a more concise way. Let Ψ be the column vector Ψ_0, $\Psi_1 \ldots, \Psi_n$ and let us define the $\mathscr{A}$ matrix with elements,

$$a_{00} = 0 \qquad a_{nn} = s_n \qquad a_{nm} = \tfrac{1}{2} t_{n-m}^{-1} (n \neq m) \tag{9.108}$$

then the system (9.106) can be written as

$$\frac{d\Psi}{dz} = 2\pi i \mathscr{A} \Psi. \tag{9.109}$$

This system of equations was derived by Howie and Whelan[227] from Schrödinger's equation in a different way.

(c) Two beam approximation. The system of Eqs. (9.106) or (9.109) will now be simplified further by adopting the 2-beam approximation: the directly transmitted beam is ψ_0 and the only diffracted beam is ψ_n. We then obtain

$$\frac{\partial \psi_0}{\partial z} = 2\pi i \frac{V_{-n}}{2k_0} \psi_n \tag{9.110}$$

$$\frac{\partial \psi_n}{\partial z} = 2\pi i s_n \psi_n + 2\pi i \frac{V_n}{2k_0} \psi_0. \tag{9.111}$$

[230d] One should in fact write

$$\frac{V_{n-m}}{k_0} = t_{n-m}^{-1} e^{i\phi_{n-m}}$$

where the ϕ_{n-m} are the phase factors of the Fourier coefficients. If the origin is placed at a center of symmetry if present the phases reduce to 0 or π and the exponential factors reduce to plus or minus unity.

With the substitution

$$\psi_0 = T'' \quad \text{and} \quad \psi_n = S'' \, e^{2\pi i s z} \tag{9.112}$$

which preserves the boundary conditions and does not change the intensities, this system of equations becomes

$$\frac{dT''}{dz} = \frac{\pi i}{t_e} S'' \, e^{2\pi i s z} \tag{9.113}$$

$$\frac{dS''}{dz} = \frac{\pi i}{t_e} T'' \, e^{-2\pi i s z} \tag{9.114}$$

with

$$s = s_n; \quad \frac{|V_{-n}|}{k_0} = \frac{|V_n|}{k_0} = 1/t_e. \tag{9.115}$$

This is the system of equations given by Howie and Whelan.[230e,231a]

The relation between the functions T', S' defined in the previous paragraph (9.96, 9.97) and T'' and S'' defined here is following:

$$T_{\bullet} = T'; \; S' = S'^{\prime} \, e^{2\pi i s z}. \tag{9.116}$$

The following substitution

$$\psi_0 = T \, e^{\pi i s z}; \, \psi_n = S \, e^{\pi i s z} \tag{9.117}$$

preserves again the boundary conditions and intensities, and also the phase relationship. The system of Eqs. (9.110) and (9.111) now becomes

$$\frac{dT}{dz} + \pi i s T = \frac{i\pi}{t_e} S \tag{9.118}$$

$$\frac{dS}{dz} - \pi i s S = \frac{i\pi}{t_e} T. \tag{9.119}$$

[230e] A. Howie and M. J. Whelan, *Proc. European Reg. Conf. on Electron Microscopy*, *Delft, 1960*, Vol. 1, p. 194 (1961).

[231a] Notice the difference in the sign of s in this system of equations and in the system as given by Howie and Whelan.[230e] Our Eqs. (9.110) and (9.111) are however identical to Eqs. (11a) and (11b) of their paper.[227] The same difference in the sign of s occurs between the system (11a, 11b) in Howie and Whelan[227] and in Howie and Whelan.[230e] Apart from this sign difference the system of Eq. (11a, b), can be reduced to (1) of paper by Howie and Whelan[230e] by the substitution

$$T' = T \quad S' = S e^{2\pi i s z}.$$

We shall make use of this system later. The relation between S, T and S'', T'' is

$$T = T'' e^{-\pi i s z} \qquad S = S'' e^{\pi i s z}. \qquad (9.120)$$

In the kinematical limit $T'' = 1$ and the system of Eqs. (9.113) and (9.114) reduces to

$$\frac{dS''}{dz} = \frac{\pi i}{t_e} e^{-2\pi i s z}$$

which immediately leads to

$$S'' = \frac{\pi i}{t_e} \int_0^t e^{-2\pi i s z} dz \qquad (9.121)$$

which can be compared with formula (9.16).

(2) *Transmitted and scattered amplitudes for a perfect crystal foil.* The system of Eqs (9.118) and (9.119) is solved by eliminating respectively S and T from the two equations. One obtains the same second-order equation with constant coefficients for S and T.

$$\frac{d^2 T}{dz^2} + (\pi\sigma)^2 T = 0. \qquad (9.122)$$

The solution, taking into account the boundary conditions $T(0) = 1$ and $S(0) = 0$ for $z = 0$, is:

$$T(z) = \cos \pi\sigma z - i\frac{s}{\sigma} \sin \pi\sigma z \qquad (9.123)$$

$$S(z) = \frac{i}{\sigma t_e} \sin \pi\sigma z. \qquad (9.124)$$

These expressions are evidently identical to Eqs. (9.96) and (9.97) obtained directly from Schrödinger's equation since the following relation exists

$$T' = T e^{\pi i s z} \qquad S' = S e^{\pi i s z}. \qquad (9.125)$$

In the kinematical approximation $\sigma \to |s|$ and they reduce to

$$T(z) = e^{-\pi i s z} \qquad S(z) = \frac{i}{s t_e} \sin \pi s z. \qquad (9.126)$$

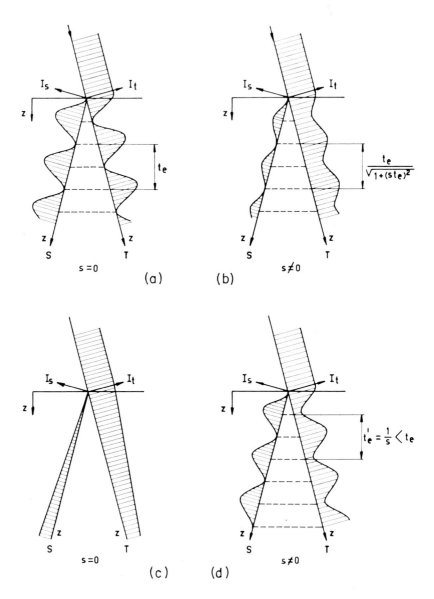

Fig. 104. Schematic representation of the periodic behavior of the intensity of the scattered and transmitted beam for a perfect crystal foil according to the kinematical and the dynamical theory. (a) Dynamical: $s = 0$. (b) Dynamical: $s \neq 0$. (c) Kinematical: $s = 0$. (d) Kinematical: $s \neq 0$.

The expressions for the scattered intensity

$$I_s = SS^* = \frac{\sin^2 \pi \sigma z}{(\sigma t_e)^2} \qquad (9.127)$$

can be compared directly with expression (9.17) for the same quantity, but in the kinematical approximation. Formally the expression is the same, but s has to be replaced by σ. For large s values such that $st_e \gg 1$, σ reduces to $|s|$ and both expressions are identical. The depth periodicity which was $t'_e = 1/s$ in the kinematical case, is now

$$t''_e = \frac{1}{\sigma} = t_e \frac{1}{\sqrt{1 + (st_e)^2}}.$$

For $s = 0$ on the other hand, the difficulty encountered in the kinematical approach is avoided:

$$I_s = \sin^2 \frac{\pi}{t_e} z. \qquad (9.128)$$

The expression for the transmitted intensity is

$$I_t = TT^* = 1 - I_s \qquad (9.129)$$

and this is complementary to the scattered intensity. This is evident since absorption has been neglected. For $s = 0$ it becomes

$$I_t = \cos^2 \frac{\pi}{t_e} z. \qquad (9.130)$$

Figure 104 is a schematic representation of the transmission and scattering of electrons through a crystal foil, while Fig. 105 shows

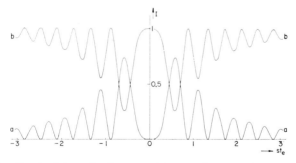

FIG. 105. Graph of intensity I_s scattered by a perfect crystal versus st_e according to the dynamical approximation, *not* taking into account absorption ($t = 3t_e$). This graph can be compared with Fig. 71 valid for the kinematical approximation.

a typical graph of I_s versus st_e. This curve can be compared with Fig. 71 valid for the kinematical theory.

(3) *Anomalous absorption.* Apart from "normal" absorption, which can be accounted for by a factor of the form $e^{-\mu z}$ where μ is the absorption coefficient, there is a so-called "anomalous" absorption, which is due to the loss of electrons by inelastic scattering processes.

The normal absorption is taken into account formally by replacing the inverse of the extinction distance corresponding to the mean inner potential

$$\frac{1}{t_0} = \frac{(2me/h^2)\, V_0}{K} \tag{9.131}$$

by

$$\frac{1}{t_0} + \frac{i}{\tau_0} \tag{9.132}$$

where $(2\pi/\tau_0) = \mu$ is the mean absorption coefficient. The quantity K then becomes complex and in Eqs. (9.89) and (9.94) a real decreasing exponential factor comes in front. This attenuation factor describes normal absorption. An explicit expression for this complex quantity K^* can easily be deduced from Eq. (9.60) in the following way

$$K = \sqrt{\frac{2meE}{h^2}} \left(1 + \frac{V_0}{E} \right)^{1/2} \simeq x \left(1 + \frac{V_0}{2E} \right) \text{ with } x^2 = \frac{2meE}{h^2}. \tag{9.133}$$

Using the expression (9.131) for t_0 this can be transformed into

$$K = x \left(1 + \frac{Kh^2}{4meE} \frac{1}{t_0} \right). \tag{9.134}$$

Performing now the substitution Eq. (9.132) in (9.134) K becomes K^*

$$K^* = x \left[1 + \frac{Kh^2}{4meE} \left(\frac{1}{t_0} + \frac{i}{\tau_0} \right) \right] = K + ix \frac{Kh^2}{4meE} \frac{1}{\tau_0}. \tag{9.135}$$

Taking into account that $eE = K^2h^2/2m$ yields:

$$K^* = K + (i/2\tau_0). \tag{9.136}$$

When substituting K^* for K in Eqs. (9.88) and (9.94) for the amplitudes, a factor $\exp(-\pi z/\tau_0)$ arises in both expressions. The absorption coefficient for intensities consequently becomes $\mu = 2\pi/\tau_0$.

In a similar way the anomalous absorption is introduced phenomenologically by replacing

$$\frac{1}{t_e} \rightarrow \frac{1}{t_e} + \frac{i}{\tau_e} \tag{9.137}$$

where τ_e has the dimensions of a length and is called the "absorption length."[227,230d]

Making this substitution in the expressions for transmitted and scattered intensity (9.129) for (9.127), we obtain, leaving out the attenuation factor describing normal absorption[231b]:

$$I_t^{(a)} = [\cosh u + (s/\sigma_r) \sinh u]^2 - [1/(\sigma_r t_e)^2] \sin^2 v \tag{9.138}$$

where $u = \pi\sigma_i z$; $v = \pi\sigma_r z$; $1/\sigma_i = \sigma_r t_e \tau_e$; $\sigma_r = [1 + (st_e)^2]^{1/2}/t_e$.
Hereby the quantity

$$\frac{\alpha}{1 + (st_e)^2} = \frac{s}{\sigma_r} \sin 2\pi\sigma_r z$$

and higher powers of α were neglected, with respect to the other terms; $\alpha = t_e/\tau_e \simeq 0.1$. The case of no absorption can be found by taking the limit for $\tau_e \rightarrow \infty$.

$I_t^{(a)}$ then reduces to the expression for I_t (9.129) as it should. The squared quantity therefore replaces 1 in the nonabsorption case. A typical curve representing $I_t^{(a)}$ versus st_e is shown in Fig. 106.

The scattered intensity becomes[231b]:

$$I_s^{(a)} = \frac{1}{(\sigma_r t_e)^2} \sinh^2 u + \sin^2 v \tag{9.139}$$

which again in the limit of $\tau_e \rightarrow \infty$ reduces to the previously found formula (9.127) for nonabsorption. Physically this corresponds to the fact that a large absorption length τ_e means little absorption.

It should be noted that the expression for the transmitted intensity is asymmetrical in s:

$$I_t^{(a)}(s) \neq I_t^{(a)}(-s) \tag{9.140}$$

while

$$I_s^{(a)}(s) = I_s^{(a)}(-s). \tag{9.141}$$

[231b] R. Gevers, *Phil. Mag.* [8], **7**, 1681 (1962).

This effect, which is known in X-ray diffraction as the Borrmann effect,[232a] is responsible for the difference in transmission for $s > 0$ and $s < 0$. Figure 106b shows the scattered intensity as a function of st_e when absorption is taken into account; $\tau_e = 10t_e$; $t = 3t_e$. It is important to note that the location of the extrema is not changed appreciably as a consequence of the absorption.

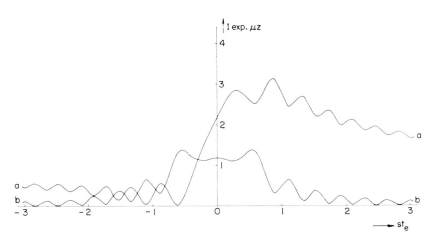

FIG. 106.   Graph of scattered and transmitted intensity versus st_e for a perfect foil, taking into account anomalous absorption. The thickness of the foil is $t = 3t_e$; the absorption length is taken as $\tau_e = 10t_e$. Curve (a) $I^{(a)}$ versus st_e. Curve (b) $I_s^{(a)}$ versus st_e.

The anomalous absorption results in a selective absorption of one of the waves, the interference of which causes the periodic character of T and S. This will now be demonstrated for the transmitted beam ψ_T. From the formulas (9.86) and (9.88) one deduces that ψ_T can be rewritten as

$$\psi_T = [A^{(+)} e^{2\pi i[\bar{K}-(\sigma\bar{e}/2)+(s\bar{e}/2)]\cdot\bar{r}} + A^{(-)} e^{2\pi i[\bar{K}+(\sigma\bar{e}/2)+(s\bar{e}/2)]\cdot\bar{r}}] e^{-\mu_a z} \quad (9.142)$$

where $\bar{e}$ is the unit vector along the z axis, and where the $A^{(\pm)}$ were defined as $A^{(\pm)} = (1/2)\,[1 \pm (s/\sigma)]$ [eq. (9.85)]. From this form of the expression it is clear that the final transmitted wave results from the interference between 2 waves with slightly different wave vectors.

$$\bar{K} + \tfrac{1}{2}\,(\sigma + s)\,\bar{e} \quad \text{and} \quad \bar{K} - \tfrac{1}{2}\,(\sigma - s)\,\bar{e}. \quad (9.143)$$

[232a] G. Borrmann, *Physik Z.* **42**, 157 (1941); *Z. Physik* **127**, 297 (1950).

The effect of including the anomalous absorption is to make σ [defined in (9.78)] complex: for instance $\sigma = \sigma_r + i\sigma_i$. For $s = 0$ one has $\sigma_r = 1/t_e$ and $\sigma_i = 1/\tau_e$ ($\sigma_i < \sigma_r$; $\sigma > 0$). The normal absorption is accounted for by the attenuation factor $e^{-\mu_a z}$ (μ_a = amplitude absorption coefficient).

One has then for the simple case, $s = 0$,

$$\psi_T = (\tfrac{1}{2})\, e^{-\mu_a z} \{ e^{\pi\sigma_i z}\, e^{2\pi i[\bar{K}-(\sigma_r/2)\bar{e}]\cdot\bar{r}} + e^{-\pi\sigma_i z}\, e^{2\pi i[\bar{K}+(\sigma_r/2)i]\cdot\bar{r}} \}. \qquad (9.144)$$

From this it is clear that the second of these waves will be attenuated much more rapidly in depth than the other one because of the factor $\exp[-(\mu_a + \sigma_i)z]$. Divergence of the amplitude of the first wave is suppressed by the factor $e^{-\mu_a z}$.

It is noteworthy that in the kinematical limit $|s| = \sigma_r$, either $A^{(+)}$ or $A^{(-)}$ is zero and only one wave is left. An estimate of τ_e can be obtained from a quantitative study of the thickness fringes in a wedge-shaped crystal. The method is discussed in Section 9,g,(3).

Just recently some light has been shed on the mechanism of anomalous absorption, by showing that the dark absorption bands observed in thin metal foils emit more X-rays than the other regions.[232b,c] This phenomenon can be understood if one notices that the rapidly attenuated Bloch wave has its maxima at the nuclear sites in the exact Bragg position, while the "passing" wave has its maxima in between the nuclear positions. Since the probability for exciting X-ray emission must be larger close to the nucleus than elsewhere it is to be expected that the wave which has its maxima at the nuclear positions will be more rapidly attenuated because it spends its energy in exciting X-rays.

(4) *Classical derivation of the system of equations of Darwin-Howie-Whelan.* The system of first-order differential equations can also be derived by the use of purely physical arguments, based on a classical picture. The reasoning is not completely rigorous in several respects, but it gives direct physical insight.

Consider a slab of crystal dz on a crystal of thickness z (Fig. 107) on which a beam is incident with the Bragg angle ϑ. The plane of the drawing contains the incident wave vector and the normal to the diffracting planes, which are themselves normal to the foil plane. We

[232b] P. B. Hirsch, A. Howie, and M. J. Whelan, *Phil. Mag.* [8] **7**, 2095 (1962).
[232c] P. Duncumb, *Phil. Mag.* [8] **7**, 2101 (1962).

limit ourselves to the symmetrical Laue case; it is also assumed that the incident beam has the same amplitude over the part of the entrance face considered. The amplitude of the wave transmitted in a point like Q will now be calculated. This amplitude results from interference between: (1) the wave transmitted by the slab (Oz) in the point P, and again transmitted through the slab dz, and (2) the

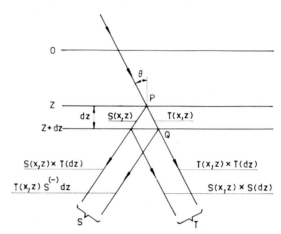

FIG. 107.   Illustrating the classical derivation of the system of differential equations of Darwin, Howie, and Whelan.

wave scattered by the slab (Oz) in the point P and scattered back into the original direction by the slab dz. One obtains

$$T(x + dz \, \text{tg } \vartheta, z + dz) = T(x, z) \, T(dz) + S(x, z) \, S(dz). \qquad (9.145)$$

Now we take into account that

$$T(dz) = 1 \qquad (9.146)$$

since the slab dz is very thin (eventually of atomic dimensions), and the kinematical approximation is certainly valid. According to the same approximation $S(dz)$ is proportional to dz. We can put

$$S(dz) = \frac{i\pi}{t_e} \, e^{2\pi i s z} \, dz. \qquad (9.147)$$

This is justified by the following argument. The coefficient of proportionality has to have the dimensions of an inverse length $1/t_e$.

The factor $i\pi$ accounts for the phase shift of $\pi/2$ on diffraction; the exponential factor is the kinematical phase factor. It assures that the amplitudes are added with the correct phase relationship.

From Eq. (9.145), (9.146), and (9.147) follows immediately, if $\operatorname{tg} \vartheta \simeq \vartheta$

$$T(x + \vartheta dz, z + dz) - T(x, z) = \frac{i\pi}{t_e} e^{2\pi i sz} S(x, z)\, dz. \quad (9.148)$$

On developing the left-hand side in a Taylor expansion and keeping only first-order terms:

$$\frac{\partial T}{\partial z} + \vartheta \frac{\partial T}{\partial x} = \frac{i\pi}{t_e} e^{2\pi i sz} S. \quad (9.149)$$

The second equation can be derived by using a completely similar argument; one finds:

$$S(x - dz \operatorname{tg} \vartheta, z + dz) = T(x, z) S^{(-)}(dz) + S(x, z) T(dz). \quad (9.150)$$

The factor $S^{(-)}dz$ represents the amplitude of the wave scattered by the slab dz; the minus sign indicates that scattering is in the reverse sense of that considered above. This can be accounted for by changing the sign of s. Taking into account Eqs. (9.146) and (9.147) and changing the sign of s gives, to the same approximation as above,

$$\frac{\partial S}{\partial z} - \vartheta \frac{\partial S}{\partial x} = \frac{i\pi}{t_e} e^{-2\pi i sz} T. \quad (9.151)$$

(5) *Imperfect crystals.* The assumption is made that the deformation is not too severe, so that it is still possible to consider a reciprocal lattice. The deformation can be described by means of a potential

$$V(\bar{r} - \bar{u}) \quad (9.152)$$

where $\bar{u}(\bar{r})$ is the displacement function. The approximation is equivalent to identifying the lattice potential in a part $\bar{r}$ of the deformed crystal with the potential in the perfect crystal but taken at a point $\bar{r} - \bar{u}$.

As above we can expand V

$$V(\bar{r} - \bar{u}) = \sideset{}{'}\sum_{n} V_n\, e^{2\pi i \bar{n} \cdot \bar{r}}\, e^{-2\pi i \bar{n} \cdot \bar{u}} = \sideset{}{'}\sum_{n} V_n\, e^{2\pi i \bar{n} \cdot \bar{r}}\, e^{-2\pi i \alpha'_n}$$

$$= \sum_{n} (V_n\, e^{-2\pi i \alpha'_n})\, e^{2\pi i \bar{n} \cdot \bar{r}}, \qquad (9.153)$$

where $\alpha'_n = \bar{n} \cdot \bar{u}$.[232d]

The equations corresponding to (9.106) now become

$$\frac{\partial \psi_n}{\partial z} + \vartheta_n \frac{\partial \psi_n}{\partial x} = 2\pi i s_n \psi_n + \sum_{m} \frac{\pi i}{t_{n-m}} e^{i\phi_{n-m}} \psi_m\, e^{-2\pi i (\alpha'_n - \alpha'_m)} \qquad (9.154)$$

where use was made of Eq. (9.107).

If we limit ourselves to the 2-beam column approximation this set of equations reduces to

$$\left. \begin{aligned} \frac{\partial \psi_0}{\partial z} &= \frac{\pi i}{t_{-n}} e^{i\phi_{-n}} \psi_n\, e^{-2\pi i \alpha'_{-n}} \\[2mm] \frac{\partial \psi_n}{\partial z} &= 2\pi i s_n \psi_n + \frac{\pi i}{t_n} e^{i\phi_n} \psi_0\, e^{-2\pi i \alpha'_n}. \end{aligned} \right\} \qquad (9.155)$$

With a proper choice of the origin $e^{i\phi_{-n}}$ and $e^{i\phi_n}$ reduce to unity and we have further $t_n = t_{-n} = t_e$. It is noted that, since $\alpha'_n = \bar{n} \cdot \bar{u}$, we also have $\alpha'_n = -\alpha'_{-n} = \alpha'$; $s_n = s$.

With these simplifications, the equations finally reduce to

$$\left. \begin{aligned} \frac{d\psi_0}{dz} &= \frac{\pi i}{t_e} \psi_n\, e^{2\pi i \alpha'} \\[2mm] \frac{d\psi_n}{dz} &= 2\pi i s \psi_n + \frac{\pi i}{t_e} \psi_0\, e^{-2\pi i \alpha'}. \end{aligned} \right\} \qquad (9.156)$$

Performing the substitution Eq. (9.112) the system (9.156) reduces to

$$\left. \begin{aligned} \frac{dT''}{dz} &= \frac{\pi i}{t_e} S''\, e^{2\pi i (sz + \alpha')} \\[2mm] \frac{dS''}{dz} &= \frac{\pi i}{t_e} T''\, e^{-2\pi i (sz + \alpha')} \end{aligned} \right\} \qquad (9.157)$$

which is the generalization of Eqs. (9.113) and (9.114).

[232d] From this derivation it is clear that the Fourier coefficient V_n of the unde-formed crystal becomes $V_n e^{-i\alpha_n}$ in the deformed crystal. This means that effectively $1/t_{e,n} = (|V_n|/k_0)$ becomes in the deformed crystal $1/t_{e,n} e^{-i\alpha_n}$. This substitution formally takes into account the deformation. This is, e.g., evident in Eq. (9.154).

The equations corresponding to the system [(9.118), (9.119)] for the perfect crystal can be obtained by means of the following substitution:

$$T'' = T\, e^{\pi i(sz + \alpha')}$$
$$S'' = S\, e^{-\pi i(sz + \alpha')}.$$

$$\left.\begin{array}{c}\\[-1ex]\\\end{array}\right\} \quad (9.158)$$

Equations (9.157) then become

$$\frac{dT}{dz} + \pi i\left(s + \frac{d\alpha'}{dz}\right) T = \frac{i\pi}{t_e}\, S$$

$$\frac{dS}{dz} - \pi i\left(s + \frac{d\alpha'}{dz}\right) S = \frac{i\pi}{t_e}\, T.$$

$$\left.\begin{array}{c}\\[-1ex]\\\end{array}\right\} \quad (9.159)$$

From these equations it is quite clear that the effect of the deformation is to replace s by $(s + d\alpha')/dz$, i.e., a function of z. This can be interpreted to mean that the deformation results in a local change of s. The physical meaning of this is obvious, small regions of the deformed crystal are rotated with respect to the orientation of the perfect crystal; therefore the local s value is also different.

In the kinematical limit we have again from Eqs. (9.157)

$$\frac{dS''}{dz} = \frac{i\pi}{t_e}\, e^{-2\pi i sz}\, e^{-2\pi i \alpha'}$$

or

$$S'' = \frac{i\pi}{t_e} \int_0^t e^{-2\pi i sz}\, e^{-i\alpha}\, dz \qquad (9.160)$$

which can be compared with Eq. (9.22).

g. Experimental Use of the Intensity Profiles

(1) *The observation of* $I_s(s, t)$. When studying carefully the fine structure of diffraction spots it was found that the intensity distribution $I_s(s, t)$ can be observed directly. The Kikuchi lines also exhibit a fine structure. There corresponds to each extremum in I_s a Kikuchi line, black or white, parallel to the main Kikuchi line; we call these supplementary lines "satellites." They provide an easy measurement of the spacing between minima and maxima. Such an observation is shown in Fig. 108 where a large number of satellites can easily be recognized. The acting spot is (10$\bar{1}$0); the specimen is graphite.

FIG. 108. Observation of the fine structure of the diffraction spots as described by Eq. (9.127). The incident beam was made slightly divergent. The specimen is graphite and the most intense spot is ($10\bar{1}0$). [After R. Siems, P. Delavignette, and S. Amelinckx, *Phys. Stat. Solidi* **2**, 421 (1962).]

The observation can be made most conveniently with the Elmiskop I by using a slightly divergent incident beam. This is simply achieved by taking out the condensor diaphragm. This observation suggests that under normal operating conditions the inelastically scattered electrons, which have different incident directions, are responsible for the appearance of the satellites. The admission of the slightly divergent beam allows us in fact to explore the intensity distribution along a reciprocal lattice rod. The relation (9.19) can then be used to measure s values directly on the plate. The foil thickness can be deduced from the difference in s between minima; this is discussed in the next paragraph.

(2) *Determination of the foil thickness.* For many purposes it is of importance to know the foil thickness t, e.g., if one wishes to measure the concentration of dislocation loops introduced by irradiation.

(a) Geometrical methods. For metals one can make use of slip traces left by moving dislocations (Fig. 197). Such slip traces are localized at the upper and lower face of the foil. If one knows the orientation of the foil, and the glide plane, one can deduce the thickness t from the width of the slip trace w. If i is the angle between the normal to the foil and the normal to the slip plane one has

$$t = w \operatorname{tg} i. \tag{9.161}$$

(b) Wedge fringes: extinction distance determination. If the crystal has a wedge shape, thickness extinction contours are visible. The thickness change in going from one extinction contour to the next is $t'' = 1/\sigma$, which can be calculated for a given reflection from the known structure using the formula:

$$\sigma \simeq \sigma_r = \frac{1}{t_e} \sqrt{1 + (st_e)^2}. \tag{9.162}$$

For $s = 0$ this is simply t_e. The value of t_e for the reflection $\mathbf{n}$ can be computed in the following way[229]

$$t_{e,\mathbf{n}} = K/|V_{\mathbf{n}}| \tag{9.163}$$

where K is given by Eqs. (9.60) and $V_{\mathbf{n}}$ is defined in Eq. (9.59).

$$V_n = \frac{2me}{\hbar^2} V'_{\mathbf{n}} \tag{9.164}$$

where V'_n is the nth Fourier coefficient in the Fourier expansion of $V(\bar{r})$. One can write this also

$$t_{e,n} = \lambda E/V'_n$$

since

$$\lambda \simeq h/\sqrt{2meE} \quad (V_0 \ll E)$$

where λ is the wavelength of the electrons used and Ee their initial energy (for $E = 80$ kv: $\lambda = 0.042$ A); V'_n can be expressed as a function of the structure[229]

$$V'_n = \frac{300ed_n^2}{\pi V} \sum_j (Z_j - f_j) \exp(-2\pi i \bar{n} \cdot \bar{r}_j) = \frac{300ed_n^2}{\pi V} F_n \qquad (9.165)$$

e is the electronic charge expressed in esu; $d_{\bar{n}}$ is the interplanar spacing of the planes perpendicular to $\bar{n}$; V is the volume of the unit cell; Z_j is the atomic number; and f_j is the atomic scattering factor for the jth atom at $\bar{r}_j$ in the unit cell. The summation is over all atoms in the unit cell. The factor F_n is measured directly in a diffraction experiment. It can be calculated if the structure is known. The factor in front of F_n can also be written[233a]

$$\frac{600h^2d_n^2 \sin^2 \vartheta}{\pi me V \lambda^2} \qquad (9.166)$$

where m is the mass of the electron.

Conversely one can in principle determine the crystal structure from a measurement of the extinction distances, provided the phase problem can be solved. In the same way fringes at stacking faults, which extend through the whole thickness of the foil, can be used to derive the thickness, provided the extinction distance is known. If the foil plane and the stacking fault plane are known the width of the fault plane allows an independent thickness determination and hence an experimental measurement of the extinction distance t_e.

(c) Diffraction method. For layer structures the previous method cannot be applied so easily, since glide planes are strictly parallel to the foil plane. There is a purely electron optical method which may be useful in such cases. The method is due to Kossel and

[233a] R. M. Fisher and M. J. Marcinkowski, *Phil. Mag.* [8] **6**, 1385 (1961).

Möllenstedt[218,226]; it makes use of the fine structure of the diffraction spots described in Eqs. (9.127) and (9.17). The intensity extrema of I_s are given by the zero's of $\partial I/\partial s$. This leads, in the kinematical region, to $\sin \pi ts = 0$ and

$$\text{tg } \pi ts = \pi ts. \tag{9.167}$$

The first condition, which is equivalent to

$$\pi ts = k\pi \tag{9.168}$$

(k is an integer) determines the minima, which are at the same time zeros of I_s. The second condition determines the maxima. We prefer to make use of the minima since in this approximation they are equally spaced and also because they are better visible. The Δs corresponding to 2 successive minima is

$$\Delta s = 1/t \tag{9.169}$$

and a determination of Δs is then equivalent to a thickness determination.

For the dynamical 2-beam case, the formula for $I_{(s)}$ corresponding to Eq. (9.17) is now from (9.127)

$$I_s(st) = \sin^2 \pi \frac{t}{t_e} \sqrt{1 + (st_e)^2}/[1 + (st_e)^2]. \tag{9.170}$$

Moreover a similar approach can be used which will yield an experimental value for the extinction distance too. The two sets of extrema of $I(s, t)$ are given by

$$\sin \pi \frac{t}{t_e} \sqrt{1 + (st_e)^2} = 0 \tag{9.171}$$

and

$$\text{tg } \pi \frac{t}{t_e} \sqrt{1 + (st_e)^2} = \pi \frac{t}{t_e} \sqrt{1 + (st_e)^2}. \tag{9.172}$$

A typical graph of $I_s(s, t)$ versus st_e for a typical value $t = 3\ t_e$ is shown in Fig. 105. The zeros of this function which are also minima are given by Eq. (9.171); the corresponding s values are:

$$s_k = \pm \left(\frac{k^2}{t^2} - \frac{1}{t_e^2} \right)^{1/2} \tag{9.173}$$

where k is an integer; they are no longer equally spaced and a determination of k becomes necessary. One can use 3 successive minima s_1, s_2, and s_3 and obtain:

$$
\left.
\begin{aligned}
1 + (s_1 t_e)^2 &= (k-1)^2 (t_e/t)^2 \\
1 + (s_2 t_e)^2 &= k^2 (t_e/t)^2 \\
1 + (s_3 t_e)^2 &= (k+1)^2 (t_e/t)^2.
\end{aligned}
\right\}
\qquad (9.174)
$$

From this set of 3 equations one obtains:

$$
t = \sqrt{2}\,(s_1^2 + s_3^2 - 2s_2^2)^{-1/2}. \qquad (9.175)
$$

Further the value of the integer k can be found from relations like:

$$
2k + 1 = (s_3^2 - s_2^2)\,t^2; \quad 2k - 1 = (s_2^2 - s_1^2)\,t^2; \quad 4k = (s_3^2 - s_1^2)\,t^2. \qquad (9.176)
$$

With the knowledge of k and t one can further deduce t_e from one of the relations (9.174). In practice one may try to find first a value for t by the use of relation (9.169) by determining the spacing, Δs, far enough from the center of gravity of the reciprocal lattice point so that the kinematical theory should be applicable. This is verified experimentally if a uniform spacing is found. The effect of anomalous absorption on the position of the extrema is small and can be neglected for this problem as shown by computation of some profiles.

(3) *Measurement of the absorption length* τ_e. An experimental value for τ_e can be deduced from a quantitative study of the thickness fringes of a wedge-shaped crystal in the special case $s = 0$. Formula (9.138) for the intensity of the transmitted wave then reduces to

$$
^{(a)}I_t = I_0\, e^{-\mu t} \left(\cosh^2 \pi\, \frac{t}{\tau_e} - \sin^2 \pi\, \frac{t}{t_e} \right)
$$

$$
= \frac{1}{2} I_0\, e^{-\mu t} \left(\cosh 2\pi\, \frac{t}{\tau_e} + \cos 2\pi\, \frac{t}{t_e} \right) \qquad (9.177)
$$

where now the normal absorption has been taken into account as well. The problem is to evaluate μ, and τ_e; t_e is assumed to be known from formulas (9.163) and (9.165). Also it can be deduced directly from the fringe profile if the wedge angle is known.

The maxima of $^{(a)}I_t$ are reached approximately for $t = kt_e$ and they are then given by

$$^{(a)}I_t^{\max} = \frac{1}{2} I_0 e^{-\mu t} \left(\cosh 2\pi \frac{t}{\tau_e} + 1 \right) \qquad (9.178)$$

whereas the minima are obtained approximately for $t = (k + 1/2)t_e$

$$^{(a)}I_t^{\min} = \frac{1}{2} I_0 e^{-\mu t} \left(\cosh 2\pi \frac{t}{\tau_e} - 1 \right). \qquad (9.179)$$

Considering t as a continuous variable Eqs. (9.178) and (9.179) represent curves passing through the maxima and the minima respectively. Such curves are obtained experimentally by drawing a smooth line through the experimental points corresponding to maxima and minima. By subtracting one obtains

$$^{(a)}I_t^{\max} - {}^{(a)}I_t^{\min} = I_0 e^{-\mu t}$$

and hence

$$\ln \frac{{}^{(a)}I_t^{\max} - {}^{(a)}I_t^{\min}}{I_0} = -\mu t. \qquad (9.180)$$

On a semilog plot μ can be derived as the slope of a straight line. One has further

$$\frac{{}^{(a)}I_t^{\max} + {}^{(a)}I_t^{\min}}{{}^{(a)}I_t^{\max} - {}^{(a)}I_t^{\min}} = \cosh 2\pi \frac{t}{\tau_e}. \qquad (9.181)$$

These relations (9.180) and (9.181) allow us to compute τ_e and μ. The I values are obtained by the use of curves passing through the minima and the maxima.

As a rough rule of thumb one can say that τ_e is twice as many times t_e (for $s = 0$) as the number of fringes that one can distinguish for a wedge-shaped crystal. It should be remembered that τ_e as well as t_e depend on the diffraction vector $\bar{n}$.

Watanabe et al.[233b] have just recently performed measurements which are similar to those proposed here. However, it turns out that the method is not capable of very high accuracy because of the presence of inelastically scattered electrons whose contribution is

[233b] H. Watanabe, H. Fukuhara, and K. Kohra, J. Phys. Soc. Japan 17, 195 (1962).

completely neglected in the formula given above. Also the effect of weak reflections is neglected in the theory, since this is a 2-beam theory.

h. Stacking Fault Contrast: Dynamical Theory

(1) *Amplitude of transmitted and scattered beam of crystal with a single fault.*[231b,234] The scattered and transmitted amplitudes for a crystal containing a stacking fault as in Fig. 109, parallel to the foil plane, can be calculated from the set of Eq. (9.159). In this calculation we will neglect anomalous absorption. We put $\alpha = 2\pi\bar{n} \cdot \bar{R} = 2\pi\alpha'$ where $\bar{R}$ is the displacement vector of the fault. One can proceed in two equivalent ways. It is possible to use the expressions for $T(t_1)$ and $S(t_1)$ for a perfect crystal of thickness t_1 and use these as the initial values $T(0)$ and $S(0)$ when solving the equation for the second crystal. For the second crystal of thickness t_2 the set (9.159) should be used.

It is also possible to write the set of Eq. (9.157) or (9.159) at once in the form (the subindex s means stacking fault):

$$\left.\begin{array}{l} \dfrac{dT_s}{dz} + i\pi s T_s = \dfrac{i\pi}{t_e}[1 - (1 - e^{i\alpha})\,\epsilon]\,S_s \\[3mm] \dfrac{dS_s}{dz} - i\pi s S_s = \dfrac{i\pi}{t_e}[1 - (1 - e^{-i\alpha})\,\epsilon]\,T_s \end{array}\right\} \quad (9.182)$$

where $\epsilon = \epsilon(z - t_1)$ is the step function defined by $\epsilon = 0$ for $z < t_1$ and $\epsilon = 1$ for $z > t_1$.

Use was made of the substitution

$$T''_s = T_s\, e^{\pi i s z - i\alpha}$$
$$S''_s = S_s\, e^{-\pi i s z + i\alpha}$$

$$(9.183)$$

taking into account that for a stacking fault α is constant. This system of equations can be reduced to the following set of second-order equations, in the same way as for the perfect crystal:

$$\left.\begin{array}{l} \dfrac{d^2 T_s}{dz^2} + (\pi\sigma)^2 T_s = -i\,\dfrac{\pi}{t_e}\,(1 - e^{i\alpha})\,\delta(z - t_1)\,S_s \\[3mm] \dfrac{d^2 S_s}{dz^2} + (\pi\sigma)^2 S_s = -\dfrac{i\pi}{t_e}\,(1 - e^{-i\alpha})\,\delta(z - t_1)\,T_s \end{array}\right\} \quad (9.184)$$

[234] The treatment followed here differs considerably from the one given by Whelan and Hirsch[229]; it is an adaption of the method followed by Gevers.[231b]

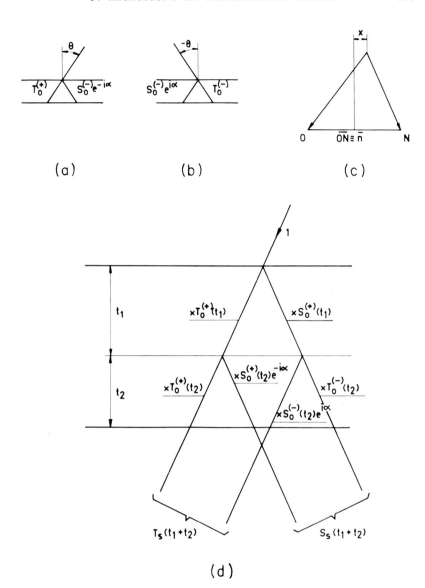

(a)   (b)   (c)

(d)

FIG. 109. Transmitted and scattered beam for a slab of crystal containing a stacking fault. (a) Incident beam under an angle $+\vartheta$ to the normal. (b) Incident beam under an angle $-\vartheta$ to the normal. (c) Reciprocal lattice construction. (d) Interference between the different beams.

where $\delta(z - t_1)$ is Dirac's delta function. The general solution of the homogeneous set of Eq. (9.184) without the right-hand side is given by (9.123) and (9.124). The general solution of an inhomogeneous equation of the form

$$\frac{d^2y}{dx^2} + \gamma^2 y = f(x)$$

can be written as

$$y = y_0(x) + \frac{1}{\gamma} \int_0^x \sin \gamma(x - \tau) f(\tau) \, d\tau \tag{9.185}$$

where y_0 is the solution of the homogeneous equation. Applying this to the system (9.184) leads to:

$$T_s = T_0(z) - \frac{i}{\sigma t_e} (1 - e^{i\alpha}) \int_0^z \sin \pi\sigma(z - \tau) \, \delta(\tau - t_1) \, S_0(\tau) \, d\tau \tag{9.186}$$

or

$$T_s = T_0(z) - \frac{i}{\sigma t_e} (1 - e^{i\alpha}) \sin \pi\sigma(z - t_1) \, S_0(t_1) \tag{9.187}$$

and

$$S_s = S_0(z) - \frac{i}{\sigma t_e} (1 - e^{-i\alpha}) \sin \pi\sigma(z - t_1) \, T_0(t_1) \tag{9.188}$$

where T_0 and S_0 are the solutions of the homogeneous equation, given by Eqs. (9.123) and (9.124). More explicitly we can write

$$T_s(z) = \cos \pi\sigma z - i \frac{s}{\sigma} \sin \pi\sigma z + \frac{1}{(\sigma t_e)^2} (1 - e^{i\alpha}) \sin \pi\sigma t_1 \sin \pi\sigma(z - t_1) \tag{9.189}$$

$$S_s(z) = \frac{i}{\sigma t_e} \left\{ \sin \pi\sigma z - (1 - e^{-i\alpha}) \left[\cos \pi\sigma t_1 - i \frac{s}{\sigma} \sin \pi\sigma t_1 \right] \sin \pi\sigma(z - t_1) \right\}. \tag{9.190}$$

At the back surface $z = t = t_1 + t_2$; t is the total thickness. Interchanging t_1 and $t_2 = z - t_1$ in formula (9.189) does not change T, which means that I_t, the intensity distribution in the bright field fringe pattern, is symmetrical with respect to the foil center. In nonabsorbing crystals this is also true for the dark field image since $I_s = 1 - I_t$.

(2) *Alternative formulation.*[231b] The formulas (9.189) and (9.190) can be transformed by introducing the expressions for the amplitudes scattered and transmitted by the two parts of the crystal separately. Let us call $S_0^{(\pm)}(t_1)$ and $S_0^{(\pm)}(t_2)$ the amplitudes of the waves scattered by perfect crystals of thickness t_1 and t_2 as given by Eq. (9.124) and $T_0^{(\pm)}(t_1)$ and $T_0^{(\pm)}(t_2)$ the amplitudes of the transmitted waves by crystals with thicknesses t_1 and t_2; $T_0(t_1 + t_2)$ and $S_0(t_1 + t_2)$ have obvious meanings. The plus sign means the expression as given by Eqs. (9.123), (9.124), or by the corresponding formula with absorption, and the minus sign the expression after $+s$ is changed to $-s$.

With this notation the relation (9.189) can be rewritten as

$$T_s(t_1 + t_2) = T_0(t_1 + t_2) - (1 - e^{i\alpha}) S_0(t_1) S_0(t_2) \qquad (9.191)$$

or

$$T_s(t_1 + t_2) = T_0^{(+)}(t_1) T_0^{(+)}(t_2) + S_0^{(+)}(t_1) S_0^{(-)}(t_2) e^{i\alpha} \qquad (9.192)$$

and similarly

$$S_s(t_1 + t_2) = S_0(t_1 + t_2) - (1 - e^{-i\alpha}) S_0(t_2) T_0(t_1) \qquad (9.193)$$

or

$$S_s(t_1 + t_2) = S_0^{(+)}(t_1) T_0^{(-)}(t_2) + T_0^{(+)}(t_1) S_0^{(+)}(t_2) e^{-i\alpha}. \qquad (9.194)$$

This shows that we can add up amplitudes scattered by two separate parts of the crystal provided we introduce the right phase factor $e^{i\alpha}$ or $e^{-i\alpha}$.

This remark can be extended and developed into a formalism which allows us to write down immediately scattered and diffracted amplitudes for crystals containing any combination of fault planes. Let us consider again the system of Eq. (9.182) valid for a part of a crystal bounded by a stacking fault

$$\left. \begin{aligned} \frac{dT_s}{dz} + \pi i s T_s &= \frac{i\pi}{t_e} S_s e^{i\alpha} \\[2mm] \frac{dS_s}{dz} - \pi i s S_s &= \frac{i\pi}{t_e} T_s e^{-i\alpha}. \end{aligned} \right\} \qquad (9.195)$$

By a suitable substitution one can reduce this set to the one valid for a perfect crystal and for which we know the solution $T_0(z)$ and

$S_0(z)$. The substitution, which at the same times preserves the boundary conditions $T(0) = 1$ and $S(0) = 0$, is

$$S_s \to S_0 \, e^{-i\alpha} \qquad T_s \to T_0. \tag{9.196}$$

The expressions for the amplitudes of the partial waves, after passage through a fault plane and crystal slab of thickness z, can therefore be written down immediately as shown in Fig. 109d.

If the incident wave has the direction $-\vartheta$, as in Fig. 109b, the origin of the reflecting sphere comes in the point N and vice versa the origin of the reciprocal lattice O becomes the diffracting node point. Evidently this means that $\bar{n}$ becomes $-\bar{n}$ and hence (since $\alpha = 2\pi\bar{n}, \bar{R}$) that α becomes $-\alpha$. Furthermore the deviation from the exact reflecting position given by x(or s) (Fig. 109c) changes its sign as well, i.e., s becomes $-s$. The wave progressing further in the direction $-\vartheta$ is effectively a transmitted wave and the expression T_0 has to be taken with $-s$ also $\alpha \to -\alpha$. The wave propagating in the direction ϑ is now a diffracted wave (with $-s$ and $-\alpha$), i.e., the expression $S_0^{(-)} \, e^{i\alpha}$ has to be used.

In this way the boundary conditions are satisfied as well since the substitution effectively amounts to interchanging S and T.

It is clear from the scheme of Fig. 109a,b,d how for a crystal containing a single fault the total amplitudes of the transmitted and diffracted waves can be written down immediately in the form (9.192) and (9.194). The transmitted wave results from the interference between the doubly transmitted wave $T_0^{(+)}(t_1)\, T_0^{(+)}(t_2)$ and the wave scattered back into the initial direction $S_0^{(+)}(t_1)\, S_0^{(-)}(t_2)e^{i\alpha}$. The scattered wave is produced by interference between the wave scattered in the first part, but transmitted through the second $S_0^{(+)}(t_1)\, T_0^{(-)}(t_2)$ and the wave transmitted through the first part, and subsequently scattered by the second $T_0^{(+)}(t_1)\, S_0^{(+)}(t_2)e^{-i\alpha}$.

Note that, if absorption is neglected,

$$S_0^{(-)}(z) = S_0^{(+)}(z) \tag{9.197}$$

since s occurs only as s^2; however, if absorption is neglected

$$T_0^{(-)}(z) = T_0^{(+)*}(z) \tag{9.198}$$

where the asterisk denotes the complex conjugate.

We have developed the latter formalism in some detail because it may give some physical insight and also because it can easily be generalized to a number of overlapping stacking faults, and other planar defects.

From the last formalism it follows that the fringe pattern observed in transmission on inclined stacking faults can be considered to result from the interference between the directly transmitted beam and the doubly diffracted beam; the mechanism of contrast formation is therefore somewhat similar to that for moiré patterns.[231b]

(3) *Overlapping stacking faults.* Using the formalism developed in the previous paragraph it is possible to write down immediately the amplitudes of the wave transmitted and diffracted by a crystal containing an arbitrary succession of stacking faults in parallel planes as represented in Fig. 110ᵃ.

We shall first condense further the notation in such a way that it can be generalized readily. Let us consider a wave with amplitude 1

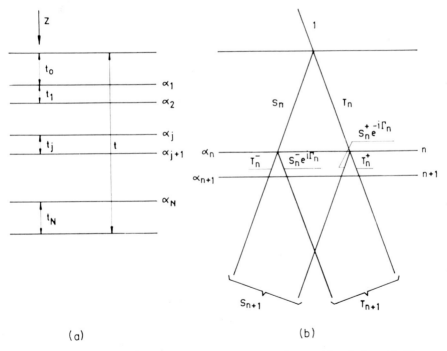

(a)   (b)

FIG. 110. Crystal slab containing several overlapping stacking faults. (a) Illustrating the notations used. (b) Derivation of the relation between S_{n+1}, T_{n+1} and S_n, T_n.

incident on the front face of the crystal. After having passed through $(n-1)$ stacking fault interfaces with total phase shift

$$\Gamma_{n-1} = \sum_{i=1}^{n-1} \alpha_i \qquad (9.199)$$

this wave emerges as two waves; one with amplitude $\mathscr{T}_n$ and which has the incident direction, and one with amplitude $\mathscr{S}_n$ and which has the direction of the Bragg diffracted beam. These two waves are now incident on the nth stacking fault interface. This stacking fault is characterized by a phase shift α_n. According to the formalism of Section 9h(2) the beam $\mathscr{T}_n$ gives rise to a transmitted beam with amplitude $T_n^{(+)}$ (per unit amplitude of the incident wave) and a scattered beam with amplitude $S_n^{(+)} e^{-i\Gamma_n}$ while the scattered beam gives rise to a transmitted beam with amplitude $T_n^{(-)}$ and a scattered beam with amplitude $S_n^{(-)} e^{i\Gamma_n}$. The total amplitude of the wave transmitted through the nth stacking fault is then evidently:

$$\mathscr{T}_{n+1} = T_n^{(+)}\mathscr{T}_n + S_n^{(-)} e^{i\Gamma_n}\mathscr{S}_n \qquad (9.200)$$

whereas the total amplitude of the scattered wave is

$$\mathscr{S}_{n+1} = S_n^{(+)} e^{-i\Gamma_n} \mathscr{T}_n + T_n^{(-)}\mathscr{S}_n. \qquad (9.201)$$

This is a recurrence relation between the $\mathscr{T}_{n+1}$, $\mathscr{S}_{n+1}$ and the $\mathscr{T}_n$, $\mathscr{S}_n$. It can be written down more concisely under matrix form

$$\begin{pmatrix} \mathscr{T}_{n+1} \\ \mathscr{S}_{n+1} \end{pmatrix} = \begin{pmatrix} T_n^{(+)} & S_n^{(-)} e^{i\Gamma_n} \\ S_n^{(+)} e^{-i\Gamma_n} & T_n^{(-)} \end{pmatrix} \begin{pmatrix} \mathscr{T}_n \\ \mathscr{S}_n \end{pmatrix} = P_n \begin{pmatrix} \mathscr{T}_n \\ \mathscr{S}_n \end{pmatrix}. \qquad (9.202)$$

It is now convenient to consider in some detail the way in which the matrix can be related directly to the stacking fault characteristics. We therefore consider P_n as the product of 3 matrices.

$$P_n = \begin{pmatrix} 1 & 0 \\ 0 & e^{-i\Gamma_n} \end{pmatrix} \begin{pmatrix} T_n^{(+)} & S_n^{(-)} \\ S_n^{(+)} & T_n^{(-)} \end{pmatrix} \begin{pmatrix} 1 & 0 \\ 0 & e^{i\Gamma_n} \end{pmatrix} = \mathscr{Q}_n^* \mathscr{Q}_n \mathscr{Q}_n \qquad (9.203)$$

where the sign* stands for the complex conjugate. The definition of the matrices $\mathscr{Q}$ and $\mathscr{Q}$ is clear from this formula.

The $\mathcal{A}_n$ have the following property:

$$\mathcal{A}_n\mathcal{A}_{n-1}^* = \begin{pmatrix} 1 & 0 \\ 0 & e^{i\Gamma_n} \end{pmatrix} \begin{pmatrix} 1 & 0 \\ 0 & e^{-i\Gamma_{n-1}} \end{pmatrix} = \begin{pmatrix} 1 & 0 \\ 0 & e^{i(\Gamma_n - \Gamma_{n-1})} \end{pmatrix} = \begin{pmatrix} 1 & 0 \\ 0 & e^{i\alpha_n} \end{pmatrix} = A_{\alpha_n}. \tag{9.204}$$

Further the A_{α_i} have the property:

$$A_{\alpha_{n+1}} A_{\alpha_n} = \begin{pmatrix} 1 & 0 \\ 0 & e^{i\alpha_{n+1}} \end{pmatrix} \begin{pmatrix} 1 & 0 \\ 0 & e^{i\alpha_n} \end{pmatrix} = \begin{pmatrix} 1 & 0 \\ 0 & e^{i(\alpha_n + \alpha_{n+1})} \end{pmatrix} = A_{\alpha_{n+1} + \alpha_n} \tag{9.205}$$

i.e., on multiplying 2 $A\alpha_i$ matrices the indices may be added.

In order to calculate now the wave transmitted and scattered by a crystal containing N stacking faults one has to apply the operation symbolized by the matrix P_n to every one of the stacking faults. One obtains

$$\begin{pmatrix} \mathcal{T}_n \\ \mathcal{S}_n \end{pmatrix} = (\mathcal{A}_N^* \mathcal{Q}_N \mathcal{A}_N)(\mathcal{A}_{N-1}^* \mathcal{Q}_{N-1} \mathcal{A}_{N-1}) \cdots (\mathcal{A}_2^* \mathcal{Q}_2 \mathcal{A}_2)(\mathcal{A}_1^* \mathcal{Q}_1 \mathcal{A}_1) \begin{pmatrix} \mathcal{T}_0 \\ \mathcal{S}_0 \end{pmatrix}. \tag{9.206}$$

The expression for $\mathcal{T}_0$ and $\mathcal{S}_0$ are those given by Eqs. (9.123) and (9.124) Making use of the property (9.204) this can be written still more concisely as

$$\begin{pmatrix} \mathcal{T}_n \\ \mathcal{S}_n \end{pmatrix} e^{+i\Gamma_n} = \mathcal{Q}_N A_{\alpha_N} \mathcal{Q}_{N-1} A_{\alpha_{N-1}} \cdots A_{\alpha_3} \mathcal{Q}_2 A_{\alpha_2} \mathcal{Q}_1 A_{\alpha_1} \mathcal{Q}_0 \begin{pmatrix} 1 \\ 0 \end{pmatrix}. \tag{9.207}$$

The phase factor in the left-hand side is of course immaterial since it disappears on calculating intensities.

From this formulation it is quite clear that expression (9.207) can be related directly to the crystal. The rule is simply this:

(1) Every time the wave passes a perfect crystal slab we multiply $\begin{pmatrix} 1 \\ 0 \end{pmatrix}$ to the left with a $\mathcal{Q}_i$ matrix.

(2) Passing a stacking fault results in a left multiplication with an A_α matrix.

One can deduce some general properties from this formulation. We give the following examples:

(a) Let t_n be an integral number of extinction distances; then one knows that $T_n^{(+)} = T_n^{(-)} = 1$ and $S_n^{(+)} = S_n^{(-)} = 0$ [see Eqs. (9.123) and (9.124)] and hence that the corresponding $\mathcal{Q}$ matrix $\mathcal{Q}_n = I$ (I: unit matrix). Making further use of the property (9.205) of the A_{α_i}

matrices one finds for the product giving the final amplitudes

$$\cdots A_{\alpha_{n+1}} \mathcal{D}_n A_{\alpha_n} \cdots = \cdots A_{\alpha_{n+1}} I A_{\alpha_n} \cdots = \cdots A_{\alpha_{n+1}+\alpha_n} \cdots. \qquad (9.208)$$

This means that the net result of the presence of the 2 faults separated by an integral number of extinction distances is the same as if only 1 fault were present between lamella t_{n-1} and t_{n+1}, with a displacement vector equal to the sum of the displacement vectors and hence producing a phase shift equal to the sum of the 2 phase shifts.

(b) If the first or the last fault is at an integral number of extinction distance from the surface we have $\mathcal{D}_0$ or $\mathcal{D}_N$ equal to I and multiplying with this $\mathcal{D}$ matrix changes nothing; the last (or first) A_{α_i} matrix changes only the phase of the outcoming wave. Such faults can therefore be ignored, they do not influence the intensities.

(4) *Discussion of stacking fault contrast.*[229]   We now discuss the formulas (9.189) and (9.190) for different cases of interest.

In the purely dynamical case, i.e., for $s = 0$, the expression (9.189) reduces to

$$T_s = \cos \pi \frac{t_1}{t_e} \cos \pi \frac{t_2}{t_e} - \sin \pi \frac{t_1}{t_e} \sin \pi \frac{t_2}{t_e} e^{i\alpha} \qquad (9.209)$$

which can also be written as

$$T_s = \frac{1}{2}\left[(1 + e^{i\alpha}) \cos \frac{\pi t}{t_e} + (1 - e^{i\alpha}) \cos \pi \frac{t_1 - t_2}{t_e}\right]$$

$$= \left(\cos \frac{\alpha}{2} \cos \pi \frac{t}{t_e} - i \sin \frac{\alpha}{2} \cos \frac{2\pi u}{t_e}\right) e^{i(\alpha/2)}. \qquad (9.210)$$

The corresponding intensity distribution is given by

$$I_t(s, t) = TT^* = \cos^2 \pi \frac{t}{t_e} \cos^2 \frac{\alpha}{2} + \cos^2 \frac{2\pi u}{t_e} \sin^2 \frac{\alpha}{2}. \qquad (9.211)$$

From the last formula (9.211) or from (9.210) it is clear that for a foil of given thickness $t_1 + t_2 = t$ the transmitted intensity is a function of $t_2 - t_1 = 2u$, i.e., the position of the fault. If the fault is in an inclined plane u varies linearly as a function of x, the distance measured perpendicular to the line of intersection of the fault plane and the foil surface. A fringe pattern of the cosine squared type parallel to the intersection line will result. The period is $1/2\, t_e$. The periodicity of the "wedge fringes," i.e., of the thickness contours can be obtained from formula (9.211) by putting $\alpha = 0$; only the $\cos^2\pi\,(t/t_e)$ term remains and the depth periodicity is t_e, i.e., the double of that of the stacking fault pattern.

For small s, the fringe profile becomes more complicated. Two kinds of maxima occur. Examples of this are shown in Fig. 111 and illustrated by Fig. 112. The periodicity is given by

$$t_1'' = \frac{1}{\sigma} = t_e/\sqrt{1 + (st_e)^2}. \tag{9.212}$$

For large values of s, i.e., for $st_e \gg 1$ the kinematical theory applies and the depth periodicity is now $t_e' = 1/|\,s\,|$. The periodicity of the

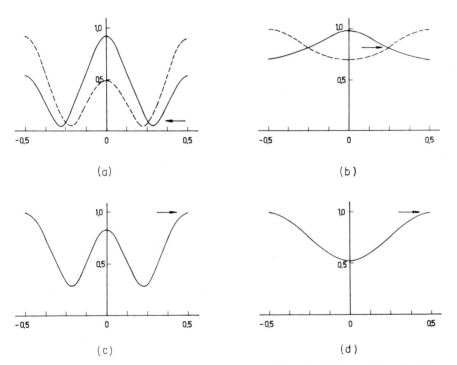

FIG. 111. Theoretical intensity profiles for stacking fault fringes. Intensity is plotted versus σz for 2 typical values of σt where t is the foil thickness, and for 2 typical values of st_e. Only one period of the pattern is shown (N: integer). (a) $\sigma t = N + 1/2$: $st_e = 1/4$. (b) $\sigma t = N + 1/2$: $st_e = 2$. (c) $\sigma t = N$: $st_e = 1/4$. (d) $\sigma t = N$: $st_e = 2$. The dotted line corresponds to negative values of st_e. In (a) and (c) there are two kinds of minima; in (b) and (d) there is but 1 minimum.

wedges fringes in the kinematical theory is also $1/|\,s\,|$; both periods are now equal.

Also the intensity distributions are different in the two cases. This is illustrated in Fig. 112. Figure 113 shows the effect of changing s on the fringe pattern of faults in stainless steel. The effect of changes in thickness is to cause a decrease in the number of fringes. Fringe reversal takes place every time the number of fringes diminishes by 1. Fringe reversal also may be caused by an overlapping fault.

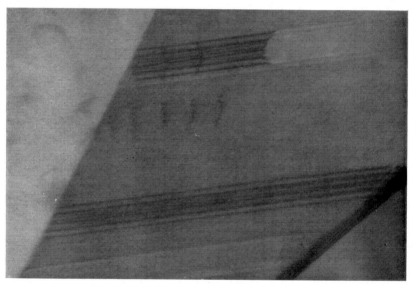

FIG. 112.   Observed example of the fringe profile of Fig. 111(a) type (a) or (c). (Courtesy of P. B. Hirsch.)

The fringe spacing is evidently closely related to the extinction distance. If the orientation of the foil plane is known, the width of a stacking fault intersecting both surfaces determines its inclination and the fringe spacing gives a direct measurement for the extinction depth. Such measurements were made[229] for stainless steel and the following values were obtained:

Reflection	t_e in A
111	228
200	263
220	413

These values are in good agreement with the theoretical values, deduced from Eqs. (9.165) and (9.166).

(5) *Graphical interpretation of the stacking fault contrast neglecting absorption.*[235] In order to facilitate the discussion of the stacking fault contrast for various values of the parameters involved it is useful

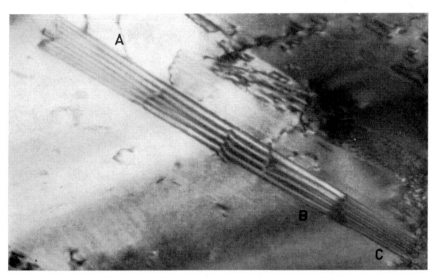

FIG. 113.   Illustrating the effect of changing s on the fringe pattern. In A there is but one maximum, i.e., in the region $st_e \gg 1$. In B the doublet structure appears, in C $st_e \ll 1$.

to write Eq. (9.189) in a somewhat different form. We call $\gamma = s/\sigma$; it is then easy to see that Eq. (9.189) can be transformed into

$$T_s(t_1 + t_2) = (\cos \pi \sigma t - i\gamma \sin \pi \sigma t)$$
$$+ i(1 - \gamma^2)\, e^{i(\alpha/2)} \sin \frac{\alpha}{2}(\cos \pi \sigma t - \cos 2\pi \sigma u) \qquad (9.213)$$

where $2u = t_1 - t_2$; u is as before the distance of the stacking fault from the middle of the foil. The quantity $|\gamma|$ varies between 0 (purely dynamical case) and 1 (purely kinematical case).

This formula can be represented graphically in the complex plane; the real axis is along the x axis and the imaginary axis along the y axis.

[235] I am very much indebted to Dr. R. Gevers for this construction.

In Fig. 114a the following numerical values were used: $\alpha = 120°$, $\gamma = 0,4$, and $\pi\sigma t = 50°$. Let the radius vector OB of magnitude 1 form an angle $\pi\sigma t$ with the real axis, then $\overline{OA_1}$ has a length $\cos \pi\sigma t$ and $\overline{A_1B}$ a length $\sin \pi\sigma t$. The quantity $-i\gamma \sin \pi\sigma t = \overline{A_1A_2}$ is found by multiplying $\overline{A_1B}$ with γ and changing the sign; hence $-i\gamma \sin \pi\sigma t = \overline{A_1A_2}$. The first part of Eq. (9.213), which is also the amplitude of the wave diffracted by the perfect crystal, is now represented by the vector $\overline{OA_2}$. To this we have to add the vector.

$$\overline{A_2A_3} = \sin \frac{\alpha}{2} (\cos \pi\sigma t - \cos 2\pi\sigma u) (1 - \gamma^2) e^{i\alpha/2}. \qquad (9.214)$$

The magnitude of this vector depends on u, i.e., on the position of the stacking fault with respect to the center of the foil. The axis $\overline{M_1M_2}$, which has the direction of this vector, is found by rotating the imaginary axis anticlockwise over an angle $\alpha/2$. The total amplitude of the transmitted wave is now $\overline{OA_3}$. As u varies, the point A_3 will describe the line M_1M_2 periodically. The extreme values for the magnitude of $\overline{A_2A_3}$ are obtained for $\cos 2\pi\sigma u = \pm 1$. The points M_1 and M_2 correspond to these extreme values; $u = 0$ in M_1 and $u = \pm(1/2) t_e''$ (mod $t_e'' = 1/\sigma$) in M_2. It is now straightforward to plot the transmitted intensity versus u for one period. Such a plot is shown in Fig. 114d (curve 1). The intensity has 2 maxima, the largest for $u = 0$ is given by $\overline{OM_1^2}$ the smaller one (the so-called subsidiary maximum) if $u = 1/2t_2''$ is given by $\overline{OM_2^2}$. In between there is a minimum of magnitude $\overline{Om^2}$ for $u = u_{min}$.

From this graph it is clear that whether a subsidiary minimum will occur or not depends on whether or not M_1 and M_2 are lying on opposite sides of m or at the same side. Figure 114b, for instance, shows a case where there is no subsidiary maximum; the values of the parameters are $\alpha = 120°$, $\gamma = 0.7$, and $\pi\sigma t = 30°$. The maximum intensity is given by $\overline{OM_1^2}$ and the minimum by $\overline{OM_2^2}$. Figure 114d (curve 2) is a plot of the intensity versus u.

The condition under which a subsidiary maximum occurs is $\overline{A_2M_2} > \overline{A_2m}$. The magnitude of $\overline{A_2m}$ can be calculated from geometrical considerations with reference to Fig. 114a; one finds

$$\overline{A_2m} = \gamma \sin \pi\sigma t \cos \frac{\alpha}{2} + \cos \pi\sigma t \sin \frac{\alpha}{2} \qquad (9.215)$$

and

$$\overline{A_2M_2} = (1 - \gamma^2) \sin \frac{\alpha}{2} (\cos \pi\sigma z + 1). \qquad (9.216)$$

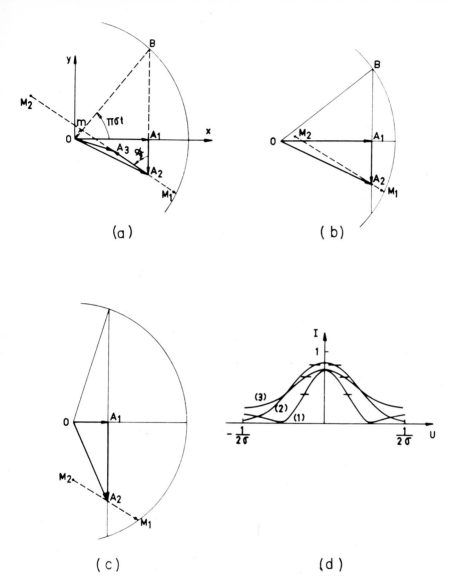

(a)

(b)

(c)

(d)

FIG. 114. Graphical construction for the amplitude of the stacking fault fringes neglecting absorption effects

(a)	$\alpha = 120°$	$\gamma = 0,4$	$\pi \sigma t = 50°.$
(b)	$\alpha = 120°$	$\gamma = 0,6$	$\pi \sigma t = 40°.$
(c)	$\alpha = 120°$	$\gamma = 0,7$	$\pi \sigma t = 70°.$

The amplitude OA_3 oscillates between OM_1 and OM_2, as the point A_3 describes the segment M_1M_2. The amplitude transmitted by the perfect crystal is given by OA_2. (d) Fringe profiles; the curves (1) (2) and (3) correspond to (a) (b) and (c), respectively; the background intensity is shown by a horizontal bar.

211

The position $u = u_{min}$ of the stacking fault for which the minimum amplitude is obtained is given by the condition:

$$\gamma \sin \pi\sigma t \cos \frac{\alpha}{2} + \cos \pi\sigma t \sin \frac{\alpha}{2} = (1 - \gamma^2) \sin \frac{\alpha}{2} (\cos \pi\sigma t - \cos 2\pi\sigma u_{min})$$

(9.217)

which, when solved for u_{min} yields

$$\cos 2\sigma\pi u_{min} = -\frac{\gamma}{1 - \gamma^2} \frac{\gamma \sin(\alpha/2) \cos \pi\sigma t + \cos(\alpha/2) \sin \pi\sigma t}{\sin(\alpha/2)}.$$ (9.218)

It is clear that the minimum will occur only if Eq. (9.218) has a solution for u_{min}, i.e., if the right-hand side is smaller than unity in absolute value.

In the dynamical case $s = 0$; γ is also zero and this condition is always satisfied. On the other hand if $\gamma \to 1$, i.e., in the kinematical case, the condition cannot be satisfied and there will be no subsidiary minimum.

The intensity at the minimum is given by $I_{min} = \overline{Om}^2$; it can again be calculated from Fig. 114a, one finds

$$I_{min} = \left[\cos \pi\sigma t \cos\left(\frac{\alpha}{2}\right) - \gamma \sin \pi\sigma t \sin\left(\frac{\alpha}{2}\right)\right]^2.$$ (9.219)

It is clear that this minimum will be zero if

$$\text{tg } \pi\sigma t = (1/\gamma) \cot\left(\frac{\alpha}{2}\right).$$ (9.220)

The contrast of the fringes will be defined as $I - I_0$, i.e., the difference between the intensity at a given point and the intensity transmitted by the perfect part of the crystal. From Fig. 114a follows that $I = \overline{OA_3^2}$ and $I_0 = \overline{OA_2^2}$, we further have

$$\overline{OA_3} = \overline{Om} + \overline{mA_2} + \overline{A_2A_3}$$

or

$$I = \overline{OA_3^2} = |\overline{Om}|^2 + |\overline{mA_2}|^2 + \overline{A_2A_3}(\overline{A_2A_3} - 2\overline{A_2m})$$

and since $|\overline{Om}|^2 + |\overline{mA_2}|^2 = I_0$

$$I - I_0 = \overline{A_2A_3}(\overline{A_2A_3} - 2\overline{A_2m}).$$

Substituting the appropriate expressions one obtains:

$$I - I_0 = (1 - \gamma^2) \sin\left(\frac{\alpha}{2}\right) (\cos \pi\sigma t - \cos 2\pi\sigma u) \left[(1 - \gamma^2) \sin\frac{\alpha}{2}\right.$$

$$\left.(\cos \pi\sigma t - \cos 2\pi\sigma u) - 2\left(\gamma \sin \pi\sigma t \cos\frac{\alpha}{2} + \cos \pi\sigma t \sin\frac{\alpha}{2}\right)\right]. \quad (9.221)$$

There will be no contrast for a stacking fault parallel to the foil plane if $\cos \pi\sigma t = \cos 2\pi\sigma u$ or $(t/2 - u = \pm k t_e'')$ where k is an integer; i.e., if the stacking fault is at an integral number of extinction distances t_e'' from one of the surfaces. Graphically this condition is equivalent to setting $A_2 A_3 = 0$. Furthermore there will be no contrast if the points A_3 come in $A_3{}^*(u^*)$ defined by $A_2 A_3 = 2 A_2 m$ since then $| OA_3{}^* | = | OA_2 |$. The condition for this to occur again can be derived explicitly from Fig. 114^d. It is

$$\cos 2\sigma\pi u^* = - \frac{(1 + \gamma^2) \sin(\alpha/2) \cos \pi\sigma t + 2\gamma \cos(\alpha/2) \sin \pi\sigma t}{(1 - \gamma^2) \sin(\alpha/2)}. \quad (9.222)$$

In the "kinematical" case, i.e., for γ close to unity there is no real value for u^* since then $A_3{}^*$ would be located outside $M_1 M_2$. In the dynamical case there may be a solution depending on whether or not the right-hand side of Eq. (9.222) is smaller than unity in absolute value. From the plot of Fig. 114d it is clear that depending on where the line representing the perfect crystal transmission I_0 comes, there will be 2 or 4 intersection points with the I versus u curve; for every one of the so obtained u values there will be absence of contrast.

For $s = 0$ the graphical construction of $| T_s |$ becomes especially simple (Fig. 114$^{(1)}$). We refer to formula (9.210) valid in this case. One draws $OM \equiv \cos(\alpha/2) \cos \pi(t/t_e)$ along the real axis, parallel to the imaginary axis, through the point M one measures $\sin(\alpha/2) \cos(2\pi u/t_e)$. The point C corresponds to $u = 0$, the point M to $u = \frac{1}{4} t_e$ and the point D to $\frac{1}{2} t_e$. As u changes the point P describes the segment CD and $OP \equiv | T_s |$. The period for $| T_s |$ or I_T is clearly $\frac{1}{2} t_e$. The diagram shows that the contrast should be optimum when $OM = 0$ i.e., for $t = (k \pm 1/2)t_e$; it is weakest for $t = k t_e$.

(6) *The effect of anomalous absorption.* In the preceding discussion the effect of absorption was neglected. As seen in Section 9f(3) the anomalous absorption can be taken into account phenomenologically by the substitution $1/t_e \to 1/t_e + i/\tau_e$ in formulas (9.189) and (9.190).

This was done by Hashimoto, Howie, and Whelan[236a,b,237a] and numerical profiles were calculated in order to explain some particular features observed in the stacking fault fringes in stainless steel. The main result of these calculations can be summarized as:

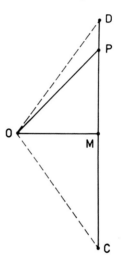

FIG. 114[(1)]. Graphical construction of the stacking fault fringes according to the dynamical theory neglecting absorption for the case $s = 0$.

(a) There are still subsidiary fringes in the middle of the foil due to the subsidiary maxima or minima in the transmitted intensity. For $s = 0$ and for small thicknesses, i.e., for $t/\tau_e \ll 1$, the spacing between the intensity maxima remains approximately $1/2\, t_e$ as in the theory without absorption.

(b) The intensity profile in the bright field image remains symmetrical about the foil center, whereas the dark field image becomes asymmetrical. The symmetry of I_t can be deduced from expression (9.192); interchanging the indices (1) and (2) does not affect $T_s(t_1 + t_2)$ since even with absorption $S^+ = S^-$. The asymmetry of $S_s(t_1 + t_2)$ follows from Eq. (9.194).

[236a] H. Hashimoto, A. Howie, and M. J. Whelan, *Proc. European Reg. Conf. on Electron Microscopy, Delft, 1960* p. 207 (1961).
[236b] A. Howie, *Met. Rev.* **6**, 467 (1961).
[237a] H. Hashimoto, A. Howie, and M. J. Whelan, *Phil. Mag.* [8] **5**, 967 (1960).

(c) For thick specimens and for $s \simeq 0$ the subsidiary extrema disappear, except near the foil center; depth periodicity of the fringes near the surface then becomes t_e.

(d) The fringes decrease in contrast near the foil center.

(e) At the top face the fringe pattern in bright and dark field images are similar, i.e., both start with a bright or a dark fringe. At the bottom face, on the other hand, the fringes are complementary. However, the sum of the intensities is smaller than unity. These remarks allow us to distinguish between the top and bottom end of the fault plane.

(f) The nature of the first fringe (next to the top surface) in the bright field image of a fault in a thick foil ($t \geqslant 6t_e$) depends only on the sign of α. A bright fringe means $\alpha = + 120°$ and a dark fringe $\alpha = - 120°$. The knowledge of the sign of α allows us to distinguish between intrinsic and extrinsic faults in face-centered cubic metals (see Appendix A for an analytical derivation of these and other results).

(7) *Antiphase boundary contrast.*[233a] In many ways antiphase boundaries behave like stacking faults with respect to contrast formation. The contrast effects have been discussed by Marcinkowski and Fisher[233a] on the basis of the dynamical theory developed for stacking faults by Whelan and Hirsch.[229] In particular we will discuss the case of $AuCu_3$, where the shear vector of the antiphase boundary is $(a/2) [110]$ since the structure is face-centered cubic. The phase angle $\alpha = 2\pi \bar{n} \bar{R}$ is $\pm 2\pi$ and the boundaries will be invisible, unless the operating reflection is a super lattice reflection, e.g., $\bar{n} = [110]$.[237b]

The phase angles then reduce to 0, or $\pm\pi$ instead of 0, $\pm 2\pi/3$ for the stacking fault. Formula (9.213) now simplifies to:

$$T(t_1 + t_2) = (\cos \pi \sigma t - i\gamma \sin \pi \sigma t) - (1 - \gamma^2) (\cos \pi \sigma t - \cos 2\pi \sigma u)$$

$$(9.223)$$

where the antiphase boundary is at t_1 and t_2 from the 2 surfaces of the foil with thickness $t = t_1 + t_2$; $u = t_2 - t_1$.

We will not discuss in detail the fringe profiles for different diffraction conditions; this is done in Fisher and Marcinkowski.[233a]

[237b] Superlattice reflections are reflections which are only present for the ordered alloy.

Instead we will show how these profiles can be deduced graphically for any specific case, with the method outlined in Section 9h(5). We refer to Fig. 115 for this. As before O is the origin of the complex plane, $\overline{OA_1}$ is $\cos \pi \sigma t$, while $\overline{A_1 A_2} = -\gamma \sin \pi \sigma t$; $\overline{OA_2}$ then represents the amplitude of the transmitted wave for the perfect crystal. The second term of the right-hand side of Eq. (9.223) is measured along the line perpendicular to the imaginary axis. A given value of u corresponds to a point A_3 on this line. As the value of u goes through a complete period, e.g., from 0 to $1/\sigma$ the image point A_3 travels along $M_1 M_2$ and back. The center of $M_1 M_2$ is given by $\overline{A_2 C} = (1 - \gamma^2) \cos \pi \sigma t$. The extrema points M_1 and M_2 correspond to $\cos 2\pi \sigma u = -1$ and $\cos 2\pi \sigma u = +1$. The transmitted amplitude in each point is measured by the vector $|\ \overline{OA_3}\ |$.

For a situation as shown in (1) in Fig. 115, i.e., for $\sigma t = 1/3$:

(1) $\gamma = 0.2$ $\pi \sigma t = 60°.$
(2) $\gamma = 0.7$ $\pi \sigma t = 60°.$
(3) $\gamma = 0.9$ $\pi \sigma t = 60°.$

$\gamma = 0.2$ there are 2 maxima [curve (1) of Fig. 115b] as is typical for the dynamical case. If on the other hand $\sigma t = 1/3$ and $\gamma = 0.9$ as in (3) there is but one maximum and the fringes are very weak [curve (3) of Fig. 115b]. This situation is typical for the kinematical case. For the intermediate value of $\gamma = 0.7$ curve (2) has a small subsidiary maximum (curve (2) of Fig. 115b). For each curve the intensity transmitted by the perfect crystal is shown also as a small horizontal bar (dotted). It should be pointed out that the result of the construction and hence the magnitude of T does not depend on the sign of s. The only term that changes sign with s is $-i\gamma \sin \pi \sigma t$ (since $\gamma = s/\sigma$), but this only means that the diagram will be transformed into its mirror image with respect to the line OA_1. This was not true for a stacking fault. If absorption is taken into account this symmetry no longer holds.

As for the stacking fault an expression for the contrast $I - I_0$ can be found from Fig. 115a.

$$C = I - I_0 = \overline{OA_3^2} - \overline{OA_2^2} = \overline{A_2 A_3}(\overline{A_2 A_3} - \overline{2 OA_2})$$

and hence

$$C = (1 - \gamma^2)(\cos \pi s t - \cos 2\pi \sigma u)\ [(\gamma^2 - 1) \cos 2\pi \sigma u - (\gamma^2 + 1) \cos \pi \sigma t].$$

$$(9.224)$$

The contrast will be zero if

$$\begin{aligned} &(1) \quad \cos \pi \sigma t = \cos 2\pi \sigma u \quad \text{i.e., if } u = \pm \frac{t}{2} + k t''_e \\ &(2) \quad \cos 2\pi \sigma u = \frac{\gamma^2 + 1}{\gamma^2 - 1} \cos \pi \sigma t. \end{aligned} \tag{9.225}$$

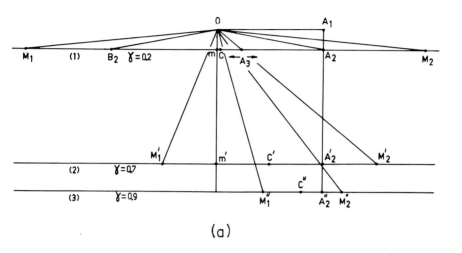

(a)

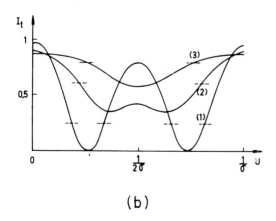

(b)

FIG. 115. Graphical construction for the antiphase boundary fringes. The amplitude of the transmitted wave OA_3 oscillates between OM_1 and OM_2 as the point A_3 travels along M_1M_2. The curves (1), (2), and (3) of (b) correspond to the 3 positions of M_1M_2.

The last condition is the analytical expression of the geometrically evident fact that there will be no contrast if A_3 comes in B_2 (Fig. 115a) defined by $\overline{mA_2} = \overline{mB_2}$.

The extinction distances for superlattice reflections were calculated by Marcinkowski and Fisher[233a] using the formulas (9.163) and (9.166) It turns out that in $AuCu_3$ they are about 4 times larger than for the ordinary strong reflections. As a consequence of this, interference patterns containing several fringes are not observed because for a thickness large enough to give rise to several fringes (several extinction distances) the beam is attenuated so much by normal absorption that no image is left. The angular setting required for good contrast also becomes very critical. This can be deduced from the graphical construction of Fig. 115. If s increases st_e becomes rapidly large (since t_e is large) and as a consequence $\gamma = st_e/\sqrt{1 + (st_e)^2}$ becomes large (close to 1) and the length of A_1A_2 in Fig. 115a increases. The contrast then becomes of the type (3), i.e., very faint.

i. Contrast at Dislocations, Taking into Account the Absorption

(1) *Principle of the calculation.* The method for calculating profiles of dislocation images is now as follows. The Eqs. (9.157) or (9.159) are integrated down a column at (x, y) and perpendicular to the foil. The displacement function of the dislocation is taken for $\bar{R}(\bar{r})$. The initial conditions are $T = 1$ and $S = 0$ at $z = 0$. The anomalous absorption is considered to be small compared to the extinction: $\tau_e \simeq 10\, t_e$. The values of SS^* and TT^* at the bottom of the column at (x, y) are identified with the scattered and transmitted intensity at the point (x, y).

The function $\alpha' = \bar{n}.\bar{R}(z)$ is relatively complicated even for the simplest case of the screw dislocation, parallel to the surface. Integration by analytical means is therefore difficult; numerical calculations have been made for different types of dislocations by Howie and Whelan.[230d] The results will be discussed here.

(2) *Screw dislocations parallel to the foil plane.*[230d] Let the dislocation be situated at a depth z_0 below the front surface; the x axis is perpendicular to the direction of the dislocation. The phase angle $2\pi\alpha'$ is then given by

$$\alpha' = \bar{n} \cdot \bar{b}\, \text{arctg}\, \frac{z - z_0}{x} \tag{9.226}$$

where $\bar{n} \cdot \vec{b} = n$ is an integer for a perfect dislocation. The intensity profiles for $n = 1$ and $n = 2$, which occur most frequently in practice, have been calculated and the results are shown in Figs. 116 and 117.

The width of the image is roughly the same as predicted by the kinematical approach, i.e., about 100 A, but the image is no longer displaced sideways with respect to the dislocation for $n = 1$. For $n = 2$ the image consists of 2 lines one on either side of the dislocation (Fig. 117). The influence of the surface is evident from Figs. 116b and 117. Near the surface the profile is a sensitive function of z_0 and large

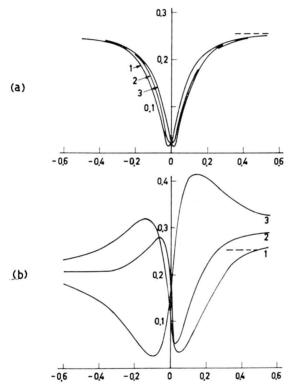

Fig. 116. Image profiles for a screw dislocation parallel to the foil plane for $s = 0$, $n = 1$, and for a foil of thickness $t = 8\,t_e$; $\tau_e = 10t_e$. (a) Curves (1), (2), and (3) correspond to dislocations at different depths below the surface, respectively, $z_0/t_e = 4$, 4.25, and 4.50. (b) Curves (1), (2), and (3) correspond to dislocations at depths $z_0/t_e = 7.25$, 7.50, and 7.75. In the first case (a) only small oscillation occurs; in the second case large oscillations are found. [After A. Howie and M. J. Whelan, *Proc. European Reg. Conf. on Electron Microscopy, Delft*, 1960, Vol. 1, p. 194 (1961).]

oscillations in the black-white contrast are to be expected for inclined dislocations. As can be seen from Fig. 116, the oscillations are far less pronounced near the middle of the foil. The influence of the surface discussed here is not a consequence of the changes in strain field near the surface, which can be accounted for by means of an image dislocation. The oscillations are due to the small length of the part of the column between the dislocation and the closest surface. Because of absorption bright and dark field images are no longer complementary.

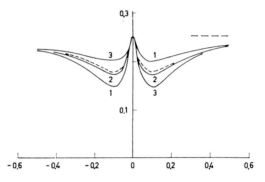

FIG. 117.   Image profiles for a screw dislocation parallel to the foil plane for $s = 0$. Curves (1), (2), and (3) correspond to dislocations at depths $z_0/t_e = 4.00$, 4.25, and 4.50. The value of $n = 2$. No large oscillations in contrast occur ($t = 8t_e$; $\tau_e = 10t_e$). [After A. Howie and M. J. Whelan, *Proc. European Reg. Conf. on Electron Microscopy, Delft,* 1960, Vol. 1, p. 194 (1961).]

An observed example of the change in profile as a function of the distance from the surface is shown in Fig. 96. A dislocation in the basal plane of SnS_2 approaches the surface which is very slightly inclined with respect to the basal plane: the dislocation finally emerges in the surface; the gradual change in profile is evident.

For inclined dislocations showing oscillating contrast it is possible to distinguish between top and bottom ends by comparing the bright and dark field images. The images are complementary at the bottom end of the dislocation. Since the period of the oscillation is t_e'' one can use the oscillating contrast to measure the distance of the dislocation from the surface.

(3) *Edge dislocations parallel to the foil plane and with a Burgers vector perpendicular to the foil.*[230d]   The profiles have been calculated[230d] for edge dislocations parallel to the foil plane and with

Burgers vector perpendicular to the foil, as in prismatic loops parallel to the foil. The intuitive picture would predict no contrast at all since $\bar{n} \cdot \bar{b} = 0$ for the reflections which are usually operative. Since the radial displacement, which is the only operative one, causes only a slight curvature of the lattice planes above and below the loop plane, the inclination being of opposite sense, the kinematical theory would also predict very little contrast. Appreciable contrast is only to be expected for very small s values where the dynamical theory can be applied.

The phase angle $2\pi\alpha'$ is now given by:

$$\alpha' = m \left\{ \ln \left[x^2 + (z - z_0)^2 \right] + 3 \frac{(z - z_0)^2 - x^2}{(z - z_0)^2 + x^2} \right\} \qquad (9.227)$$

where $-1/4 \leqslant m \leqslant 1/4$ depending on the magnitude of Poisson's ratio and the loop segment considered. At the segment perpendicular to $\bar{n}$, $m = \pm 1/4$; at the segment parallel to $\bar{n}$, $m = 0$ and there is no contrast since $\alpha' = 0$. Again z_0 is the distance of the loop plane to the front surface; x is measured radially, i.e., perpendicular to the loop segment considered. Since α' is an even function of x the image is symmetrical. For loops near the surface an image dislocation was added to take into account the presence of the surface. The image profiles obtained are shown in Fig. 118, while Fig. 119 shows examples

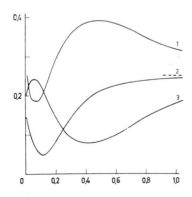

FIG. 118. Image profiles for an edge dislocation parallel to the foil plane and having its Burgers vector perpendicular to it: the following numerical values were used $t = 8t_e$; $\tau_e = 10t_e$. (a) $z_0/t_e = 7.75$; $m = 1/4$. (b) $z_0/t_e = 4$; $m = 1/4$. (c) $z_0/t_e = 7.75$; $m = -1/4$. [After A. Howie and M. J. Whelan, *Proc. European Reg. Conf. on Electron Microscopy, Delft,* 1960, Vol. 1, p. 194 (1961).]

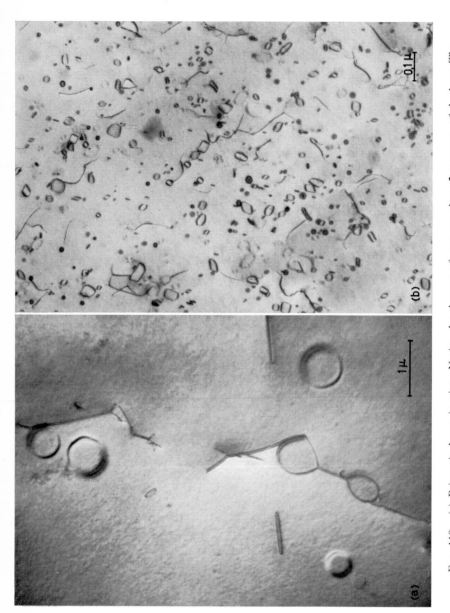

FIG. 119. (a) Prismatic loops in zinc. Notice the absence of contrast along 2 segments of the loop. The reflecting planes are parallel to the line joining these 2 segments. Although $\bar{n} \cdot \bar{b} = 0$ for these loops, contrast is visible due to the radial component of the displacement. [After A. Berghezan, A. Fourdeux, and S. Amelinckx, Acta Met. 9, 464 (1961).] (b) Prismatic loops in α-particles irradiated platinum annealed 1 hr at

observed in zinc and platinum.[215] These loops have an inclined Burgers vector, but since $\bar{n} \cdot \bar{b} = 0$ for the reflection used, the contrast is due only to the radial component. The images of Fig. 119a allow to locate immediately the orientation of the diffracting planes; they are parallel to the line connecting the 2 segments where contrast is absent. For other reflecting planes perpendicular to the loop but for which $\bar{n} \cdot \bar{b} \neq 0$ there would be no "out of contrast" segments. If "out of contrast" segments are observed for 2 different vectors parallel to the loop plane one can conclude that the Burgers vector is perpendicular to the loop plane. If the Burgers vector is inclined, "out of contrast" segments will only be found for 1 $\bar{n}$ vector; the projection of the Burgers vector is then perpendicular to $\bar{n}$. Using these rules allows us to decide whether a Burgers vector is perpendicular or inclined to the loop plane.

(4) *Edge dislocations seen end on.* (a) Simplified theory.[238a] Hashimoto and Mannami[238a] showed that the picture used in the intuitive theory outlined in Section 9d(3) can be put into a quantitative form. Using the column approximation the crystal is dissected into columns parallel to the z axis, i.e., to the dislocation and to the incident beam. The Burgers vector is along the x axis. The components of the displacement functions are then

$$u_x(x, y) = \frac{b}{2\pi} \left(\operatorname{arctg} \frac{y}{x} + \frac{1}{2(1 - \nu)} \frac{xy}{x^2 + y^2} \right) \tag{9.228}$$

$$u_y(x, y) = \frac{-b}{8\pi(1 - \nu)} \left[(1 - 2\nu) \ln (x^2 + y^2) + \frac{x^2 - y^2}{x^2 + y^2} \right]. \tag{9.229}$$

A change in lattice spacing is translated in the reciprocal lattice as a change in length of the reciprocal lattice vector. If the spacing d_n of the reflecting planes changes by Δd_n the reciprocal lattice vector changes in length by:

$$-\Delta n = \frac{1}{d_n} - \frac{1}{d_n + \Delta d_n} = \frac{\Delta d_n}{d_n(d_n + \Delta d_n)}.$$

The change in s, Δs corresponding to the change in Δn is given by

$$\Delta s = + \vartheta \Delta_n \tag{9.230}$$

when changes in Δs due to lattice rotations are neglected.

[238a] H. Hashimoto and M. Mannami, *Acta Cryst.* **13**, 363 (1960).

Applying Bragg's law $2d_n\vartheta = \lambda$ one can write

$$\Delta s = -\vartheta\frac{\Delta d_n}{d_n}\cdot\frac{1}{d_n + \Delta d_n} = -\frac{\lambda}{2d_n^2}\frac{\Delta d_n}{d_n}\frac{1}{1 + (\Delta d_n/d_n)} \simeq \frac{\lambda}{2}\frac{\Delta d_n}{d_n^3}. \quad (9.231)$$

Along a column Δs is a constant, i.e., it is independent of z; but it does depend on x and y.

$$s(x, y) = s_\infty + \Delta s(x, y). \quad (9.232)$$

The local value of s, $s(x, y)$ determines the local intensity of the scattered and transmitted wave by the relations (9.127) and (9.129).

In order to calculate $\Delta s(x, y)$ it is sufficient to express $\Delta d_n/d_n$ as a function of the lattice displacements. Hashimoto and Mannami[238a] limited themselves to the simple case $n = [h00]$ in a cubic crystal. One then has $d_{h00} = a/h$ where a is the lattice parameter

$$\Delta d_{h00} = \frac{\partial u_x}{\partial x}d_{h00} \quad (9.233)$$

and hence

$$\Delta s = -\frac{\lambda h^2}{2a^2}\frac{\partial u_x/\partial x}{1 + (\partial u_x/\partial x)} \simeq -\frac{\lambda h^2}{2a^2}\frac{\partial u_x}{\partial x}. \quad (9.234)$$

Since $\Delta d_{h00} \ll d_{h00}$ one can neglect $\partial u_x/\partial x$ with respect to 1. Contours of equal scattered intensity obtained by Hashimoto and Mannami are reproduced in Fig. 120.

(b) General theory.   This result can be generalized for an arbitrary nonorthogonal lattice. Let the direct lattice be determined by the 3 vectors $\bar{a}_i$ ($i = 1, 2, 3$) and the reciprocal lattice by the vectors $\bar{A}_i$ ($i = 1, 2, 3$) defined by $\bar{a}_i \cdot \bar{A}_j = \delta_{ij}$. The change in lattice vector $\overline{\Delta a}_i$ at a given point $\bar{r}$, due to the strain field of the dislocation, is the difference in displacement vector between 2 points separated by $\bar{a}_i$:

$$\overline{\Delta a}_i = \bar{R}(\bar{r} + \bar{a}_i) - \bar{R}(\bar{r})$$

where $\bar{r}$ is a lattice vector. The vectors $\bar{r}$ and $\bar{R}$ can be expressed in trimetric coordinates as

$$\bar{r} = \sum_i x_i\bar{a}_i \text{ and } \bar{R} = \sum_i u_i\bar{a}_i.$$

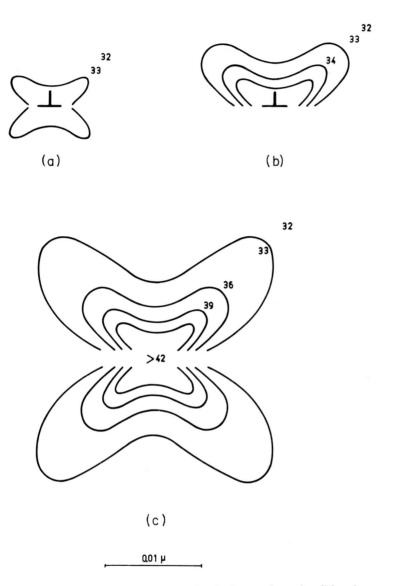

(a)

(b)

(c)

0,01 μ

FIG. 120. Contours of equal intensity for the image of an edge dislocation seen end-on. The reflections operating are ($h00$) while the slip plane is as indicated. (a) $s_{100}(\infty) = 0$. (b) $s_{100}(\infty) = 2 \times 10^3 \, \text{cm}^{-1}$. (c) $s_{100}(\infty) = 0$. [After H. Hashimoto and M. Mannami, *Acta Cryst.* **13**, 363 (1960).]

Using only the first term in the Taylor expansion:

$$\overline{\Delta a_i} = \sum_j \left[u_j(\bar{r} + \bar{a}_i) - u_j(\bar{r}) \right] \bar{a}_j$$

reduces to

$$\overline{\Delta a_i} = \sum_j (\bar{a}_i \cdot \overline{\text{grad}\, u_j})\, \bar{a}_j.$$

Making use of the definition of the reciprocal lattice vector one has:

$$\bar{a}_j \cdot \overline{\Delta A_i} = -\bar{A}_i \cdot \overline{\Delta a_j} = -\bar{A}_i \cdot \left(\sum_k (\bar{a}_j \cdot \overline{\text{grad}\, u_k})\, \bar{a}_k \right) = -\bar{a}_j \cdot \overline{\text{grad}\, u_i}$$

or finally

$$\overline{\Delta A_i} = -\overline{\text{grad}\, u_i}. \tag{9.235}$$

We make use of this relation to derive an expression for $\Delta d_n / d_n^3$ which enters in formula (9.231). First of all we notice that

$$\bar{n} = \sum_i \bar{h}_i \bar{A}_i$$

where the h_i are the indices of the reflecting lattice planes (perpendicular to $\bar{n}$). From this we deduce immediately:

$$\overline{\Delta n} = \sum_i h_i \overline{\Delta A_i} = - \sum_i h_i \overline{\text{grad}\, u_i}.$$

One now has $n^2 = 1/d_n^2$ and hence $\bar{n} \cdot \Delta \bar{n} = -\Delta d_n / d_n^3$. This leads immediately to

$$\frac{\Delta d_n}{d_n^3} = -\bar{n} \cdot \Delta \bar{n} = \sum_{ij} h_i h_j \bar{A}_j \cdot \overline{\text{grad}\, u_i}. \tag{9.236}$$

By definition

$$\text{grad}\, f = \sum_k \bar{A}_k\, \partial f / \partial x$$

and therefore we find

$$\frac{\Delta d_n}{d_n^3} = \sum h_i h_j \bar{A}_j \cdot \bar{A}_k \frac{\partial u_i}{\partial x_k}. \tag{9.237}$$

Substituting this into Eq. (9.231) gives the required expression for Δs. If the projection C of the center of Ewald's sphere on the reciprocal lattice plane parallel to the foil plane is *not* on the reciprocal lattice vector $\bar{n}$ the deviation from the exact Bragg condition has to be specified by means of 2 parameters.

In general one can specify the position of C by giving the vector $\overline{CN}$ = projection of $(\bar{k}_0 + \bar{n})$ on the reciprocal lattice plane. One can always express $\overline{CN}$ in terms of reciprocal lattice vectors:

$$\overline{CN} = \sum_i \rho_i \bar{A}_i.$$

A more general expression for Δs can now be obtained in the following way. From Fig. 103a for instance one can deduce

$$k_n^2 = (\bar{k}_0 + \bar{n})^2 = k_0^2 + s^2 + 2\bar{k}_0 \cdot \bar{s}.$$

Neglecting s^2 and solving for s one finds

$$s = \frac{(k_0 + \bar{n})^2 - k_0^2}{2k_0}$$

and hence

$$\Delta s = \frac{(\bar{k}_0 + \bar{n}) \cdot \overline{\Delta n}}{k_0}$$

and finally

$$\Delta s = -\frac{1}{k_0} \sum_{i,j,k} h_i \rho_j \bar{A}_j \cdot \bar{A}_k \frac{\partial u_i}{\partial x_k}. \tag{9.238}$$

In the special case $s_\infty = 0$ and if $\overline{CN} = \frac{1}{2}\bar{n}$ so that $\rho_j = \frac{1}{2}h_j$ the formula becomes

$$\Delta s = -\frac{1}{2k_0} \sum_{i,j,k} h_i h_j \bar{A}_j \cdot \bar{A}_k \frac{\partial u_i}{\partial x_k} \tag{9.239}$$

which reduces to Hashimoto's formula in the special case considered by him, i.e., for $n = [h00]$.

In a cubic crystal with lattice parameter a, considering reflections from $(hk0)$ planes, and taking $s_\infty = 0$ and $\overline{CN} = \frac{1}{2}\bar{n}$ one obtains:

$$\Delta s = -\frac{\lambda}{2a^2} \left[h^2 \frac{\partial u_x}{\partial x} + k^2 \frac{\partial u_y}{\partial y} + hk \left(\frac{\partial u_y}{\partial x} + \frac{\partial u_x}{\partial y} \right) \right]. \tag{9.240}$$

A similar generalization of the theory has been given recently by Mannami.[238b] Using a different approach he also arrived at the result given by Eq. (9.240). A more detailed discussion of the images is also presented in this last reference.

(c) Applications. Observed examples of dislocations seen end-on are reproduced in Fig. 121; they refer to platinum and uranium oxide. For this last substance both bright and dark field images are reproduced. Due to the absorption the images are not complementary.

Anomalous absorption can be taken into account by using formulas (9.138) and (9.139) instead of (9.127) and (9.128). By comparing the bright and dark field images of Fig. 121b,c it becomes clear that the dark field image is symmetrical, i.e., consists of 2 black regions, whereas the bright field image consists of a bright and a dark region. From the curves relating intensity (transmitted and scattered) and s for a perfect crystal, it follows that s should be negative in the region where bright and dark field images are complementary if in the perfect crystal $s_\infty = 0$.

Knowing the sign of s in the two regions the sign of an edge dislocation follows from the considerations of Section 9e(4)(c). The result is that for $s_\infty = 0$ the expanded region ($\Delta s > 0$) shows up in the same contrast, while the compressed region ($\Delta s > 0$) will exhibit complementary contrast in bright and dark field.

In discussing the contrast effects for screw dislocations seen end-on, the anomalous absorption has to be taken into account also. This changes somewhat the reasoning presented in Section III. 9. d. 3. The best contrast, like that shown in Fig. 121(a) is obtained for $s \simeq 0$. In this case the region D of Fig. 79[(1)]d corresponds to $s > 0$ and hence to the "easy" transmission side of the rocking curve (Fig. 106). This region will therefore appear bright. On the contrary the region C corresponds to $s < 0$ and will appear dark. The conclusion is therefore the same as that deduced from the naive picture given in Section III. 9. d. 3. Changing the sign of $\bar{n}$ and changing the sign of the dislocation results in inverting the black and white regions of Fig. 79[(1)].

That surface relaxation is important in producing contrast can be concluded from the fact that when the dislocations themselves are out of contrast, black or white dots may still be visible at the endpoints of slightly inclined dislocations.

[238b] M. Mannami, *J. Phys. Soc. Japan* **17**, 1160 (1962).

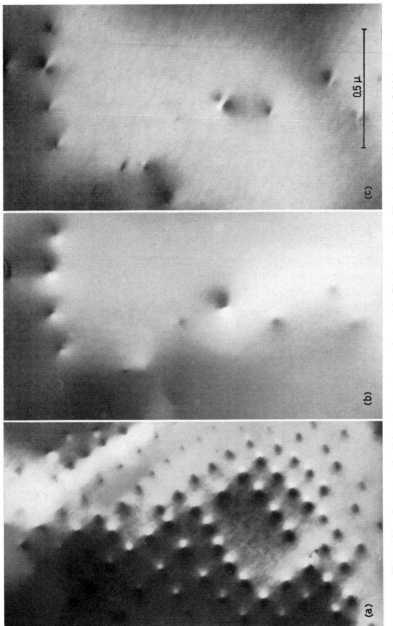

Fɪɢ. 121. Images of dislocations seen end-on. (a) Platinum. (b) Uranium oxide: bright field image. (c) Uranium oxide: dark field image. The 2 images are not complementary as a consequence of absorption.

(5) *Dislocation contrast calculated by the scattering matrix method.*

The method outlined in Section 9h(3) for the calculation of contrast due to a large number of overlapping stacking faults can be adapted to compute image profiles for dislocations. The principle of such a method has been given by Howie and Whelan[239] for the multiple beam case.

In a very recent paper Mannami[240] has presented a dynamical theory for contrast formation at dislocations, based on a model whereby the crystal is dissected in slices parallel to the foil plane. The resulting wave at the exit surface is composed of 2^n plane waves, if the crystal is dissected into n slices.

Numerical calculations have been carried out for a screw dislocation parallel to the foil plane, by dividing the foil into 21 slabs of a thickness of 50 A. Use was made of a dynamical 2-beam theory, without absorption, taking into account changes in phase shift in the successive slabs as well as changes in s.

We will use a somewhat different approach and limit ourselves to the 2-beam case. The model used is as follows: the crystal column is sliced into thin slabs of perfect crystal which are shifted according to the displacement function of the dislocation, i.e., the shift of the slab at level z is $\alpha(z) = 2\pi\bar{n} \cdot \bar{R}(z)$. To describe scattering by this model use will be made of the "*scattering*" matrix $\mathscr{D}$, which was in fact already defined in Section 9h(3) and the "*shift*" matrix $\mathscr{A}$, which is also discussed in the same paragraph. We will discuss both matrices now in somewhat more detail.

(a) Scattering matrix.   The scattering matrix relates the amplitudes of the waves entering the crystal slab considered to the amplitudes of the waves that leave this slab. It is defined as:

$$\mathscr{D} = \begin{pmatrix} T^+ & S^- \\ S^+ & T^- \end{pmatrix} \tag{9.241}$$

where T and S are given by Eqs. (9.123) and (9.124) if absorption is neglected. One then has:

$$\begin{pmatrix} T \\ S \end{pmatrix}_{out} = \begin{pmatrix} T^+ & S^- \\ S^+ & T^- \end{pmatrix} \begin{pmatrix} T \\ S \end{pmatrix}_{in} = \mathscr{D} \begin{pmatrix} T \\ S \end{pmatrix}_{in}. \tag{9.242}$$

[239] A. Howie and M. J. Whelan, *Proc. Roy. Soc.* **A263**, 217 (1961).
[240] M. Mannami, *J. Phys. Soc. Japan* **17**, 1423 (1962).

It is easy to show that $\mathscr{Q}$ is a unitary matrix:

$$\begin{pmatrix} T^+ & S^- \\ S^+ & T^- \end{pmatrix}\begin{pmatrix} T^{+*} & S^{+*} \\ S^{-*} & T^{-*} \end{pmatrix} = \begin{pmatrix} T^+T^{+*} + S^-S^{-*} & T^+S^{+*} + S^-T^{-*} \\ S^+T^{+*} + T^-S^{-*} & S^+S^{+*} + T^-T^{-*} \end{pmatrix} = \begin{pmatrix} 1 & 0 \\ 0 & 1 \end{pmatrix} = I.$$

From this derivation it is clear that this property is a manifestation of the physically obvious fact that since absorption is neglected all electrons that enter the crystal also leave it either in the transmitted beam or in the diffracted beam, since only 2 beams are allowed.

If the slab thickness $\varDelta z$ is chosen small enough one can make the following approximations

$$T^+ = 1 - \pi i s \varDelta z \qquad T^- = 1 + \pi i s \varDelta z \qquad S^+ = S^- = \frac{\pi i}{t_e} \varDelta z$$

and the $\mathscr{Q}$ matrix becomes

$$\mathscr{Q}(\varDelta z) = \begin{pmatrix} 1 - \pi i s \varDelta z & \dfrac{\pi i \varDelta z}{t_e} \\ \dfrac{\pi i \varDelta z}{t_e} & 1 + \pi i s \varDelta z \end{pmatrix}. \tag{9.243}$$

It is interesting to note that this thin slab scattering matrix can be derived immediately from the system of Eqs. (9.118) and (9.119). Multiplying by dz or $\varDelta z$ and adding T and S, respectively, to both sides of the equation leads to:

$$T + \varDelta T = (1 - \pi i s \varDelta z) T + \frac{i\pi}{t_e} \varDelta z S$$

$$S + \varDelta S = \frac{i\pi}{t_e} \varDelta z T + (1 + \pi i s \varDelta z) S$$

which formulated in matrix form is:

$$\begin{pmatrix} T + \varDelta T \\ S + \varDelta S \end{pmatrix} = \mathscr{Q}(\varDelta z) \begin{pmatrix} T \\ S \end{pmatrix}. \tag{9.244}$$

The scattering matrix for an absorbing slab of perfect material can be obtained by the usual substitution $((1/t_e) \rightarrow (1/t_e) + (i/\tau_e)$

$$\mathscr{Q}_a(\varDelta z) \begin{pmatrix} 1 - \pi i s \varDelta z & \pi i \left(\dfrac{1}{t_e} + \dfrac{i}{\tau_e} \right) \varDelta z \\ \pi i \left(\dfrac{1}{t_e} + \dfrac{i}{\tau_e} \right) \varDelta z & 1 + \pi i s \varDelta z \end{pmatrix}. \tag{9.245}$$

It is clear that $\mathcal{Q}_a$ is no longer a unitary matrix. The scattering matrix for a shifted slab in a deformed crystal can be obtained directly from the set of equations (9.182), which can be written

$$T + \varDelta T = (1 - \pi i s \varDelta z)\, T + \frac{i\pi}{t_e} \varDelta z S\, e^{i\alpha}$$

$$S + \varDelta S = \frac{i\pi}{t_e} \varDelta z T\, e^{-i\alpha} + (1 + \pi i s \varDelta z)\, S \tag{9.246}$$

or

$$\begin{pmatrix} T + \varDelta T \\ S + \varDelta S \end{pmatrix} = \begin{pmatrix} 1 - \pi i s \varDelta z & \dfrac{i\pi}{t_e} \varDelta z\, e^{i\alpha} \\ \dfrac{i\pi}{t_e} \varDelta z\, e^{-i\alpha} & 1 + \pi i s \varDelta z \end{pmatrix} \begin{pmatrix} T \\ S \end{pmatrix} = P \begin{pmatrix} T \\ S \end{pmatrix}. \tag{9.247}$$

This relation also follows immediately by noting that the deformation can be taken into account formally by the substitution $(1/t_e) \to (1/t_e)e^{\pm i\alpha}$ (see footnote [232d]). This matrix is the product of 3 matrices:

$$P = \begin{pmatrix} 1 & 0 \\ 0 & e^{-i\alpha} \end{pmatrix} \begin{pmatrix} 1 - \pi i s \varDelta z & \dfrac{i\pi}{t_e} \varDelta z \\ \dfrac{i\pi}{t_e} \varDelta z & 1 + \pi i s \varDelta z \end{pmatrix} \begin{pmatrix} 1 & 0 \\ 0 & e^{i\alpha} \end{pmatrix} = \mathcal{A}^* \mathcal{Q}(\varDelta z)\, \mathcal{A} \tag{9.248}$$

where the $\mathcal{A}$ matrices are "shift" matrices. Sandwiching $\mathcal{Q}(\varDelta z)$ between $\mathcal{A}^*$ and $\mathcal{A}$ results in a shift of the slab considered over α. Absorption can, of course, be taken into account easily. The central matrix does not depend on the level z, the shift matrices do.

(b) Amplitudes of scattered and transmitted waves, for an imperfect crystal. It is now possible to formulate a simple rule according to which a numerical calculation of the scattered and transmitted amplitudes can be performed, taking into account anomalous absorption. One has:

$$\begin{pmatrix} T \\ S \end{pmatrix} = P_n P_{n-1} \cdots P_3 P_2 P_1 \begin{pmatrix} 1 \\ 0 \end{pmatrix} = \mathcal{T} \begin{pmatrix} 1 \\ 0 \end{pmatrix} \tag{9.249}$$

where

$$P_n = \begin{pmatrix} 1 & 0 \\ 0 & e^{-i\alpha(n\Delta z)} \end{pmatrix} \begin{pmatrix} 1 - \pi i s \Delta z & i\pi\left(\dfrac{1}{t_e} + \dfrac{i}{\tau_e}\right)\Delta z \\ i\pi\left(\dfrac{1}{t_e} + \dfrac{i}{\tau_e}\right)\Delta z & 1 + \pi i s \Delta z \end{pmatrix}$$

$$\begin{pmatrix} 1 & 0 \\ 0 & e^{i\alpha(n\Delta z)} \end{pmatrix}. \qquad (9.250)$$

P_n refers to the slab at level $n\Delta z$ from the entrance face. The initial values of T and S are evidently 1 and 0. If the dislocation is along the y axis the calculation has to be made for columns with a sufficient number of points along the x axis to obtain an intensity profile.

One could also multiply 2 consecutive shift matrices and obtain:

$$A_n = \begin{pmatrix} 1 & 0 \\ 0 & e^{i\alpha(n\Delta z)} \end{pmatrix} \begin{pmatrix} 1 & 0 \\ 0 & e^{-i\alpha[(n-1)\Delta z]} \end{pmatrix} = \begin{pmatrix} 1 & 0 \\ 0 & e^{i(d\alpha/dz)_{n\Delta z}\,\Delta z} \end{pmatrix}. \qquad (9.251)$$

The calculation can then be formulated as a product of $\mathcal{Q}_a$ and A_n matrices.

As yet no results of computations along this line are available since Howie and Whelan used the system of differential Eq. (9.157) for their machine computation.

(6) *Symmetry properties of dislocation images.* If one wants to compute dislocation image profiles it is of importance, in order to reduce the work to a minimum, to make use of possible symmetry properties. Such symmetry properties can be deduced from the matrix formulation for the calculation of the amplitudes. The properties follow directly from the symmetry properties of the scattering and shift matrices. Howie and Whelan[239] have treated the theory for the multiple beam case. We will again restrict ourselves to the 2-beam case, but the conclusions and their derivations are valid for the multiple beam case.

(a) Symmetry properties of the $\mathcal{Q}$ (and $\mathcal{Q}(\Delta z)$) matrix. In the nonabsorption case one has, from Eqs. (9.123) and (9.124)

$$S^+ = S^-; \quad (S^+)^* = -S^+; \quad (S^-)^* = -S^- \qquad (9.252)$$

and $(T^+)^* = T^-$ and also $(T^-)^* = T^+$.

These properties are true for a slab of finite thickness as well as for a slab of infinitesimal thickness. From this it can be concluded[240a]:

(1) $$\mathscr{D} = \tilde{\mathscr{D}} \tag{9.253}$$

Proof: $$\tilde{\mathscr{D}} = \begin{vmatrix} T^+ & S^+ \\ S^- & T^- \end{vmatrix} = \begin{vmatrix} T^+ & S^- \\ S^+ & T^- \end{vmatrix} = \mathscr{D}$$

(2) $$\mathscr{D}^{-1} = \mathscr{D}^+. \tag{9.254}$$

This property follows immediately from the unitarian character of $\mathscr{D}$, i.e., from $\mathscr{D}\mathscr{D}^+ = I$

(3) $$\mathscr{D}^* = \mathscr{D}^{-1}. \tag{9.255}$$

Since $\mathscr{D}^* = \tilde{\mathscr{D}}^* = \mathscr{D}^+$ and using the just proved property $\mathscr{D}^* = \mathscr{D}^{-1}$. In an absorbing crystal part of the relations (9.252) remains valid

$$(T^+)^* = T^-; (T^-)^* = T^+; S^+ = S \tag{9.256}$$

however

$$(S^-)^* = \left[\pi i \left(\frac{1}{t_e} + \frac{i}{\tau_e} \right) \Delta z \right] = -\pi i \left(\frac{i}{t_e} - \frac{i}{\tau_e} \right) \Delta z \neq -S^-.$$

The only symmetry property of the $\mathscr{D}$ matrix is now

$$\mathscr{D} = \tilde{\mathscr{D}}. \tag{9.257}$$

(b) Symmetry properties of the shift matrices. If

$$^+\mathscr{A}_n = \begin{vmatrix} 1 & 0 \\ 0 & e^{i\alpha_n} \end{vmatrix}$$

describes the shift resulting from the displacement $\bar{R}(n\Delta z)$ the matrix

$$^-\mathscr{A}_n = \begin{pmatrix} 1 & 0 \\ 0 & e^{-i\alpha_n} \end{pmatrix}$$

evidently describes the displacement $-R(n\Delta z)$.

[240a] The symbol $\sim$ means transposing.

The matrices have the following properties:

$$^-\alpha^{-1} = \begin{pmatrix} 1 & 0 \\ 0 & e^{i\alpha} \end{pmatrix} = {}^+\alpha = {}^-\alpha^*$$

$$^+\alpha^{-1} = \begin{pmatrix} 1 & 0 \\ 0 & e^{-i\alpha} \end{pmatrix} = {}^-\alpha = {}^+\alpha^*.$$

(9.258)

Since the α matrices are diagonal they do not change on transposing:

$$^+\tilde{\alpha} = {}^+\alpha \quad \text{and} \quad {}^-\tilde{\alpha} = {}^-\alpha. \tag{9.259}$$

(c) The symmetry properties of the P_n matrices. We now consider 2 deformed crystals related to each other by an inversion through a center. These crystals will be called plus and minus crystals.

It is clear that a translation in the plus crystal becomes a translation in the opposite sense in the minus crystal.

For the plus crystal the P matrix of the nth slab is:

$$^+P_n = {}^+\alpha_n^{*+}2^+\alpha_n = {}^+\alpha_n^{-1+}2^+\alpha_n \tag{9.260}$$

whereas for the minus crystal, i.e., after inversion, the matrix becomes

$$^-P_n = {}^-\alpha_n^{*-}2^-\alpha_n = {}^-\alpha_n^{-1-}2^-\alpha_n. \tag{9.261}$$

If we assume the perfect crystal to be centrosymmetrical one has furthermore:

$$^+2 = {}^-2 = 2$$

because the inversion with respect to the center of symmetry does not change the structure. This statement is in fact equivalent to Friedel's law which says that the diffraction phenomenon is invariant under the inversion. Making use of these remarks one can prove that

$$^-\tilde{P}_n = {}^+P_n. \tag{9.262}$$

In fact one has

$$^-\tilde{P}_n = ({}^-\alpha_n^{-1}2^-\alpha_n)^\sim = ({}^+\alpha_n 2^+\alpha_n^{-1})^\sim$$

by making use of Eqs. (9.258) and (9.259) and further by noting that on transposing the order of multiplication reverses while the α matrices do not change

$$-\tilde{P}_n = {}^+\alpha_n^{-1}\tilde{\mathscr{D}}{}^+\alpha_n = {}^+\alpha_n^{-1}\mathscr{D}{}^+\alpha_n = {}^+P_n.$$

Similarly one has

$$+\tilde{P}_n = {}^-P_n. \tag{9.263}$$

These properties apply to absorbing as well as to nonabsorbing crystals since use was made of Eqs. (9.258) and (9.259) only.

(d) Symmetry property of the total $\mathscr{P}$ matrix. The total scattering matrix for the deformed plus crystal is given by

$$+\mathscr{P} = {}^+P_1{}^+P_2 \cdots {}^+P_n$$

whereas the matrix for the minus crystal is:

$$-\mathscr{P} = {}^-P_n{}^-P_{n-1} \cdots {}^-P_2{}^-P_1$$

since the same electron wave passes the slabs of the minus crystal in the inverse order.

One can prove that

$$-\mathscr{P} = {}^+\tilde{\mathscr{P}}. \tag{9.264}$$

Proof:

$$+\tilde{\mathscr{P}} = ({}^+P_1{}^+P_2 \cdots {}^+P_n)^\sim = {}^+\tilde{P}_n{}^+\tilde{P}_{n-1} \cdots {}^+\tilde{P}_2{}^+\tilde{P}_1.$$

Making use of the property (9.263)

$$+\tilde{\mathscr{P}} = {}^-P_n{}^-P_{n-1} \cdots {}^-P_2{}^-P_1 = {}^-\mathscr{P}.$$

(e) Symmetry properties of the images.[239] A point at level z in a given column of the plus crystal is transformed by inversion into a point at level $t - z$ in the corresponding column of the minus crystal. Since the inversion also changes the sense of the displacement one has the correspondence:

$$\bar{R}(z) \xrightarrow{\text{inversion}} -\bar{R}(t - z).$$

For 2 such columns we have shown that the total transmission matrices $\mathscr{P}$ are transposed one with respect to the other. On calculating the transmitted and scattered intensities one uses the formula

$$\begin{pmatrix} T \\ S \end{pmatrix} = \mathscr{P} \begin{pmatrix} 1 \\ 0 \end{pmatrix} = \begin{pmatrix} \mathscr{P}_{11} & \mathscr{P}_{12} \\ \mathscr{P}_{21} & \mathscr{P}_{22} \end{pmatrix} \begin{pmatrix} 1 \\ 0 \end{pmatrix}.$$

From this it is clear that the transmitted intensity will be the same for both columns since on calculating T only the diagonal elements $\mathscr{P}_{11}$ are used, and these are the same in both $\mathscr{P}$ matrices.

Since a constant displacement R_0 of the whole column does not change the intensity of the diffracted and scattered waves, one can state quite generally that in a centrosymmetrical crystal columns with displacement functions $\bar{R}(z)$ and $\bar{R}_0 - \bar{R}(t - z)$ will produce the same bright field intensity.

The same property does not hold in general for the dark field image because S is obtained from the matrix element $\mathscr{P}_{21}$. This element is in general different from the corresponding one in the transposed matrix except in the case of the 2-beam theory without absorption.

From these general principles it follows, for instance, that the stacking fault fringes have a bright field image which is symmetrical with respect to the center of the foil, even in an absorbing crystal. The dark field image is asymmetrical.[236b]

(f) Matrix formulation of the kinematical theory. The transmitted and scattered amplitude for a perfect slab of thickness z is given by

$$T = e^{-\pi i s z}, \qquad \text{i.e., } |\bar{T}| = 1$$

$$S = \frac{i}{st_e} \sin \pi s z.$$

These amplitudes can be derived from the initial values at the entrance face $T = 1$, $S = 0$ by means of the following matrix

$$\begin{pmatrix} T \\ S \end{pmatrix} = \begin{pmatrix} e^{-\pi i s z} & 0 \\ \dfrac{i \sin \pi s z}{st_e} & e^{\pi i s z} \end{pmatrix} \begin{pmatrix} 1 \\ 0 \end{pmatrix}. \qquad (9.265)$$

The zero in the matrix, according to the kinematical theory, means physically that no electrons are scattered back into the incident direction.

The amplitude for a scattered wave for a crystal containing a stacking fault can be found by the following matrix multiplication:

$$\begin{pmatrix} T_s \\ S_s \end{pmatrix} = \begin{pmatrix} e^{-\pi i s t_1} & 0 \\ \dfrac{i \sin \pi s t_1}{s t_e} e^{\pi i s t_1} & e^{\pi i s t_1} \end{pmatrix} \begin{pmatrix} 1 & 0 \\ 0 & e^{i\alpha} \end{pmatrix} \begin{pmatrix} e^{-\pi i s t_2} & 0 \\ \dfrac{i \sin \pi s t_2}{s t_e} e^{\pi i s t_2} & e^{\pi i s t_2} \end{pmatrix} \begin{pmatrix} 1 \\ 0 \end{pmatrix}.$$

The result given in formula (9.24) or (9.29) is found back in this way

$$T = e^{-\pi i s (t_1 + t_2)}$$

$$S = \frac{i}{s t_e} \{ e^{-\pi i s (t_1 + t_2)} \left[e^{\pi i s t_1} \sin \pi s t_1 + e^{\pi i s (t_1 + 2 t_2)} e^{i\alpha} \sin \pi s t_2 \right] \}. \tag{9.266}$$

It is obvious that this result can be generalized to an arbitrary succession of stacking faults and also to dislocations in the same way as for the dynamical theory.

It is worthwhile pointing out that the thin slab transmission matrix in the kinematical approximation is

$$\mathscr{Q}^{\mathrm{kin}}(\Delta z) = \begin{pmatrix} 1 - \pi i s \Delta z & 0 \\ \dfrac{\pi i}{t_e} \Delta z & 1 + \pi i s \Delta z \end{pmatrix} \tag{9.267}$$

as compared to Eq. (9.243) for the dynamical case. The only difference is that one element is zero. This element is in the dynamical theory S^-, i.e., the amplitude of the wave scattered back into the incident direction. Since this process is not taken into account in the kinematical approximation this element is evidently zero.

(7) *Screw dislocations inclined with respect to the foil plane.* A more detailed account of the machine calculations by Howie and Whelan[241] has now become available. In particular the oscillating contrast is studied for screw dislocations inclined with respect to the foil plane. Two different aspects are distinguished. The contrast is said to be dotted if the image remains in place but fluctuates periodically in darkness. It is called alternating when the black and white parts of the image are interchanged, i.e., if the black line goes alternatively left and right of the dislocation. On consulting the curves of Fig. 8 (for $n = 1$) of their paper, it becomes apparent that the period is the

[241] A. Howie and M. J. Whelan, *Proc. Roy. Soc.* **A267**, 206 (1962).

same for both types of contrast and about equal to the extinction distance.

Depending on the total thickness of the foil, the bright and dark field images may have different character. For instance for a foil of thickness $t = 3t_e$ with $s = 0$, and $\tau_e = 10t_e$ the bright field image is dotted and the dark field image alternating, whereas for a foil of thickness $t = 3.5 \, t_e$ the reverse is true. For thicker crystals the oscillations are confined to the region near the surface; they die away in the middle of the foil. The alternating effects essentially result from absorption since they even occur for $s = 0$, where the nonabsorption dynamical theory predicts symmetrical images (for $n = 1$) at all depths and for all values of n (see Appendix B).

There is a simple way of deriving intuitively that dislocation contrast is oscillating in certain cases, with the same period of the thickness fringes, at least for the part of the image at the position of the dislocation, i.e., for $x = 0$. Let us consider first the simple case of the screw dislocation with the displacement function given by Eq. (9.226). For a reflection with $\bar{n} \cdot \bar{b} = n$ there is a sudden phase shift of $\alpha = n\pi$ between the upper and lower part of the column, passing through the dislocation ($x = 0$).

This can be deduced immediately as follows; for $z > z_0$

$$\lim_{\substack{x \to 0 \\ >}} \alpha = \lim_{\substack{x \to 0 \\ >}} n \, \text{arctg} \frac{z - z_0}{x} = n \frac{\pi}{2}$$

while for $z < z_0$, $\lim_{\substack{x \to 0 \\ >}} \alpha = -n(\pi/2)$. The phase difference is clearly $n\pi$. The same result holds for $x \to 0$.

For odd values of n the intensity transmitted and scattered by the column at $x = 0$ is the same as if there were an antiphase boundary at the level of the dislocation, the rest of the column being perfect. Therefore one can apply the theory of Sections 9h(7) and (4) to predict the variation of intensity at $x = 0$ if the dislocation changes its level, i.e., if z_0 varies. For the case $s = 0$ the formula (9.211) simply reduces to

$$I_t = \cos^2 \frac{\pi}{t_e} (t_2 - t_1). \tag{9.268}$$

Crystals of layer structures grown by sublimation often have top and bottom faces gradually sloping with respect to the c plane, in which the dislocations are situated. If one side of the platelet is perfectly parallel to the c plane, while the other is sloping, the image of a dislocation in the c plane will be periodic with period t_e the same as for the thickness fringes. These conditions were satisfied in Fig. 96.

For even values of n there is a phase difference of $\pm 2k\pi$, i.e., no net phase difference. The dislocation image at $x = 0$ has now the same intensity as the perfect crystal of the same thickness whatever the position of the dislocation is. The part of the image at $x = 0$ is never periodic.

For partial dislocations, n is fractional and there will again be periodic behavior at $x = 0$.

For the edge dislocation with slip plane parallel to the foil plane the part of the displacement function to be considered is

$$\frac{b}{2\pi} \left[\phi + \frac{\sin 2\phi}{4(1 - \nu)} \right].$$

For $x = 0$; $\phi = \pm (\pi/2)$; $\sin 2\phi = 0$ and the phase difference between upper and lower part of the column is the same as for the screw. If valid for edge and screw dislocations, the conclusion is also valid for mixed dislocations.

(8) *Partial dislocations.* In Howie and Whelan[241] the bright field images of partial screw dislocations for which $\bar{n} \cdot \bar{b} = \pm 1/3, \pm 2/3$, and $\pm 4/3$ are computed. In accord with the calculations by Gevers,[231a] based on the kinematical theory, and with observations on graphite (Fig. 100) it is found that for $\bar{n} \cdot \bar{b} = \pm 1/3$ the contrast is too weak to be visible. It would consist of a very faint white line between the background and the stacking fault intensity. For $\bar{n} \cdot \bar{b} = \pm 4/3$ the line should be visible, however. The calculated profiles are shown in Fig. 99,b. They can be compared with those of Fig. 99a calculated on the basis of the kinematical theory.

(9) *Slip trace contrast.* The contrast of slip traces also has been computed in Howie and Whelan,[241] on the assumption that it originates from the presence of a dislocation just below the surface, pinned by an oxide film. The effect of the surface is taken into account by means of an image dislocation at a distance $0.1 t_e$ from the real dislocation and symmetrically placed with respect to the surface.

For $n = \pm 1$ the image is found to be a black-white double line. For $n = + 1$ the inside of the bright field images due to top and bottom trace are both bright. The dark field image is again similar to the bright field image at the top trace, but "complementary" at the bottom trace (cf. stacking fault fringes). This remark allows us to distinguish top and bottom end of the slip trace. For $n = 1$ the situation is reversed. The inside of the image is now a dark line in the bright field image. Although the model is certainly oversimplified the agreement with observations on stainless steel is surprisingly good.

(10) *Determination of the length of the Burgers vector.* The image profile of a dislocation depends on the value of n. In particular for $s = 0$ the image for $n = 2$ consists of a double peak separated by a "line" which has the background intensity, whereas the image for $n = 1$ is a single peak. For $s \neq 0$ the image for $n = 2$ consists of a single peak.

On crossing an inclination extinction contour, along which s changes sign, the image side changes, since it is essentially determined by the sign of $(g \cdot b)s$. The behavior on crossing such a contour is shown in Fig. 121[(1)] for $n = 1$ and $n = 2$; from this it is evidently

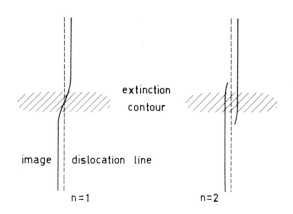

FIG. 121[(1)]. Behavior of the image of a dislocation at an inclination extinction contour for $n = 1$ and $n = 2$.

possible to conclude whether n is 1 or 2. If we know already the direction of b, and knowing the diffraction vector, it is possible to determine the length of b. The knowledge of n is essentially equivalent to knowing the length of the projection of $\bar{b}$ onto $\bar{n}$.

10. Electron Microscopy in Transmission: Applications

a. Ribbons of Partial Dislocations

A high-resolution technique is required to observe the individual partial dislocations in a ribbon. Using electron microscopy this has been possible, and detailed knowledge concerning partial dislocations and stacking faults has resulted. In particular a number of methods for estimating stacking fault energies from geometrical configurations of partials has been developed. These methods are at present the only direct ones. In the past, stacking fault energies have been deduced from the energy of a coherent twin boundary[242a] and also from the temperature dependence of the stress-strain curves.[242b]

Partials have been observed for the first time in stainless steel[243–245] and since then in a number of crystals, mainly layer structures.[202] In this section the geometry of ribbons in a number of particular structures will be discussed.

(1) *Partial dislocations in the close-packed metals.* As pointed out by Heidenreich and Shockley[246] glide along close-packed planes, e.g., on (111) planes in the face-centered cubic lattice takes place in the two steps shown in Fig. 122. This can easily be demonstrated on a ball model; on glide the spheres follow "valleys" rather than going over the "hills." In dislocation language this means that the dislocation that produces glide over $(a/2)$ [110], which is the glide vector in the f.c.c. system, separates into 2 partial dislocations according to the scheme

$$\frac{a}{2}[110] \rightarrow \frac{a}{6}[211] + \frac{a}{6}[12\bar{1}]. \tag{10.1}$$

The region between the 2 partials is faulted, as shown in Fig. 123. If the stacking in the perfect crystal is *a b c a b c a b c* ... the structure

[242a] R. L. Fullman, *J. Appl. Phys.* **22**, 448 (1950).

[242b] A. Seeger, R. Berner, and H. Wolf, *Z. Physik* **155**, 247 (1959).

[243] M. J. Whelan, P. B. Hirsch, R. W. Horne, and W. Bollmann, *Proc. Roy. Soc.* **A240**, 524 (1957).

[244] M. J. Whelan, *Proc. Roy. Soc.* **A249**, 114 (1958).

[245] M. J. Whelan, *Proc. 3rd Intern. Conf. on Electron Microscopy, London, 1954* p. 539 (1956).

[246] R. O. Heidenreich and W. Shockley, *Rept. Conf. Strength of Solids, Bristol, 1947* p. 57 (1948).

of the fault is

$$a\,b\,c\,a\,|\,c\,a\,b\,c\,a\,b\,c\cdots \qquad (10.2)$$

i.e., there are 2 layers in hexagonal close-packed stacking and this arrangement evidently will have a larger energy than the normal cubic stacking. The difference in energy per unit area between the

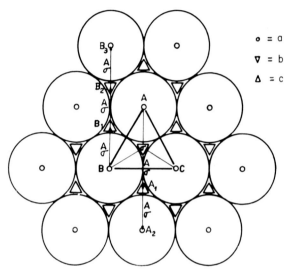

Fig. 122. Glide between close-packed planes illustrating the concept of Shockley partials. Glide from A to B proceeds in 2 steps: $A\sigma$ and σB.

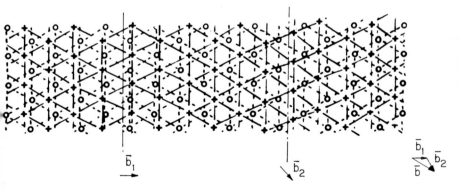

Fig. 123. Each of the 2 glide motions from Fig. 122 gives rise to a partial dislocation. The region between the 2 partials is faulted.

faulted and the normal stacking is called the specific stacking fault energy γ. The stacking fault, described by the symbol (10.2) can also be considered as resulting from the removal of one lattice plane: the plane marked by an arrow in the sequence

$$a\,b\,c\,a\,\underset{\uparrow}{b}\,c\,a\,b\,c\ldots .\tag{10.3}$$

Such a fault has been called "intrinsic" by Frank.[247] One can also imagine a stacking fault resulting from the introduction of 1 lattice plane, producing a sequence of the kind

$$a\,b\,c\,\underset{\uparrow}{b}\,a\,b\,c\,a\,b\,c\ldots .\tag{10.4}$$

This is called an "extrinsic" fault. There are again 2 infractions against the stacking rule, but they are localized in planes separated by 1 lamella in the cubic stacking. It is therefore not evident which fault will have the lowest energy. The extrinsic fault can also be generated by 2 intrinsic faults on adjacent lattice planes. Whereas an intrinsic fault can be introduced and eliminated by the motion of 1 partial, the introduction or the removal of an extrinsic fault requires the passage of 2 partials along glide planes on either side of the fault.

The dissociation into partials will actually occur, if elastic energy is gained provided the stacking fault energy is not too large. Whether or not elastic energy is gained can easily be verified by using the "square of the Burgers vector" criterion. The elastic energy of a dislocation is proportional to the square of its Burgers vector. The elastic energy at the left-hand side of Eq. (10.1), e.g., is proportional to $1/2$ as compared to $1/6 + 1/6 = 1/3$ for the right-hand side.

In the hexagonal close-packed structure $a\,b\,a\,b\,a\,b\,a\,b \ldots$ the same type of dissociation can take place, for dislocations lying in the c planes. The fault now has the structure $a\,b\,a\,b\,c\,a\,c\,a \ldots$, it consists of a lamella in the cubic close-packed arrangement.

(2) *Notations for Burgers vectors.* (a) Face-centered cubic lattice: Thompson's notation.[248] It is convenient to introduce at this point a notation for Burgers vectors in the face-centered cubic lattice, which we shall use frequently in what follows.

[247] F. C. Frank and J. F. Nicholas, *Phil. Mag.* [7] **44**, 1213 (1953).
[248] N. Thompson, *Proc. Phys. Soc.* (*London*) **66**, 481 (1953).

Consider the tetrahedron consisting of {111} planes drawn in full lines in Fig. 124. The summits are denoted by capital Latin letters *A*, *B*, *C*, and *D*; the centers of its faces by means of Greek letters; α opposite *A*, β opposite *B*, etc. The faces are denoted by means of small italic letters; *a* opposite *A*, etc.

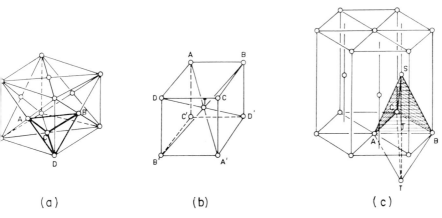

FIG. 124.   Notations for Burgers vectors in the different structures. (a) Thompson's tetrahedron for the face-centered cubic lattice. (b) Body-centered lattice. (c) Hexagonal lattice.

The glide planes are now the planes, *a,b,c,* and *d*, while the Burgers vectors of perfect dislocations [i.e., vectors of the type $(a/2)\,[110]$] are given by the edges $\overline{AB}$, $\overline{CA}$, etc. The Burgers vectors of Shockley partials are denoted by symbols of the type $A\beta$, $C\delta$, etc. whereas the Burgers vectors of stair rod dislocations [i.e., with $\overline{b} = (a/6)\,[110]$] are given by $\overline{\alpha\delta}$, $\overline{\beta\gamma}$, etc.

The vector $X\delta$ with $X = A$, *B*, or *C* is, for instance, lying in the *d* plane. The vector *AB* is lying in the *c* and *d* planes; $\alpha\beta$ is parallel to *AB*.

Frank developed a notation for the description of dislocations in networks, associated with these symbols. He proposed to put the letters denoting the Burgers vector left and right of the line as in Fig. 125 and read from left to right when looking away from the node point. In such a way it is possible to describe in a concise way the distribution of Burgers vectors in a net.

Two kinds of 3-fold nodes, satisfying the nodal sum rule, can be considered: the so-called *P* and *K* nodes. *K* nodes are based on the sum relation $AB + BC + CA = 0$, the P nodes on a relation of the type $AB + CA + BC = 0$. The lettering patterns for the two types of nodes are shown in Fig. 125a,b. If the dislocations forming these nodes dissociate into partials the K nodes become extended and the P nodes contracted. It is easy to show that in a stable hexagonal net K and P nodes have to alternate.

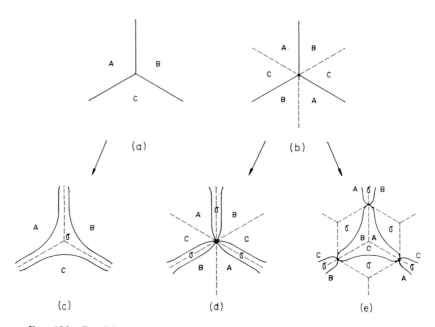

FIG. 125. Possible nodes in the face-centered cubic lattice, illustrating at the same time the use of Thompson's notation. (a) P node. (b) K node. Dissociation into partials gives rise to the configurations (c), (d), and (e), respectively. The last case is observed only if both intrinsic and extrinsic faults have small energy.

(b) Hexagonal lattice.   A notation for the hexagonal close-packed metals recently was proposed by Berghezan et al.[249]; it can be extended easily to most hexagonal structures, in particular to graphite. It makes use of the bipyramid shown in Fig. 124c. The 6 perfect basal vectors are denoted by AB, AC, BC and their negatives. The perfect dislocations perpendicular to the c plane and of magnitude c are ST and TS. The imperfect Shockley partials have Burgers vectors of the

[249] A. Berghezan, A. Fourdeux, and S. Amelinckx, Acta Met. 9, 464 (1961).

type $A\sigma$, $B\sigma$, $C\sigma$, they form by dissociation according to the relation $AB \rightarrow A\sigma + \sigma B$. The imperfect dislocations perpendicular to the c plane have vectors σS, σT, $S\sigma$ and $T\sigma$ of magnitude $c/2$. There are also imperfect dislocations with vectors AS, BS, etc. Larger Burgers vectors can be denoted as the sum of vectors of the kind shown in Fig. 124c, e.g., $1/3 \langle 11\overline{2}3 \rangle$ can be represented as $(ST + AB)$.

(c) The body-centered lattice. In the body-centered metals the simplest reference system is the centered cube itself, as proposed by Amelinckx and Dekeyser[250] and more recently used extensively by Carrington et al.[251] The reference cell is shown in Fig. 124b. The usual perfect dislocations have Burgers vectors of the type $(a/2) \langle 111 \rangle$ and they can be represented by a symbol involving T: AT, BT, ..., etc. There is a second kind of stable vector for which the symbol does not involve T, e.g., AB, $A'D'$; i.e., $a \langle 100 \rangle$. Two T vectors can react to form a vector of this last type: $AT + TB = AB$. Dislocations with vectors like DA', joining opposite corners of the side face, i.e., of the type $a \langle 110 \rangle$, are unstable against dissociation according to the reaction $DA' \rightarrow DT + TA'$.

(3) *Width of the ribbons.* Evidently the magnitude of the separation of the partials will be determined by the balance between the repulsive force between the dislocations, which diminishes (in an infinite solid) as the inverse of the distance d, and the constant attractive force due to the stacking fault with specific energy γ. The repulsive force per unit length between two parallel dislocations with Burgers vector $\bar{b}_1$ and $\bar{b}_2$ is given by

$$F = \frac{\mu}{2\pi d} \left(\frac{b_{1e} b_{2e}}{1 - \nu} + b_{1s} b_{2s} \right) \tag{10.5}$$

where the indices e and s refer to edge and screw components, respectively; μ and ν are, respectively, the shear modulus and Poisson's ratio. The attractive force due to the stacking fault is deduced in the following way. If the distance between partials increases by dx the energy of the faulted area increases by $dE = \gamma dx$ per unit length. The force per unit length is thus

$$-\frac{dE}{dx} = -\gamma. \tag{10.6}$$

[250] S. Amelinckx and W. Dekeyser, *Solid State Phys.* **8**, 325 (1959).
[251] W. Carrington, K. Hale, and D. McLean, *Proc. Roy. Soc.* **A259**, 203 (1960).

Applying Eqs. (10.5) and (10.6) to the particular problem of the equilibrium separation of partials in the ribbon of Fig. 126 leads to

$$d = d_0 \left(1 - \frac{2\nu}{2 - \nu} \cos 2\phi\right) \qquad (10.7)$$

where

$$d_0 = \frac{\mu b^2}{8\pi\gamma} \frac{2 - \nu}{1 - \nu} \qquad (10.8)$$

ν is Poisson's ratio, ϕ is the angle between the total Burgers vector and the direction of the ribbon.[252] Formula (10.7) was verified quantitatively for the first time for graphite in which the separation is large enough so that d can be measured directly. Although the structure is not close-packed, we will show below that this theory can be applied.

The way in which formulas (10.7) and (10.8) can be used is as follows. For widely curved dislocation ribbons the Burgers vector is determined, using the procedure described in Section 9d(2). From this, ϕ can be deduced for every segment, and the width d can be measured as a function of ϕ. A plot of d versus cos 2 ϕ should be a straight line. The slope of the straight line gives an effective value for ν; the intercept gives d_0, and finally Eq. (10.8), allows us to evaluate the dimensionless quantity $\gamma/\mu b$; if μ is known γ is determined.

One should also take into account the finite thickness of the foil, which tends to reduce the repulsive forces more rapidly than $1/d$.[252a] This correction is discussed further in Section 10k(3). The anisotropy, especially in layer structures, has to be taken into account also; this will be discussed in Section 10a(6).

(4) *Extended nodes.* The interactions of 2 ribbons with a common partial gives rise to an isolated extended node by the process represented in Fig. 127. The 2 meeting partials with the same Burgers vector annihilate over part of their length and the extended node results. The mechanism whereby extended nodes in networks are generated is different and will be discussed in Section 10c(5). The isolated extended nodes can be used to evaluate stacking fault energies even in certain cases where the separation of partials in the ribbons is not resolved.[244]

[252] W. T. Read, "Dislocations in Crystals." McGraw-Hill, New York, 1953.

[252a] R. Siems, P. Delavignette, and S. Amelinckx, *Phys. Stat. Solidi* 2, 636 (1962).

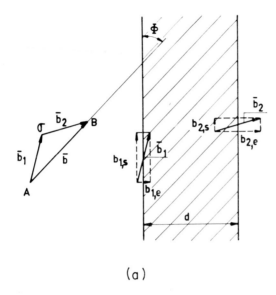

(a)

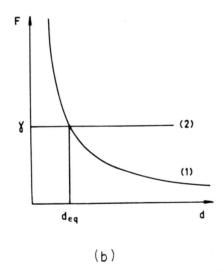

(b)

FIG. 126. (a) Model for ribbons of partial dislocations of the Shockley type. The angle between the direction of the ribbon and the Burgers vector is called ϕ. (b) The repulsive force between partials (curve 1) goes as $1/d$ whereas the force due to the stacking fault is constant (curve 2). The intersection point gives the equilibrium separation d_{eq}.

To a first approximation one can consider the partials in the node to be independent. The equilibrium curvature of one of them is then determined by the balance between the line tension, which can be taken as $1/2\,\mu b^2$ and the effective shear stress due to the stacking

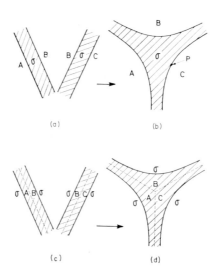

Fɪɢ. 127.   Fusion of 2 ribbons giving rise to an isolated extended node.

fault, which is given by γ/b. The equilibrium radius of curvature is then

$$R = \mu b^2/2\gamma. \tag{10.9}$$

For such crystals where ribbons and extended nodes are simultaneously resolved it was possible to investigate how good an approximation Eq. (10.9) is. It turns out that it is only a first approximation. Moreover the measurement of the radius of curvature is rather subjective. Siems et al.[252b] have proposed a more accurate procedure. The interaction between neighboring partials is taken into account as well as the character of the ribbons. The stacking fault energy is expressed as a function of y_0 (see Fig. 128) rather than R. The relation is deduced by minimizing the total energy associated with the node. The different contributions are:

(1) the line energy;

[252b] R. Siems, P. Delavignette, and S. Amelinckx, Z. Physik 165, 502 (1961).

(2) the interaction energy;

(3) the stacking fault energy.

The result is

$$\gamma = K \Psi_0 / y_0 \sqrt{3} \qquad (10.10)$$

where

$$K = \frac{\mu b^2}{8\pi} \frac{2 + \nu}{1 - \nu} \qquad (10.11)$$

for edge ribbons and

$$K = \frac{\mu b^2}{8\pi} \frac{2 - 3\nu}{1 - \nu} \qquad (10.12)$$

for screw ribbons.[253] The dimensionless quantity Ψ_0 is the solution of the following equation

$$\ln \psi_0 = \psi_0 - 1 - 0.267(\epsilon/K) \qquad (10.13)$$

ϵ is the line energy. One has $(\epsilon/K) = 8.95\ [2/(2 + \nu)]$ for edge ribbons and $(\epsilon/K) = 8.95\ [2/(2 - 3\nu)]$ for screw ribbons. For graphite and materials with a similar structure of the dislocation network, the values of $\gamma/\mu b$ deduced in this way compare very well with these derived from ribbon widths.

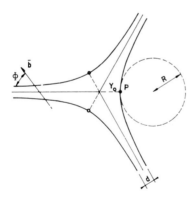

FIG. 128.   Extended node. Illustrating the notations used in deriving the expression for the stacking fault energy.

[253] A ribbon is called an edge or a screw ribbon if the total dislocation (outside of the node) is in edge or screw orientation.

(5) *Threefold ribbons.* (a) Structure and formation mechanism. In the hexagonal close-packed structure, and also in the graphite structure, the dissociation into partials in successive c planes takes place in a different succession. With reference to Fig. 124 the dissociation can be described by the reaction:

$$AB \rightarrow A\sigma + \sigma B; \ AC \rightarrow A\sigma + \sigma C; \ BC \rightarrow B\sigma + \sigma C \qquad (10.14)$$

in one lattice plane and by

$$AB \rightarrow \sigma B + A\sigma; \ AC \rightarrow \sigma C + A\sigma; \ BC \rightarrow \sigma C + B\sigma \qquad (10.15)$$

in the next lattice plane. With the notation used here the sum sign is usually *not* commutative. As a consequence of the different dissociation schemes single ribbons in neighboring glide planes or planes separated by a distance $(2n + 1)\,(c/2)$ can combine to form more complex, and wider, ribbons. In particular 3-fold ribbons result from the fusion of 2 single ribbons according to the relation

$$(A\sigma + \sigma B) + (\sigma C + A\sigma) \rightarrow A\sigma + A\sigma + A\sigma. \qquad (10.16)$$

It is clear that the 3 Burgers vectors are the same; this is confirmed by observations [see Section 10a(1) (c)]. Further the ribbon is symmetrical since the stacking faults in both halves have the same specific energy, being related by a symmetry operation. This becomes evident if it is realized that by changing the notation the stacking symbols can be made identical. A model for the arrangement of layers within such a ribbon is presented in Fig. 129.

Two ribbons in glide planes separated by an integral number of repeat distances can combine and build an asymmetrical triple ribbon according to the reaction

$$AB + AC = (A\sigma + \sigma B) + (A\sigma + \sigma C) = A\sigma + A\sigma + (\sigma B + \sigma C)$$
$$= A\sigma + A\sigma + A\sigma. \qquad (10.17)$$

Since this formation process requires the crossing of $A\sigma$ by σB, these 2 partials repel, but under a suitable shear stress the process can take place. A cross section through such a ribbon is represented in Fig. 129b. It is clear that the stacking fault energies in both halves are different; this causes the difference in width between the 2 halves. Whereas in one half there are 2 layers in the cubic arrangement, there are 4 of these in the other half.

(b) Equilibrium configuration. For an infinite isotropic medium the triple ribbons should be about 5 times wider than single ones; therefore their width can be measured with a better precision. Moreover the one-sided nature of dislocation contrast Section 9d(1) does not disturb the measurements; since the 3 partials have the same Burgers vector the sidewise displacement is the same for all 3. The stacking fault energy can again be obtained from the equilibrium separation, which we now calculate.

By putting the resultant force on each partial of Fig. 129c equal to zero one obtains

$$\left\{ \begin{array}{l} A\left(\dfrac{1}{x} + \dfrac{1}{x+y}\right) = \gamma_1 \\[2mm] A\left(\dfrac{1}{y} + \dfrac{1}{x+y}\right) = \gamma_2 \\[2mm] A\left(\dfrac{1}{x} - \dfrac{1}{y}\right) = \gamma_1 - \gamma_2 \end{array} \right\} \qquad (10.18)$$

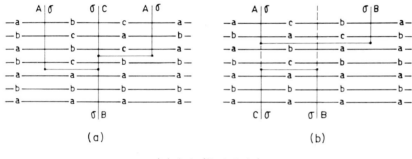

(a)   (b)

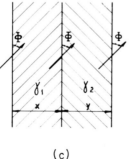

(c)

FIG. 129. Arrangement of layers in threefold ribbons in the hexagonal structure, in particular in graphite; the 3 partials have parallel Burgers vectors. (a) Symmetrical. (b) Asymmetrical. (c) Notations used in calculating the equilibrium configuration.

where

$$A = \alpha \sin^2 \phi + \beta \cos^2 \phi; \; \alpha = \mu b^2/2\pi(1 - \nu); \; \beta = \mu b^2/2\pi. \quad (10.19)$$

The solution of this set of compatible equations is

$$\left. \begin{array}{l} x = 3A/[2\gamma_1 - \gamma_2 + g] \\ y = 3A/[2\gamma_2 - \gamma_1 + g] \end{array} \right\} \quad (10.20)$$

with

$$g = \sqrt{\gamma_1^2 + \gamma_2^2 - \gamma_1\gamma_2}.$$

In the symmetrical case $\gamma_1 = \gamma_2 = \gamma$ and one finds $x = y$ and

$$w = x + y = \frac{3A}{2\gamma} = w_0 \left(1 - \frac{\nu}{2 - \nu} \cos 2\phi\right) \quad (10.21)$$

where

$$w_0 = \frac{3\mu b^2}{4\pi\gamma} \frac{2 - \nu}{1 - \nu}. \quad (10.22)$$

The same functional dependence of w on ϕ as for the single ribbon is still valid.

In the asymmetrical case one finds

$$w_0 = \frac{3\mu b^2}{4\pi} f(\gamma_1, \gamma_2) \frac{2 - \nu}{1 - \nu} \quad (10.23)$$

where

$$f(\gamma_1, \gamma_2) = \frac{\gamma_1 + \gamma_2 + 2g}{4\gamma_1\gamma_2 - \gamma_1^2 - \gamma_2^2 + g(\gamma_1 + \gamma_2)}. \quad (10.24)$$

In the latter case it is possible to express the ratio of the stacking fault energies γ_1/γ_2 directly as a function of $r = y/x$. One finds

$$\gamma_1/\gamma_2 = r(r + 2)/(2r + 1). \quad (10.25)$$

The ratio of the widths of the single and triple ribbons with the same ϕ depends only on ν, and can therefore be used to determine ν $(w/d) = [12/(2+\nu)]$ for edge ribbons and $(w/d) = [12(1 - \nu)/(2 - 3\nu)]$ for screw ribbons. An example of an asymmetrical ribbon is visible in Fig. 82; from the widths of the 2 halves we can conclude $\gamma_1/\gamma_2 = 5/8$. The energy of the superposed faults is less than twice the energy of a single fault which suggests that there is either some interaction between faults in neighboring planes, or that the influence of the surface on the stress field is important. This will naturally decrease, especially the width of the wide half, i.e., decrease γ_1/γ_2.

(6) *The width of ribbons in anisotropic substances.* The width d of ribbons in the basal plane of an anisotropic hexagonal crystal is given by formula (10.7) provided one uses the appropriate expression for the effective shear modulus μ and the effective Poisson's ratio ν

$$\mu_{\text{eff}} = \sqrt{\tfrac{1}{2}\, C_{44}(C_{11} - C_{12})} \tag{10.26}$$

$$\frac{\mu_{\text{eff}}}{1 - \nu_{\text{eff}}} = \frac{C_{11}C_{33} - C_{13}^2}{C_{33}(2\delta_1 + \delta_2)^{1/2}} \tag{10.27}$$

where

$$\delta_1 = \sqrt{\frac{C_{11}}{C_{33}}} \tag{10.28}$$

and

$$\delta_2 = \frac{C_{11}C_{33} - C_{13}^2 - 2C_{13}C_{44}}{C_{33}C_{44}} \tag{10.29}$$

the C_{ik} are the elastic constants in a system with the x_3 axis along the hexagonal axis.

For triple ribbons the anisotropy can be taken into account by using the same substitution (10.26, 10.27) in the expression (10.21).

If the complete set of elastic constants is known, which is unfortunately not often the case, the formulas (10.26) to (10.29) in conjunction with (10.7) and (10.8) allow us to deduce the stacking fault energy. For very anisotropic substances like graphite, taking into account the anisotropy, this leads to 0.6 ergs/cm² against 3.5×10^{-2} ergs/cm² derived from the isotropic theory. Graphite is one of the few substances for which the required elastic constants are known.

For a ribbon in the (111) plane of a cubic crystal it is possible to give an approximate expression for μ_{eff} and ν_{eff} as a function of the elastic constants

$$\mu_{\text{eff}} = \sqrt{\tfrac{1}{2}\, C_{44}(C_{11} - C_{12})} \tag{10.30}$$

$$\frac{\mu_{\text{eff}}}{1 - \nu_{\text{eff}}} = \frac{1}{3}(C + C_{12}) \sqrt{\frac{C_{44}(C - C_{12})}{C_{11}(C_{11} + C_{12} + 2C_{44})}} \left(1 + 2\,\frac{C_{11}}{C}\right) \tag{10.31}$$

with

$$C = \sqrt{\tfrac{1}{2}(C_{11} + C_{12} + 2C_{44}) C_{11}} \tag{10.32}$$

For dislocation lines along [110] directions the equations are then exact. For silicon[210] the formula leads to: $15 < \gamma < 75$ ergs/cm^2.

(7) *Observation of ribbons in particular structures.* Dislocations[7] which could be resolved into partials, were first observed in stainless steel[243,244] and since then in the alloys: Cu-Al[254-256] and Ni-Co[25], and in many layer structures. Polytypism is a frequent occurrence in this last class of substances; this indicates that the energy difference between different stackings is small or, in other words, that the stacking fault energy γ is small. This is, of course, the reason why the basal dislocations usually are widely split into partials. Further the ribbons are exactly parallel to the close-packed basal plane, and as a consequence they will more likely adopt their equilibrium separation, the influence of the surface being relatively small. Moreover it is possible to correct the observed width for the influence of the finite foil thickness.[252c]

(a) Stainless steel. Extended nodes have been observed (see, for instance, Fig. 130) and from them the stacking fault energy was estimated using formula (10.9): the value of γ is 13 ergs/cm^2. The equilibrium separation of partials in a ribbon has not been measured. Often widely separated segments of partials connecting upper and lower surfaces of the foil, and kept together by a stacking fault, are found (see Fig. 81). Presumably the partials were separated under high local stress and were subsequently pinned at the surface. Therefore their distance of separation is not a measure for the stacking fault energy.

Reactions between ribbons on intersecting glide planes have been observed and analyzed in detail by Whelan.[244] Two important interactions will be discussed here.

(i) Lomer-Cottrell barriers. Consider 2-ribbon dislocations, one in the glide plane (111) and with total vector $b_1 = (a/2) [10\bar{1}] =$

[254] P. R. Swann and J. Nutting, *J. Inst. Metals* **88**, 478 (1960).

[255] A. Howie, *in* "Direct Observation of Imperfections in Crystals" (J. B. Newkirk and J. H. Wernick, eds.), p. 283. Wiley (Interscience), New York, 1962.

[256] A. Howie and P. R. Swann, *Phil. Mag.* [8] **6**, 1215 (1961).

[257] S. Mader and E. Simsch, *Proc. European Reg. Conf. on Electron Microscopy Delft, 1960* **1**, 379 (1961).

$(a/6)$ [11$\bar{2}$] + $(a/6)$ [2$\bar{1}\bar{1}$] and the second one in the glide plane (11$\bar{1}$) and with total vector $(a/2)$ [011] = $(a/6)$ [112] + $(a/6)$ [$\bar{1}$21]. The two glide planes intersect each other along the line [1$\bar{1}$0] and the parts that contain the ribbons form an acute angle. On meeting along the

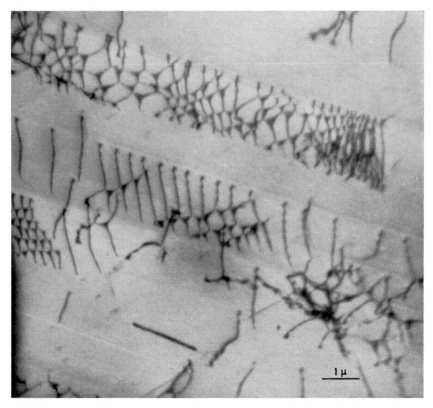

FIG. 130. (a) Network of extended and contracted nodes observed in stainless steel. The network lies in a plane inclined with respect to the foil plane. (Courtesy of Whelan.)

line of intersection, as shown in Fig. 131a the following reaction can take place between the first partials of the two ribbons

$$\frac{a}{6}[11\bar{2}] + \frac{a}{6}[2\bar{1}\bar{1}] + \frac{a}{6}[\bar{1}21] + \frac{a}{6}[112] \rightarrow \frac{a}{6}[11\bar{2}] + \frac{a}{6}[110] + \frac{a}{6}[112].$$

In Thompson's notation the glide planes would be, e.g., d and a,

Fig. 130.   (b) Network in Cu-8%Al alloy. (Courtesy of Art.)

and the reacting ribbons would be $AC = A\delta + \delta C$ and $CD = C\alpha + \alpha D$. The reaction can then be noted

$$A\delta + \delta C + C\alpha + \alpha D = A\delta + \delta\alpha + \alpha D. \qquad (10.33)$$

Two of the partials combine into the new dislocation $\delta\alpha = (a/6)[110]$; the reaction liberates energy as can be judged from the square of the

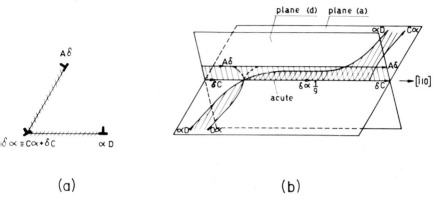

(a)   (b)

Fɪɢ. 131.   Lomer-Cottrell lock. (a) Seen in cross section, perpendicular to the line of intersection of the 2 glide planes. (b) Seen in space. [After M. J. Whelan, *Proc. Roy. Soc.* **A249**, 114(1958).]

Burgers vector criterion. The new type of dislocation is called a "stair rod" dislocation and it has pure edge character. However, its Burgers vector is not in a plane of easy slip and it is therefore sessile; furthermore it is bound to the intersection line. The 2 other partials are repelled by the stair rod and they take up equilibrium positions in their respective glide planes since they are bound to the stair rod by the stacking faults. This reaction is shown in space in Fig. 131b and in cross section in Fig. 131a. The configuration is sufficiently characteristic so that it can be recognized in micrographs; examples are visible in Fig. 132. Configurations of this type are believed to play an important role in work-hardening since they constitute effective barriers against which dislocations can pile up. Evidence for the formation of barriers from etching experiments has been presented in Section 3j(2) (9).

(*ii*) The formation of contracted and extended nodes.   The 2 intersecting dislocations have different glide planes and Burgers vectors enclosing an angle of 120°. Let the intersected and the intersecting dislocation have, respectively, Burgers vectors $DC = D\alpha + \alpha C$ and $CB = C\delta + \delta B$. The first stage in the interaction is represented in Fig. 133a. The partials $C\delta$ and δB of the intersecting dislocation

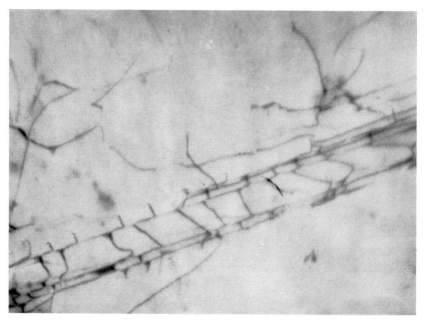

FIG. 132.   (a) Lomer-Cottrell interactions in stainless steel. (Courtesy of Whelan.)

combine, with a gain in elastic energy, with αC of the intersected dislocation according to the scheme

$$\alpha C + C\delta + \delta B \rightarrow \alpha B.$$

The partial αB will be pulled down by the line tension of $D\alpha\alpha C$ into the plane (a) as shown in Fig. 133b where intersections of several dislocations of the type $\alpha CD\alpha$ with the 1 dislocation $C\delta\delta B$ have taken place. The contracted nodes indicated with arrows in Fig. 133b will migrate along the line of intersection of the slip planes and cause $C\delta\delta B$ to cross slip from (d) onto (a). Finally one will obtain the planar configuration of Fig. 133c which consists of an alternation of contracted and extended nodes. Presumably the extended nodes contain an intrinsic fault; the contracted nodes could then possibly

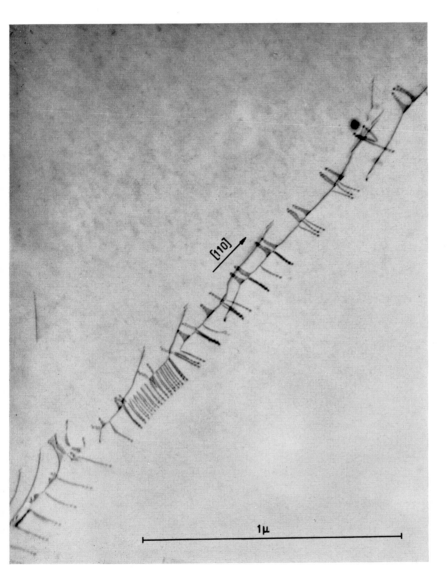

FIG. 132.   (b) Lomer-Cottrell interaction in Cu-8% Al alloy. The geometry is particularly well defined in this case. (Courtesy of Art.)

dissociate also with formation of an extrinsic fault. Apparently this does not take place.

Figure 130 shows an example of an hexagonal network containing extended and contracted nodes in stainless steel.

(b) Metals and alloys. (*i*) Presence of stacking faults. As described in Section 9h(4) stacking faults can be recognized easily in

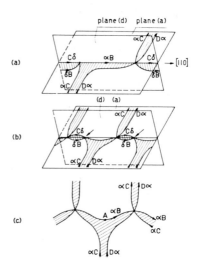

FIG. 133. Formation of contracted and extended nodes on intersection. (a) The first stage of the interaction: $C\delta$ and δB are absorbed by αC. (b) In (b) αB pulls down and the nodes indicated with arrows are driven along by the line tension of $D\alpha\, \alpha C$. (c) Final configuration. (After Whelan.)

the electron microscope. If they are situated in inclined planes the characteristic fringe system appears. If they lie in planes parallel to the foil plane, as in the layer structures, only a shade difference becomes visible under suitable conditions; the observation is somewhat more difficult. The fault vector can be determined in the same way as the Burgers vector (Section 9d(2)).[257a]

It should be kept in mind that twin boundaries may give rise to a fringe system similar in most respects so that produced at a stacking fault.

Stacking faults have now been identified in a large number of

---

[257a] Fringes disappear for $\bar{n} \cdot \bar{R} = 0$ or an integer.

metals and alloys. Very often they occur as a consequence of high local stresses and then the separation between partials is not its equilibrium value. This was presumably the case for the stacking faults observed in zinc,[249,258]copper,[259] gold,[260] silver,[213] cobalt,[261,262] and silicon.[263]

In niobium Fourdeux and Berghezan[264] observed wide stacking faults on (112) planes as well as networks with extended and contracted nodes. The observations were taken as evidence that the stacking fault energy should be small in this metal. On the other hand the observations of Segall[265] seem to suggest that the presence of stacking faults in niobium is a consequence of the segregation of impurities on dislocations. In pure niobium cross slip is very frequent, resulting in wavy slip markings. However this cannot be taken as evidence that the dislocations are not dissociated, because the Burgers vector of the 2 partials may be parallel. The Burgers vector of the perfect dislocation could dissociate according to

$$\frac{a}{2}[111] \rightarrow \frac{a}{6}[111] + \frac{a}{3}[111]. \tag{10.34}$$

Both partials adopt the screw orientation simultaneously so that cross slip is not inhibited. Further Segall established that stacking faults lie on (310) planes more often than on (112) planes; they only occur in annealed and hence impure specimens. Coherent twin boundaries formed on annealing also lie in (310) and have a [310] twin axis.

In tungsten the visible splitting also was attributed to lowering of the stacking fault energy by impurity segregation.[266]

(ii) Stacking fault energies.   Using the method outlined in Section 10a(4) the stacking fault energy has been measured in a number of alloys by Howie and Swann.[256] The authors measured the radius of curvature of extended nodes and applied formula (10.9) in a somewhat

[258] A. Fourdeux and A. Berghezan, *Compt. Rend.* **250**, 3019 (1960).

[259] J. Fourie and R. Murphy, *Phil. Mag.* [8] **6**, 1069 (1961).

[260] D. Pashley, *Phil. Mag.* [8] **4**, 324 (1959).

[261] E. Votava, *J. Inst. Metals* **90**, 129 (1961, 1962).

[262] E. Votava, *Acta Met.* **8**, 901 (1960).

[263] H. Queisser, R. H. Finch, and J. Washburn, *J. Appl. Phys.* **33**, 1536 (1962).

[264] A. Fourdeux and A. Berghezan, *J. Inst. Metals* **89**, 31 (1960, 1961).

[265] R. L. Segall, *Acta Met.* **9**, 975 (1961).

[266] A. Keh and S. Weissmann, *in* "Electron Microscopy and Strength of Crystals" (G. Thomas and J. Washburn, eds.), p. 231. Wiley (Interscience), New York, 1963.

improved form. In this way $\gamma/\mu b$ values between 0.1×10^{-3} and 1.9×10^{-3} can be measured. Still smaller values can be measured if the equilibrium separation of ribbons can be used [see Section 10a(3)]. Their results for the systems Cu + Zu; Cu + Al, Ag + Zu and Ag + Al are shown in Fig. 134.

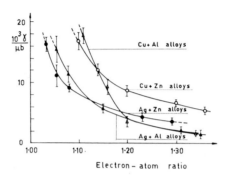

FIG. 134. Stacking fault energies as a function of the composition for a number of alloys. The stacking fault energies were deduced from the geometry of extended nodes. (a) Cu + Zn. (b) Cu + Al. (c) Ag + Zu. (d) Ag + Al. [After A. Howie and P. R. Swann, *Phil. Mag.* [8] **6**, 1215 (1961).]

They also investigated the system Ni + Co and this result is plotted in Fig. 135. The stacking fault energy goes through zero at about 75% Co (at room temperature). The transition f.c.c. → h.c.p. occurs at this composition.

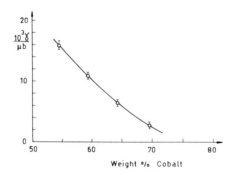

FIG. 135. Stacking fault energy as a function of composition in the alloy Ni + Co. The stacking fault energy goes through zero at about 75% Co. [After A. Howie and P. R. Swann, *Phil. Mag.* [8] **6**, 1215 (1961).]

From these results the authors obtained estimates of the stacking fault energies for the pure metals by extrapolation:

Cu :   40 ergs/cm²
Ag :   25 ergs/cm²
Ni :   150 ergs/cm²

Mader and Simsch[257] also examined the nickel-cobalt system. Estimates of the stacking fault energy could be made from the quenching behavior. In Ni + 20% Co loops were found on quenching; on the other hand in Ni + 60% Co tetrahedral defects were produced.

Prior to the use of extended nodes or ribbons stacking fault energies were deduced from the temperature and strain rate dependence of the stress at the onset of stage III of the stress-strain curve. The theory[267] is based on the thermally activated formation of a constriction in the leading dislocations in a pile-up, resulting in cross-slip. Thornton, Mitchell, and Hirsch[268] have compared γ-values, for a number of copper based alloys, obtained by means of the two methods. It turns out that the "stage III" values are too high in general. Sometimes they become negative. The disagreement is most pronounced for the alloys with an electron/atom ratio in excess of 1.10. The extrapolated values for the pure metals copper and silver are respectively 70 ± 10 ergs/cm² and 25 ± 5 ergs/cm²; they are presumably lower limits.

Measurements on particularly well-developed stacking fault ribbons in (112) planes in quenched tungsten, deformed above 2500° C, lead to the value $\gamma = 14.5$ ergs/cm² ($\mu = 1.54 \times 10^{12}$ dynes/cm²; $\nu = 0.28$).[266]

The stacking fault energy in f.c.c. thorium seems to be intermediate. No visible splitting of dislocations occurs, but on the other hand the slip traces give no evidence for cross-slip.[269]

(iii) The variation of γ with temperature. Swann and Nutting[270] have measured the stacking fault energy of the copper- 7% aluminum alloy as a function of temperature. The radius of curvature of the extended nodes does not change very much up to 275° C but at

[267] A. Seeger, R. Berner, and H. Wolf, Z. Physik 155, 247 (1959).
[268] P. R. Thornton, T. E. Mitchell, and P. B. Hirsch, Phil. Mag. [8] 7, 1349 (1962).
[269] J. O. Stiegler and C. J. McHargue, Acta Met. 11, 225 (1963).
[270] P. Swann and J. Nutting, J. Inst. Metals 90, 133 (1961, 1962).

$340°$ C a sharp increase in stacking fault energy is apparently observed. However, on cooling down the node does not expand again; an irreversible change has apparently taken place.

It was found that in the alloy, quenched from about $400°C$, the stacking fault energy was between 5 and 18 ergs/cm² against 2.6 ergs/cm² at room temperature, after deformation.

(c) Graphite. The structure of graphite is shown schematically in Fig. 136. It consists of a stacking of hexagonally linked layers of carbon atoms as shown in projection in Fig. 136c. The normal stacking can be described by the symbol *a b a b a b* (Fig. 136a). Occasionally rhombohedral graphite with a stacking symbol *a b c a b c* (Fig. 136b) is found, especially in deformed samples. The *c* plane is glide and cleavage plane. In cleaved specimens all dislocation arrangements

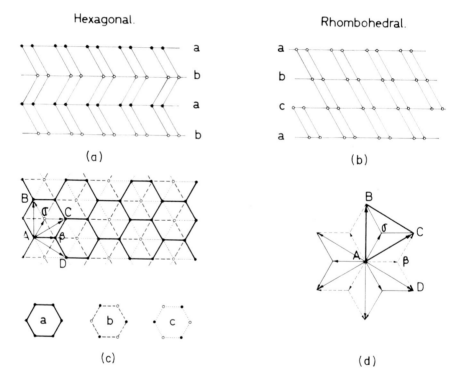

Fig. 136. Structure of graphite. (a) Hexagonal modification seen in cross section. (b) Rhombohedral modification seen in cross section. (c) Projection on the *c* plane. (d) Key for denoting directions of Burgers vectors in the basal plane.

will therefore lie in the plane of observation. It is found that all basal dislocations are split into 0.1μ wide ribbons; this can be understood in the following way. The vectors in the c plane which connect one atom to the nearest crystallographically equivalent one are AB, AC, and BC (Fig. 136c) as well as their negatives. These are the potential glide vectors. They can be decomposed into 2 partial vectors according to the reaction $AB \rightarrow A\sigma + \sigma B$, in other words the perfect dislocations can split into 2 partials having Burgers vectors enclosing an angle of $120°$. The stacking fault between the 2 partials is 1 lamella in rhombohedral stacking with 2 violations against the stacking rule as visualized in Fig. 137.

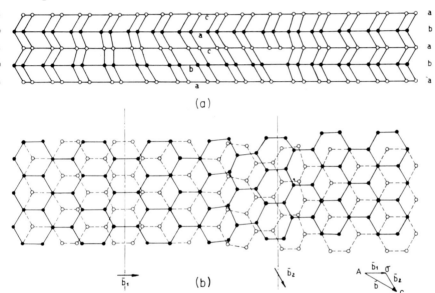

FIG. 137. Schematic view of a dislocation ribbon in the graphite structure. No C-C bonds are broken; they are only deformed. The fault between the 2 partials is equivalent to a lamella in rhombohedral stacking. This can be seen in the cross section (a). A projected view is shown in (b). [After P. Delavignette and S. Amelinckx, *J. Nucl. Mater.* **6**, 17 (1962).]

Geometrically a second type of dissociation is possible $AC \rightarrow A\beta + \beta C$ as shown in Fig. 136c. The corresponding first partial would bring one layer on top of the second one. The energy associated with such an "a over a" stacking is however considerably higher than in the first case. The second type of dissociation probably does not take place for this reason.

The Burgers vectors of dislocations were determined using the method outlined in Section 9d(2). Since the partials go out of contrast for reflections of the type $(11\bar{2}0)$, they are of the type $A\sigma$, $B\sigma$, $C\sigma$. A model of an extended dislocation having the proposed Burgers vector is represented in Fig. 137; no C-C bonds are broken, only deformation of the hexagons is required to take up the strain. Other models, differing considerably from the one given here, were proposed by Grenall and Sosin.[271]

Graphite was the first substance for which a visible equilibrium separation of partials into ribbons was found. This allowed verification of the relation (10.26)[272-274]; Figure 138 shows the curved ribbons with a radius of curvature large as compared to the ribbon widths which were used for this purpose. The direction of the total Burgers vector is indicated on the photograph. A plot of d versus $\cos 2\phi$ is shown in Fig. 139; as predicted by formula (10.7) a straight line is found. From this we conclude $d_0 = 0.85\,\mu$ and $\nu = 0.25 \pm 0.05$ and consequently using formulas (10.7) and (10.8) $\gamma = 3.5 \times 10^{-2}\,\mathrm{erg/cm^2}$ with $\mu = 2.3 \times 10^{10}\,\mathrm{dynes/cm^2}$ and $b = 1.42 \times 10^{-8}$ cm. The value of γ becomes 0.6 erg/cm when the anisotropy is properly taken into account (see Section 10a(6).

Threefold ribbons are frequently observed in graphite (Fig. 140). As can be judged from Fig. 140b the 3 partials go out of contrast simultaneously which shows that they have the same Burgers vector; this is indicated also in the same figure. Therefore the model described in Section 10a(5) is adequate. The total width of the symmetrical ribbon of Fig. 140 is about 0.5 μ and it has approximately screw character. Formulas (10.21) and (10.22) then allow deduction of the stacking fault energy. The value found is in agreement with the one deduced from the single ribbons.

Baker et al.[274a] deduced independently the stacking fault energy of graphite, using anisotropic elasticity theory. They measured the change in width of a circular ribbon as a function of orientation. From the ratio of the width of edge and screw parts, the unknown

[271] A. Grenall and A. Sosin, Proc. 4th Conf. Carbon Buffalo, 1959 p. 371 (1960).
[272] S. Amelinckx and P. Delavignette, J. Appl. Phys. 31, 2126 (1960).
[273] P. Delavignette and S. Amelinckx, J. Nucl. Mater. 5, 17 (1962).
[274] G. K. Williamson, Proc. Roy. Soc. A257, 457 (1960).
[274a] C. Baker, Y. T. Chou, and A. Kelly, Phil. Mag. [8] 6, 1305 (1961).

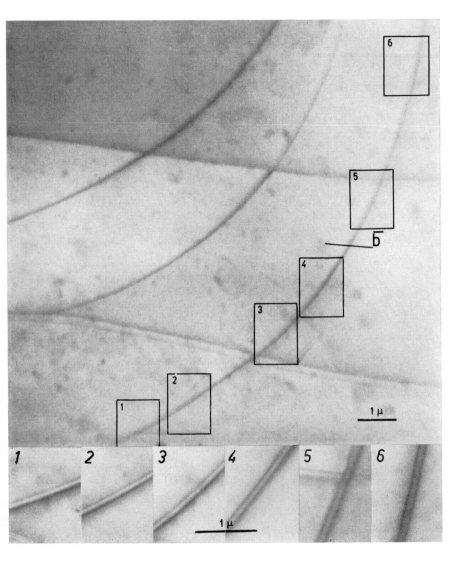

Fig. 138.   Curved ribbon in graphite showing the change in width with orientation. The insets show enlarged sections of the ribbons. The Burgers vector is indicated. [After P. Delavignette and S. Amelinckx, *J. Nucl. Mater.* **5**, 17 (1962).]

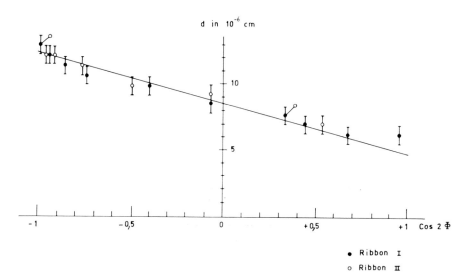

FIG. 139. Plot of ribbon width w versus cos 2ϕ, where ϕ is the angle between Burgers vector and the ribbon. The slope of the straight line allows deducing an effective Poisson's ratio while the intercept allows to calculate γ. [After P. Delavignette and S. Amelinckx, *J. Nucl. Mater.* **5**, 17 (1962).]

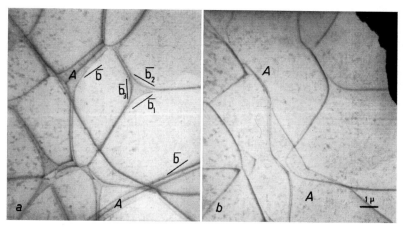

FIG. 140. Dislocation pattern in graphite. (a) Isolated extended node and threefold ribbon. All partials are in contrast. (b) The same region as (a) but with some of the lines out of contrast; particularly the threefold ribbon is out of contrast. The direction of Burgers vectors is indicated.

elastic constant C_{13} was determined. From the absolute value of the ribbon width the stacking fault energy is found: 0.58 erg/cm^2, in accord with the value found by Siems *et al.*[274b]

(d) The cadmium iodide structure (AX$_2$).   The cadmium iodide structure can be described by the stacking symbol

$$a\gamma ba\gamma ba\gamma b... . \qquad (10.35)$$

The italic letters represent the anion or X ions while the Greek letters represent the cation or A ions. The close-packed layers of cations X are sandwiched between 2 close-packed layers of anions; they occupy the octahedral interstices. The binding between 2 successive X layers is probably weaker than between A and X layers. This results in easy cleavage and glide between the 2 close-packed layers. As a consequence partials with Burgers vectors of the type $A\sigma$, $B\sigma$, or σA, σB are to be expected. Indeed they have been found in CdI$_2$, SnS$_2$, SnSe$_2$, and SnSSe.[275] The ribbons are particularly wide in SnS$_2$; an example is shown in Fig. 141. However, the situation is somewhat complicated by the presence of 2 possible (0001) glide planes either between X and X or between X and A; we call them X — X or A — X glide planes.

(*i*) X — X glide planes.   On glide along a X — X glide plane a stacking fault of the type

$$a\gamma ba\gamma bc\beta ac\beta a \qquad (10.36)$$

results. This can be considered as 1 lamella in the cadmium chloride structure for which the stacking symbol is

$$a\gamma bc\beta ab\alpha c. \qquad (10.37)$$

The sandwiches X-A-X remain unsheared in this process. The glide vectors are of the type $A\sigma$, $B\sigma$, $C\sigma$.

(*ii*) X-A glide.   Glide along the second type of glide plane generates a fault of the type

$$a\gamma ba\gamma \mid cb\alpha cb\alpha c. \qquad (10.38)$$

This process would bring 1 layer of A atoms into tetrahedral interstices. Therefore it is reasonable to assume that a synchro-shear

---

[274b] R. Siems, P. Delavignette, and S. Amelinckx, *Z. Physik* **165**, 502 (1961).

[275] P. Delavignette and S. Amelinckx, unpublished work, 1962.

motion will take place bringing these A atoms again into octahedral sites.

The stacking symbol would then become:

$$a\gamma ba\beta \mid cb\alpha cb\alpha c \cdots \qquad (10.39)$$

and again 1 lamella in the cadmium chloride structure results.

A second synchro-shear motion is required to complete the glide movement. These glide motions are indicated in Fig. 142. The direction of movement of the A atoms makes an angle of 60° with the direction of displacement of the upper part of the crystal, i.e., with the Burgers vector. This type of motion and the term "synchro-shear" were first introduced by Kronberg[276] to describe glide motions in aluminum oxide.

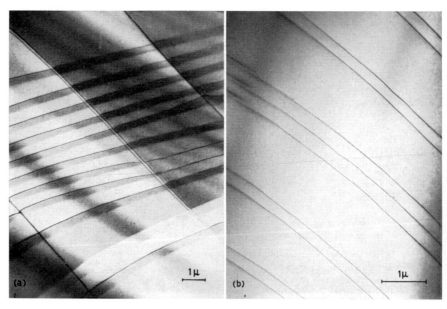

Fig. 141. (a) Dislocation ribbons in molybdenum sulfide observed with stacking fault contrast. Notice the extinction contours. In the contours the dislocation contrast is most pronounced. Some of the ribbons become undissociated presumably as a consequence of jogging into a different type of glide plane. [After P. Delavignette and S. Amelinckx, unpublished work, 1962.] (b) Wide dislocation ribbons in tin disulfide. A threefold ribbon is also visible. Notice that the observed width of the threefold ribbon is smaller than the theoretical width deduced from the width of the single ribbon.

[276] M. Kronberg, *Acta Met.* **9**, 970 (1961).

There is an alternative possibility. If glide is considered, e.g., between a and γ without synchro-shear taking place, one could imagine the formation of a fault of the type

$$a\gamma ba \mid \beta ac\beta ac \tag{10.40}$$

which could be considered as 1 lamella of the molybdenum sulfide structure $a\beta ab\alpha ba\beta ab\alpha b$ (10.41). However a consideration of the direction of the Burgers vectors associated with the different possible glide motions suggests that this is less probable.[275] We refer to Fig. 142 for the discussion. For X-X glide of an a layer over a b layer the vectors are of the type $A\sigma + \sigma B$. On the other hand, for the X-A glide accompanied by synchro-shear the glide vectors are of the type $\sigma A + B\sigma$, ..., etc., whereas the vectors leading to 1 layer in the molybdenum structure are again of the first type. This remark

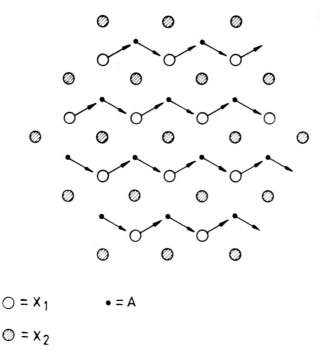

$$\bigcirc = X_1 \qquad \bullet = A$$

$$\oslash = X_2$$

FIG. 142. Synchro-shear movement on glide along the A-X glide planes in the cadmium iodide AX_2 structure. The synchro-shear movement takes place because the A atoms have a tendency to stay in octahedral environment.

can be combined with the frequent observation in tin disulfide of ribbons consisting of 3 partials with the same Burgers vector, resulting from the fusion of 2 single ribbons. This observation proves that the two types of ribbons are present with vector $A\sigma + \sigma B$ and $\sigma C + A\sigma$ and it provides therefore direct evidence that two types of glide planes operate in this substance.

Price and Nadeau[207] have studied nickel bromide which has the cadmium chloride structure. In this structure the anions form a close-packed cubic arrangement. The stacking fault energy is small and wide ribbons are observed. Occasionally asymmetrical triple ribbons are observed; they are attributed to the combination of single-ribbons in the way described in Section 10a(5). One of the ribbons contains a stacking fault consisting of 1 layer in the cadmium iodide structure; the other contains a stacking fault consisting of 1 layer in the molybdenum sulfide structure.

(e) Molybdenum sulfide.   Dislocations in this substance have been studied by a number of investigators[277-280a]; however, their behavior is not completely understood. Not less than three different kinds of dislocations have been found, undissociated dislocations,[202,278] widely dissociated ribbons,[202,278] and ribbons of an intermediate width[202] (Fig. 143).

The structure of molybdenum sulfide can be represented by the stacking symbol

$$a\beta ab\alpha ba\beta ab\alpha b \ldots \tag{10.41}$$

The italic letters refer to sulfur and the Greek letters to molybdenum atoms.

The binding energy between two sulfur layers is mainly of the van der Waal's type and hence considerably weaker than the covalent bonds between molybdenum and sulfur. It is therefore reasonable to expect that S/S glide, i.e., glide between 2 sulfur layers, will be

[277] S. Amelinckx and P. Delavignette, in "Direct Observation of Imperfections in Crystals" (J. B. Newkirk and J. H. Wernick, eds.), p. 295. Wiley (Interscience), New York, 1962.

[278] D. W. Pashley and A. E. B. Presland, Proc. European Reg. Conf. on Electron Microscopy, Delft, 1960 1, 417 (1961).

[279] M. Gillet, Bull. Microscop. Appl. [2] 10, 83 (1960).

[280a] A. Yeada and co-workers, cited after Pashley and Presland[278]; F. C. Boswell, Proc. European Reg. Conf. on Electron Microscopy, Delft, 1960 1, 409 (1961); Y. Kamiya, K. Ando, M. Nonoyama, and R. Uyeda, J. Phys. Soc. Japan 15, 2025 (1960).

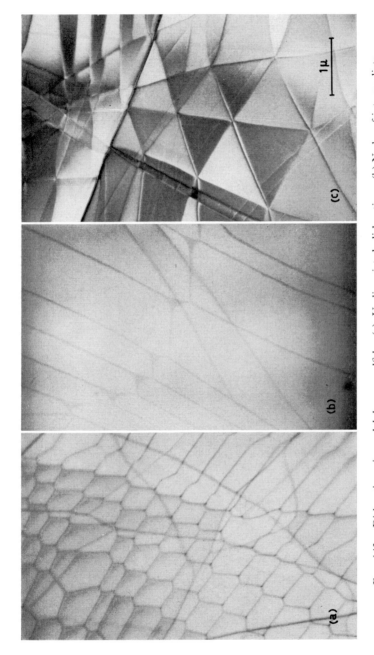

FIG. 143. Dislocations in molybdenum sulfide. (a) Undissociated dislocations. (b) Nodes of intermediate width. (c) Widely dissociated nodes.

easier than Mo/S glide, and *a priori* it might seem unlikely that glide on Mo/S glide planes would take place at all. However the observations are consistent with the two types of glide planes.

The complete Burgers vectors are of the type AB (Fig. 144).

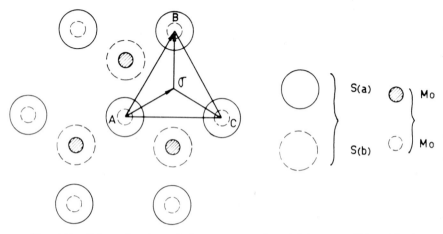

FIG. 144. Schematic view of the structure of molybdenum sulfide with the Burgers vectors indicated.

For both assumptions concerning the location of the slip plane in the structure, glide takes place between close-packed layers, and on a purely geometrical basis dissociation into partials is possible in either case. Whether a visible dissociation occurs or not will depend on the value of the stacking fault energy.

A large fraction of the dislocations (type I) looks simple, apparently undissociated, the stacking fault energy being large. In the same specimen one finds an almost equally large fraction of very wide ribbons (type II) and even isolated partials. Finally, ribbons having an intermediate width comparable to that found for ribbons in graphite, are also found. The Burgers vectors of the three types of dislocations are drawn with respect to the crystal structure in Fig. 144. The γ/μ values for the three types of dislocations are summarized in the following table:

Type	γ/μ
I	$\gamma/\mu > 10^{-11}$ cm
II	0.26×10^{-12} cm $< \gamma/\mu < 0.75 \times 10^{-12}$ cm
III	3×10^{-12} cm $< \gamma/\mu < 5 \times 10^{-12}$ cm

There is no obvious explanation for the occurrence of three types of dislocations or stacking faults in the structure (10.41). It is possible to account for two kinds[202,278] by assuming that Mo/S glide as well as S/S glide takes place. It is clear that Mo/S glide over a vector like $A\sigma$ would give rise to a stacking

$$a\beta ab\alpha ba \mid \gamma bc\beta c \cdots \qquad (10.42)$$

whereby the environment of the molybdenum is wrong in 1 layer. The covalent bonds which are normally oriented along the diagonals of a triangular prism, as shown in Fig. 145a, would have to reorient

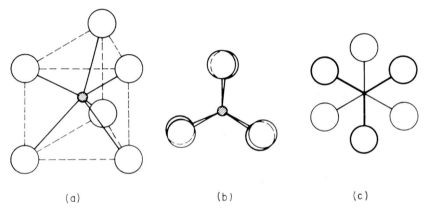

(a)   (b)   (c)

FIG. 145.   Molybdenum sulfide structure. (a) Environment of molybdenum in the normal molybdenum sulfide structure. (b) In projection on the c plane. (c) Wrong environment as would occur on Mo/S glide.

into the configuration shown by Fig. 145c. It is then reasonable to consider this as a high energy stacking fault, and hence one would assimilate this type of glide with the undissociated dislocations. On the other hand, glide between two S layers, i.e., S/S glide, would give rise to a stacking fault, which can be represented by the symbol:

$$a\beta ab\alpha ba\beta a \mid c\beta cb\gamma b \cdots \qquad (10.43)$$

and which is such that covalent bonds within the MoS_2 layers are conserved. It is therefore reasonable to assimilate this with the small energy stacking fault.

However, the difficulty with this scheme is that there is no room for a third kind of stacking fault. The complex behavior is probably related to the occurrence of polytypism, which itself is a consequence of a small stacking fault energy.[280b]

(f) The diamond structure; silicon and germanium. Dislocations in silicon and germanium were first observed in the electron microscope by Geach et al.[209] in indented specimens, but no detailed experimental study was made. A theoretical study was presented by Hornstra[281] and by Haasen and Seeger.[282] Alexander[283] studied the plastic deformation of germanium single crystals. In foils cut parallel to active glide planes he found numerous dislocation dipoles left in the wake of moving screws. Occasionally wide stacking faults were found. In epitaxially grown silicon, Queisser et al.[263] found wide stacking faults also. However the stacking fault energy cannot be deduced from such observations.

The diamond structure can be represented by the symbol

$$a\alpha b\beta c\gamma a\alpha b\beta c\gamma \cdots. \tag{10.44}$$

Glide is presumably easiest between the planes a and α (b and β, or c and γ) because this involves breaking the smallest number of bonds. However, dissociation into partials would take place between 2 close-packed layers, i.e., between α and b, β and c, or γ and a. The stacking fault then would have the structure

$$a\alpha b\beta c\gamma \mid b\beta c\gamma a\alpha b\beta c\gamma a\alpha \cdots \tag{10.45}$$

i.e., it would consist of 1 lamella of the wurtzite structure. An undissociated dislocation is shown in Fig. 146a; a dissociated dislocation containing an intrinsic fault is represented in Fig. 146b. Dissociation into a ribbon containing an extrinsic fault leads to the structure shown in Fig. 146c. Extended dislocations have been observed in silicon[210] as well as in germanium.[284a] Booker and Stickler, however, found no evidence for dissociation.[284b]

[280b] P. Braun, private communication, 1962.

[281] J. Hornstra, *Phys. Chem. Solids*, **5**, 129 (1958).

[282] P. Haasen and A. Seeger, "Halbleiterprobleme," Vol. IV, p. 68. Vieweg, Braunschweig, 1958.

[283] H. Alexander, *Z. Metallk.* **52**, 344 (1961).

[284a] A. Art, E. Aarts, P. Delavignette, and S. Amelinckx, *Appl. Phys. Letters* **2** 40 (1963).

[284b] G. R. Booker and R. Stickler, *Acta Met.* **10**, 993 (1962). G. R. Booker and R. Stickler, *Proc. 5th Intern. Conf. on Electron Microscopy, Philadelphia, 1962* **1**, B-8 (1962).

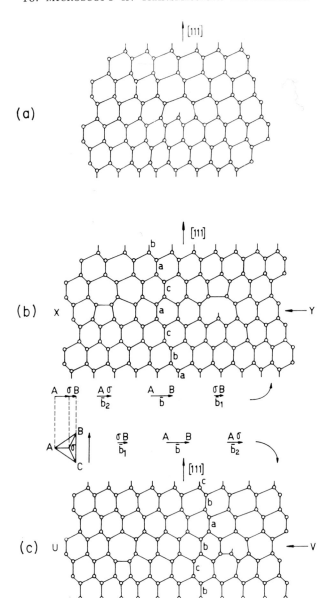

FIG. 146.   The diamond structure. (a) Undissociated dislocations in the diamond lattice. (b) Cross section of a ribbon containing an intrinsic fault. (c) Cross section of a ribbon containing an extrinsic fault. The Burgers vectors are indicated for all partials.

A considerable amount of attention has recently been devoted to defects in epitaxially grown silicon.[284c-f] On etching one reveals triangular features on (111) surfaces of epitaxial layers. In thin films the same features give rise to bands cutting through the foil and which exhibit fringes of the same type as those due to stacking faults. The bands are in (111) planes, suggesting that they are either stacking faults or micro twins. There is evidence that both are present.[284c,d] Presumably the faults are not generated by deformation but result from growth accidents at the interface substrate-epitaxial layer. In layers grown on clean surfaces the concentration of faults is very small. It has also been suggested that the faults result from the contact between two different nuclei, e.g., one in register with the substrate and one nucleated in faulted position at the interface.[284e] The nucleation in faulted position has been attributed to the presence of oxygen.[284f]

(g) The wurtzite, zinc sulfide, and related structures. Observations are available for aluminum nitride (wurtzite structure)[285] and for zinc sulfide[286] as well as for gallium sulfide and gallium selenide.[287].

(i) Wurtzite and sphalerite structures. The wurtzite structure in general and the aluminum nitride structure in particular can be represented by the stacking symbol

$$a\alpha b\beta a\alpha b\beta. \tag{10.46}$$

The Greek letters denote nitrogen and the italic letters aluminum. A schematic view, as seen along the c direction is shown in Fig. 147a while Fig. 147b gives a cross section. The main glide plane is found to be the c plane. The complete glide vectors are AB, AC, AD, and their negatives (Fig. 147a).

There are two possible locations for the glide planes within the structure: either between β and a (or α and b) or between b and β

[284c] J. M. Charig, B. A. Joyce, D. J. Stirland, and R. W. Bicknell, *Phil. Mag.* [8] **7**, 1847 (1962).

[284d] R. H. Finch, H. J. Queisser, J. Washburn, and G. Thomas, *J. Appl. Phys.* **34**, 406 (1963).

[284e] G. R. Booker and R. Stickler, *J. Appl. Phys.* **33**, 3281 (1962).

[284f] H. J.Queisser, R. H. Finch, and J. Washburn, *J. Appl. Phys.* **33**, 1536 (1962).

[285] P. Delavignette, H. B. Kirkpatrick, and S. Amelinckx, *J. Appl. Phys.* **32**, 1098 (1961).

[286] H. Blank, P. Delavignette, and S. Amelinckx, *Phys. Stat. Solidi* **2**, 1660 (1962).

[287] Z. S. Basinski, D. B. Dove, and E. Mooser, *Helv. Phys. Acta* **34**, 373 (1961).

(or α and a). We will call them type I and type II glide planes; they are indicated in Fig. 147b. The perfect 60° dislocation in a type II glide plane can be represented as in Fig. 148a. The number of Al-N bonds to be broken on motion is smaller for a dislocation in this glide plane than for one in a type I glide plane. On the other hand it is not obvious how such a dislocation would dissociate into partials

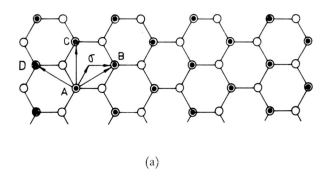

(a)

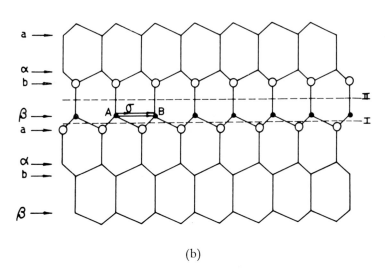

(b)

FIG. 147. Schematic view of the wurtzite structure. (a) As projected on the c plane; Burgers vectors are indicated. (b) In cross section. The two types of glide planes I and II are shown.

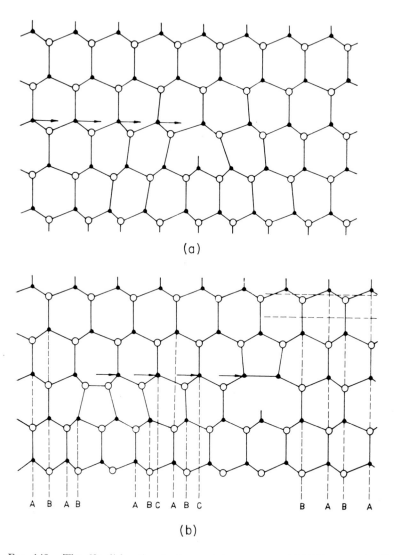

(a)

(b)

Fig. 148.   The 60° dislocation in the wurtzite structure. (a) With a type I glide plane there is no dissociation into partials. (b) Dissociation into partials along a type II glide plane. The fault between the partials consists of 1 lamella in the sphalerite structure.

by glide on a type II glide plane. For type I glide planes, on the other hand, the extended 60° dislocation would look as represented in Fig. 148b. Dissociation into partials is now possible, at least geometrically because glide takes place between 2 close-packed layers. The glide vectors of the partials are $A\sigma$, $B\sigma$, $C\sigma$, etc. The fault within the ribbon has the structure $b\beta a\alpha b\beta c\gamma a\alpha c\gamma a\alpha$ ... and it is equivalent to a lamella having the sphalerite structure

$$a\alpha b\beta c\gamma a\alpha b\beta c\gamma \cdots \qquad (10.47)$$

as is evident also from Fig. 148b.

Extended dislocations were indeed found. This suggests strongly that glide takes place on type I glide planes. The value of γ/μ deduced from extended nodes is $\gamma/\mu = 8.1 \times 10^{-12}$ (Fig. 149). Occasionally undissociated dislocations were also found; they may be due to glide on type II glide planes.

Sublimed crystals of zinc sulfide were studied by Blank et al.[286]. The crystals were in the wurtzite (high-temperature) form when grown. On cooling the stacking fault energy becomes effectively negative and all dislocations which are present in the basal plane split into infinitely wide stacking fault ribbons, transforming in this way 1 lamella into the sphalerite structure. Faults are also present in prism planes, presumably $(12\bar{3}0)$; they are always connected to basal plane faults. Closed loops of stacking faults are sometimes formed in this way.

Rather similar observations were made by Chadderton et al.[287a] According to these authors the non basal plane faults should lie on $(11\bar{2}0)$ and $(1\bar{2}10)$ planes. They also concluded that the basal plane faults are not single layers, but rather thin lamella of sphalerite.

(ii) Gallium sulfide and selenide.[287] The structure of gallium sulfide (GaS) can be described by the stacking symbol

$$a\beta\beta ab\alpha\alpha ba\beta\beta a \cdots \qquad (10.48)$$

while gallium selenide occurs in 2 modifications with structures

$$a\beta\beta ab\gamma\gamma ba\beta\beta a \cdots \text{ (hexagonal)} \qquad (10.49)$$

$$c\alpha\alpha cb\gamma\gamma ba\beta\beta a \cdots \text{ (rhombohedral).} \qquad (10.50)$$

[287a] L. T. Chadderton, A. G. Fitzgerald, and A. D. Yoffe, *Phil. Mag.* [8] **8**, 167 (1963).

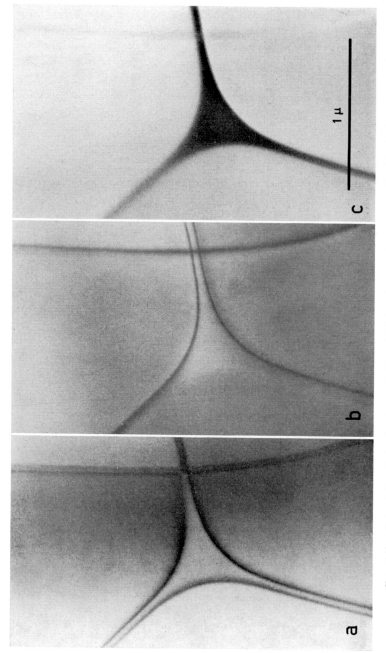

FIG. 149. Extended node in aluminum nitride, which has the wurtzite structure. Stacking fault contrast and 2 different line contrasts.

The italic letters represent sulfur or selenium; while the Greek letters represent gallium. The 3 structures in fact result from different stackings of "4 layers," of the type $a\beta\beta a$ as shown in Fig. 150. Within one 4-layer the binding is covalent, whilst between 4 layers the binding is of the van der Waal's type and weak; glide and cleavage, therefore, take place between 2 close-packed anion layers. It is clear that there is again the possibility of dissociation into Shockley partials.

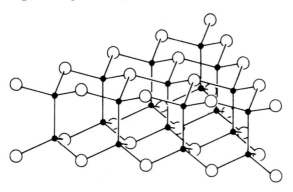

Fig. 150.   Four-layer lamella forming the basic unit of the gallium sulfide and gallium selenide structures.

It is found that the stacking fault energy is very small in GaSe while it is larger in GaS; in both cases a visible dissociation is observed.

The stacking fault in gallium sulfide resulting from glide of a partial can be represented as

$$a\beta\beta a \mid c\beta\beta c b\gamma\gamma b \tag{10.51}$$

while the corresponding stacking fault in the hexagonal modification of gallium selenide is

$$a\beta\beta a \mid c\alpha\alpha c b\gamma\gamma b. \tag{10.52}$$

It is clear that the nearest neighbor relationship between 4 layers in Eq. (10.52) is the same as that in (10.49), i.e., the stacking fault energy is only due to a wrong stacking of a second nearest neighbor 4 layers, and this is certainly very small. On the other hand, the difference between Eq. (10.51) and (10.48) is more pronounced since gallium ions come into the same position in adjacent 4-layers. These considerations explain qualitatively the observed differences in stacking fault energy. No quantitative measurements were made.

(h) Fourfold ribbons in talc and aluminum oxide. In a number of structures the total glide vector can split into 4 partial dislocations. Such behavior was predicted by Kronberg[276] for glide on the c plane in aluminum oxide and in the related spinel structure.[288,289] However, for these substances observations are not yet available. Fourfold ribbons were first observed in talc[290] and more recently in chromium chloride[291a] and chromium bromide[291] where also 6-fold ribbons were found.

(i) Talc. Talc is a silicate layer structure; it can be considered as a stacking of multilayers in which the succession of layers is (Fig. 151):

$$\cdots O - Si - (O - OH) - Mg - (O - OH) - Si - O \cdots$$

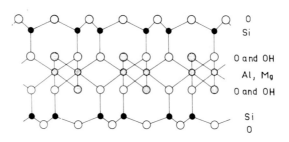

FIG. 151.   The crystal structure of talc; cross section of the multilayers.

The (O — OH) layers are close-packed, and magnesium is in the octahedral interstices. The silicon atoms are tetrahedrally surrounded by oxygen atoms. The structure of the limiting oxygen layers is represented in Fig. 152. Glide and also cleavage take place between 2 such oxygen sheets. The stacking of these sheets is also shown in Fig. 152. It is now clear that the shortest glide vector that will reproduce the same configuration is, for instance, AA_1. Using a ball model it will be realized that glide proceeds in at least 4 steps which are indicated in Fig. 152. In dislocation language this means that the dislocation with total Burgers vector AA_1 splits into 4 partial dis-

[288] J. Hornstra, *Phys. Chem. Solids* **15**, 311 (1960).
[289] J. Hornstra, *Proc. 4th Symp. on Reactivity of Solids, Amsterdam, 1960*. p. 563.
[290] S. Amelinckx and P. Delavignette, *J. Appl. Phys.* **32**, 341 (1961).
[291] P. Delavignette and S. Amelinckx, *Trans. Brit. Ceram. Soc.* **62**, 687 (1963).
[291a] S. Amelinckx and P. Delavignette, *J. Appl. Phys.* **33**, 1458 (1962).

locations forming a ribbon of which a model is shown in Fig. 153a,b. The dissociation will be observable provided the energy of the 3 stacking faults concerned is small enough; this is the case here, as can be judged from Fig. 154.

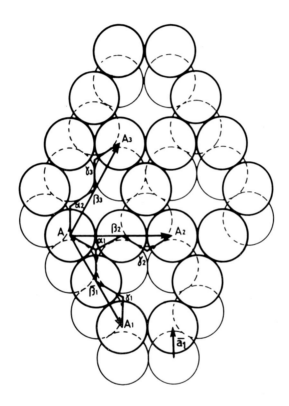

FIG. 152.   Stacking of oxygen layers (between which glide takes place in talc) and possible Burgers vectors with their dissociation are shown.

The energies of the stacking faults can be deduced from the equilibrium configuration, which we now formulate. Let the positions of the dislocations be indicated by the x_i and the stacking fault energies by γ_i. The angle between the *total* Burgers vector and the direction of the ribbon is called ϕ (see Fig. 153a,b). By setting the resultant force on each dislocation equal to zero we obtain the following set of four compatible equations

$$B\left(\frac{1}{x_1} + \frac{1}{x_3}\right) - \frac{C}{x_2} + \gamma_1 = 0$$

$$-B\left(\frac{1}{x_1} - \frac{1}{x_2 - x_1}\right) - D\left(\frac{1}{x_3 - x_1}\right) + \gamma_2 - \gamma_1 = 0$$

$$-B\left(\frac{1}{x_2 - x_1} - \frac{1}{x_3 - x_2}\right) + \frac{C}{x_2} + \gamma_3 - \gamma_2 = 0$$

$$-B\left(\frac{1}{x_3} + \frac{1}{x_3 - x_2}\right) + \frac{D}{x_3 - x_1} - \gamma_3 = 0$$

$$(10.53)$$

where the symbols have the following meaning[291b]

$$B = \tfrac{1}{2}(\alpha - \beta)\cos 2\phi - \tfrac{1}{4}(\alpha + \beta)$$
$$C = \beta\cos^2(\phi - 30°) + \alpha\sin^2(\phi - 30°)$$
$$D = \beta\cos^2(\phi + 30°) + \alpha\sin^2(\phi + 30°).$$

$$(10.54)$$

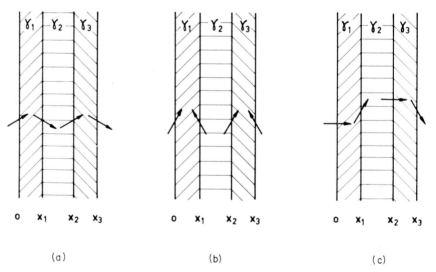

(a)   (b)   (c)

FIG. 153.   Models for fourfold ribbons, also illustrating the notation used for calculating the equilibrium conditions.(a) (1) Edge ribbon in talc; (2) Screw ribbon in talc. (b) Aluminum oxide edge ribbon.

[291b] α and β have the meanings defined in eq. (10.19).

(a)

(b)

FIG. 154. Fourfold ribbons in talc: all partials are in contrast.

By inserting the observed x_i values it is easy to deduce the γ_i values provided the elastic constants are known. For talc this leads to 6×10^{-10} cm $< \gamma_i/\mu < 9 \times 10^{-10}$ cm. More complicated glide paths are possible in talc which has in fact monoclinic symmetry. The $(a\,b)$ plane of the unit cell is outlined in Fig. 155. In the b direction one expects, for instance, the possibility of dissociation into 8 partials.

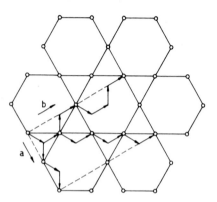

FIG. 155.   More complicated glide paths are possible in the talc structure. The projection of the unit cell is outlined. Apart from fourfold ribbons eightfold ribbons are also to be expected.

(ii) Aluminum oxide structure.   In aluminum oxide the configuration of Burgers vectors is different. A model for edge ribbons in this substance is shown in Fig. 153c.

The equilibrium conditions are now

$$\gamma_1 - \frac{A_1}{x_1} - \frac{B_1}{x_2} - \frac{C_1}{x_3} = 0$$

$$\gamma_2 - \gamma_1 + \frac{A_1}{x_1} - \frac{A_1}{x_2 - x_1} - \frac{D_1}{x_3 - x_1} = 0$$

$$\gamma_3 - \gamma_2 + \frac{B_1}{x_2} + \frac{A_1}{x_2 - x_1} - \frac{C_1}{x_3 - x_2} = 0$$

$$- \gamma_3 + \frac{C_1}{x_3} + \frac{D_1}{x_3 - x_1} + \frac{C_1}{x_3 - x_2} = 0$$

(10.55)

where

$$A_1 = \alpha \sin \phi \sin (\phi - 60°) + \beta \cos \phi \cos (\phi - 60°)$$
$$B_1 = \alpha \sin^2 \phi + \beta \cos^2 \phi$$
$$C_1 = \alpha \sin \phi \cos (\phi - 30°) - \beta \cos \phi \sin (\phi - 30°)$$
$$D_1 = \alpha \sin (\phi - 60°) \cos (\phi - 30°) - \beta \cos (\phi - 60°) \sin (\phi - 30°).$$

$$(10.56)$$

Ribbons of this type have been observed in chromium chloride and chromium bromide[291a] (see also next paragraph).

According to Kirkpatrick[291c] the dislocations in aluminum oxide itself do not appear to be dissociated.

(i) Ribbons in the chromium chloride and the chromium bromide structures. Particularly instructive examples of multiple ribbons are found in these substances. Observations were made on thin single crystal flakes prepared by sublimation. The ribbons consist either of 4 or of 6 partials. This can be understood on the basis of the structure of these compounds. We first describe chromium chloride.

(*i*) Chromium chloride.   The structure of this crystal can be described as a cubic close packing of chlorine ions with the chromium ions in octahedral interstices. The chromium ions are arranged in a ring pattern similar to that in graphite leaving $1/3$ of the sites unoccupied. The layers of chromium (Greek letters) alternate with two chlorine layers (italic letters) according to the scheme

$$c\beta_1 ab\alpha_1 ca\gamma_1 bc\beta_1 ab\alpha_1 ca\gamma_1 b \cdots. \qquad (10.57)$$

The index indicates the sites that are left empty. A projection on the c plane of one Cl-Cr-Cl sandwich of this structure is shown. The unit cell is indicated by the double line, it is 3 such sandwiches high (Fig. 156).

The shortest perfect Burgers vector is of the type $X_1 X_2$. With reference to Fig. 156 it is clear that this can dissociate into 4 partials of the Shockley type according to the scheme

$$\overline{X_1 X_2} = \bar{a} + \bar{b} + \bar{c} + \bar{d}. \qquad (10.58)$$

[291c] H. B. Kirkpatrick, private communication, 1962.

The Burgers vectors of the partials were determined and found to be in accord with this model.

A crystallographically equivalent position can also be reached after

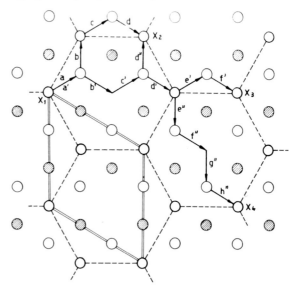

○ ⊘ : Chlorine   ○ : Chromium

FIG. 156. Projection of 1 sandwich of the chromium chloride or the chromium bromide structure on the c plane. Possible glide paths are indicated. The shortest one consists of 4 steps $\bar{a} + \bar{b} + \bar{c} + \bar{d}$. The most frequently observed one consists of 6 steps in zig-zag $a' + b' + c' + d' + e' + f'$.

6 steps in the direction X_1X_3. The large vector can be decomposed into 6 partials according to the scheme

$$\overline{X_1X_3} = \bar{a}' + \bar{b}' + \bar{c}' + \bar{d}' + \bar{e}' + \bar{f}'. \tag{10.59}$$

In both cases there is a substantial gain in energy. Whereas in the first case the glide path goes around 1 atom, it has a zig-zag form in the second case. The observations show that a majority of the ribbons are sixfold, which suggests that most glide is in X_1X_3 directions. This is probably a consequence of the fact that a zig-zag path is a better approximation to a straight line than the "round the corner" path.

An example of sixfold ribbon is visible in Fig. 157. The widths of the different stacking fault ribbons reflect the different stacking fault energies. To a rough approximation the width of each strip is inversely proportional to the energy of the stacking fault within it. It is easy to account qualitatively for the relative width of the different strips by considering the geometry of the stacking faults.[291,291a]

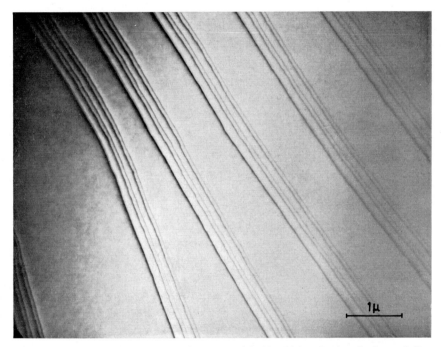

FIG. 157. Sixfold ribbons in chromium chloride. [After S. Amelinckx and P. Delavignette, *J. Appl. Phys.* **33**, 1458 (1962).]

(*ii*) Chromium bromide. In chromium bromide the bromine ions are hexagonally close packed; the chromium ions again form a ring pattern. The possible Burgers vectors are therefore the same as in the chromium chloride structure. However, on glide the succession of stacking faults is different and this is reflected in the relative widths of the fault ribbons. Sixfold as well as fourfold ribbons have been observed. In both the chromium chloride and the chromium bromide more complicated glide paths are possible, and are in fact often observed.

(j) Dislocations in ordered alloys.[292-295a,b] Consider for simplicity a dislocation in the ordered AuCu$_3$ superlattice. The passage of a dislocation with Burgers vector $(a/2)$ $\langle 110 \rangle$ restores the geometrical order, but leaves a so-called *antiphase boundary*, as represented in Fig. 158. A certain energy per unit area is associated with such an

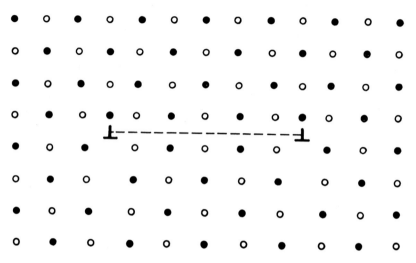

FIG. 158.   Generation of an antiphase boundary by the passage of a dislocation in an ordered alloy: two-dimensional model.

antiphase boundary. A second perfect dislocation in the same glide plane and following the first one, restores both geometrical and chemical order. From these considerations it will be evident that the 2 perfect dislocations are coupled by means of a strip of antiphase boundary. Depending on the specific energy of this boundary, closely or more widely coupled dislocations will result. Moreover since each perfect dislocation can split into Shockley partials, ribbons consisting of 4 components tend to form. A model for such a ribbon is shown in Fig. 159. The equilibrium condition is exactly the same as that derived for fourfold ribbons in talc. The significance of the γ_i is now: γ_1 and γ_3 are the energies of boundaries whose sides have

[292] M. Marcinkowski and R. Fisher, *J. Appl. Phys.* **31**, 1687 (1960).

[293] M. Marcinkowski, N. Brown, and R. Fisher, *Acta Met.* **9**, 129 (1961).

[294] N. Marcinkowski and N. Brown, *Phil. Mag.* [8] **6**, 811 (1961).

[295a] M. Marcinkowski and D. S. Miller, *Phil. Mag.* [8] **6**, 871 (1961).

[295b] M. Marcinkowski and N. Brown, *Acta Met.* **9**, 764 (1961).

undergone the antiphase- and the geometrical stacking fault movements relative to each other, while γ_2 is now the antiphase boundary energy. From observations by Marcinkowski *et al.*[293] on AuCu$_3$ and AuCu it follows that in the first alloy no visible splitting into Shockley

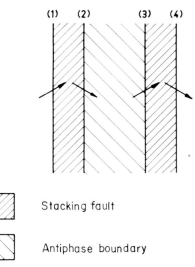

(1)   (2)   (3)   (4)

▨ Stacking fault

▨ Antiphase boundary

FIG. 159.   Model for a superlattice edge dislocation in a face-centered cubic alloy. Between dislocations 1 and 2 and between 3 and 4 there is a stacking fault; between partials 2 and 3 there is an antiphase boundary.

partial occurs; however the superlattice dislocations are coupled and the equilibrium distance is about $d = 130$ A. The equilibrium condition can then clearly be simplified to:

$$\gamma \simeq A/d \tag{10.60}$$

where A has the same meaning as in Eq. (10.19).

Using the following numerical values: $\nu = 1/3$; $b^2 = 2.34 \times 10^{-16}$ cm^2; $\mu = 4.7 \times 10^{11}$ dynes/cm^2 leads to about 13 ergs/cm^2 for the antiphase boundary energy. Marcinkowski *et al.*[293] found *no* coupling between dislocations in the disordered form of the AuCu$_3$. Theory predicts that the antiphase boundary energy should increase as the square of the ordering parameter S, and hence that the distance between coupled dislocations should be proportional to $1/S^2$. In the completely disordered alloy $S = 0$ and the distance should be infinite, as observed.

In the Fe$_3$Al type fully ordered alloy superlattice dislocations consisting of 4 $(a/2)$ $\langle 111 \rangle$ dislocations with the same Burgers vectors, connected by antiphase boundaries of two types are to be expected. For one kind of antiphase boundary only the first neighbors are of the wrong kind; for the second type the second nearest neighbors are of the wrong type. In the partially ordered Fe$_3$Al alloy the dislocations should consist of pairs of $(a/2)$ $\langle 111 \rangle$ dislocations connected by antiphase boundaries for which the nearest neighbors are wrong. All antiphase boundaries have a small energy and as a consequence, a very large separation is to be expected. It is found experimentally that $(a/2)$ $[111]$ dislocations move independently and leave long strips of antiphase boundary in the thin foil examined. The dislocations show a pronounced tendency to adopt the screw orientation; as a result they cross slip frequently. This is probably another reason why dislocations move separately rather than grouped as superlattice dislocations.

On the other hand in the ordered alloy Ni$_3$Mn dislocations are coupled as superlattice dislocations up to temperatures just below the critical temperature for long-range order.[295a] The spacing increases as the long-range order decreases; it is 180 A in perfectly ordered crystals and about 700 A below the critical ordering temperature. The increase in spacing evidently reflects a decrease in antiphase boundary energy. An example of superlattice dislocations in the ordered alloy is presented in Fig. 160.

In the disordered alloy the stacking fault energy is relatively low (~ 20 ergs/cm^2); as a result the ordinary dislocations are often widely dissociated under the applied stress. On ordering the stacking fault energy apparently increases since no separated dislocations are observed.

(k) Miscellaneous crystals. Undissociated dislocations are observed in uranium oxide[295c] and calcium fluoride[211] which have the same crystal structure as well as in magnesium oxide[145] and nickel oxide[296a] which have the sodium chloride structure.

In titanium oxide dissociation in 2 partials is found.[296b]

Ashbee and Smallmann[296c] have grown thin films of rutile (titanium dioxide) epitaxially on titanium carbide single crystals.

[295c] H. Blank and S. Amelinckx, *J. Appl. Phys.* **34**, 2200 (1963).

[296a] P. Delavignette and S. Amelinckx, *Appl. Phys. Letters* **2**, 236 (1963).

[296b] K. H. G. Ashbee, R. E. Smallmann, and G. K. Williamson, *Proc. Roy. Soc. (London)* **A276**, 524 (1963).

[296c] K. H. G. Ashbee and R. E. Smallmann, *Phil. Mag.* [8] **7**, 1933 (1962).

The epitaxial orientation relationship is given by

$$(110)_{TiO_2}//(100)_{TiC}.$$

$$[001]_{TiO_2}//[010]_{TiC}.$$

These films contain dislocation ribbons presumably with a $\frac{1}{2}(a\,0\,c)$ Burgers vector. Wide faults are sometimes observed also; it is thought that these are in fact anti-phase boundaries.

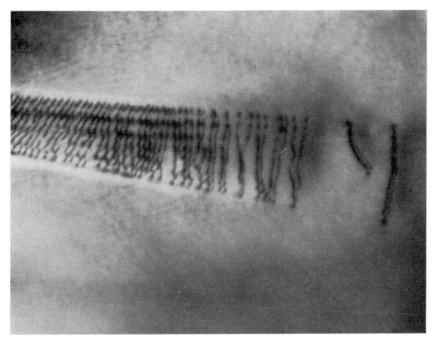

FIG. 160.   Superlattice dislocations in the ordered alloy: Ni_3Mn. The stacking fault energy is so large that splitting in partials is not observed. [Courtesy of M. Marcinkowski and D. S. Miller, *Phil. Mag.* [8] **6**, 871 (1961).]

b. Dislocation Reactions in Particular Structures.

If 2 dislocations with Burgers vectors $\bar{b}_1$ and $\bar{b}_2$ meet they may form a segment of dislocation with a Burgers vector $\bar{b}_3 = \bar{b}_1 + \bar{b}_2$. The conditions under which the reaction takes place is to a first approximation that elastic energy should be gained. Since the elastic energy of a dislocation is proportional to the square of its Burgers vector this condition can be written approximately as $\bar{b}_3^2 < \bar{b}_1^2 + \bar{b}_2^2$.[296d]

The various reactions are, of course, at the basis of network formation. We will discuss the main reactions that occur in a number of particular structures and for which there is direct experimental evidence.

(1) *Primitive cubic lattice.* The Burgers vectors of stable dislocations are a [100], a[010], and a[001]. The stability of a [110] is questionable. Its formation by means of the reaction

$$a\,[100] + a\,[010] \to a\,[110] \tag{10.61}$$

has been observed in cesium bromide by means of decoration.[132] It is the simplest reaction that can give rise to a threefold node in this structure.

(2) *Face-centered cubic lattice.* (a) Undissociated dislocations in the sodium chloride structure. In this structure the Burgers vectors are of the type $(a/2)$ [110] and the dislocations are undissociated. This Burgers vector has been confirmed by the use of contrast effects in MgO by Washburn *et al.*[145] The most frequently occurring reaction is of the form

$$\frac{a}{2}\,[110] + \frac{a}{2}\,[0\bar{1}1] \to \frac{a}{2}\,[101] \qquad \text{or} \qquad \overline{AB} + \overline{BC} \to \overline{AC}. \tag{10.62}$$

It gives rise to symmetrical threefold nodes and hexagonal networks.[297,298] It is easy to verify that elastic energy is gained. Using Thompson's notation such nodes have lettering patterns of two types, shown in Fig. 125a,b. Using this formalism it is easy to determine the sector of the intersection in which the new segment will tend to develop. It forms in the sector which has a common letter (Fig. 161a).

In some cases the following reaction is observed[126] in KCl[299] and AgCl[300] and also in aluminum[301]:

$$\frac{a}{2}\,[110] + \frac{a}{2}\,[1\bar{1}0] \to a\,[100]. \tag{10.63}$$

[296d] F. C. Frank, *Physica* **15**, 131 (1949).

[297] S. Amelinckx and W. Dekeyser, *Solid State Phys.* **8**, 325 (1959).

[298] J. Bartlett and J. W. Mitchell, *Phil. Mag.* [8] **6**, 271 (1961).

[299] S. Amelinckx, *Nuovo Cimento* [10] **7**, 569 (1958).

[300] J. Bartlett and J. W. Mitchell, *Phil. Mag.* [8] **5**, 799 (1960).

[301] P. B. Hirsch, *J. Inst. Metals* **87**, 406 (1958-59).

Although there is no gain in elastic energy, the reaction takes place if catalyzed, either by the presence of a third crossing dislocation which brings the reacting dislocations into the antiparallel configuration, or by the fact that the dislocations form only a small angle.[299]

(b) The face-centered cubic metals: dissociated dislocations. As a consequence of the dissociation the dislocation reactions in fact take place among partials with Burgers vectors of the type $(a/6)$ $[11\bar{2}]$. We have discussed a few examples of particular interest without exhausting all possibilities in Section 10a(7)(a).

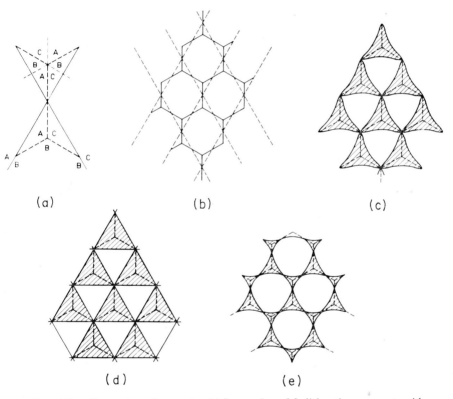

(a)   (b)   (c)

(d)   (e)

FIG. 161. Generation of networks. (a) Interaction of 2 dislocation segments with vector AB and BC, with formation of a new segment, with vector AC. (b) Formation of hexagonal meshes from the intersection of 2 families of dislocations. (c) Net after the nodes have become extended. (d) If the stacking fault energy is very small the network consists of an alternation of faulted and unfaulted triangular regions. (e) If both the intrinsic and the extrinsic fault have small energy doubly dissociated nets result.

The 2 nodes considered above (Fig. 125a,b) will become, respectively, extended and contracted if dissociation into partials takes place: Fig. 125c,d. The process whereby such nodes are formed is represented in Fig. 127. If the 2 ribbons in the same glide plane meet with a common partial, mutual annihilation takes place, resulting in the formation of an extended node containing a stacking fault.

If on the other hand 2 ribbons with vectors $A\sigma + \sigma B$ and $C\sigma + \sigma A$ are forced together in the same glide plane the 2 meeting partials will combine into a perfect dislocation CB which will however dissociate again according to the scheme $CB \rightarrow C\sigma + \sigma B$ and form an isolated contracted node as in Fig. 162. The intersection of 2 ribbons with Burgers vectors $B\sigma + \sigma A$ and $C\sigma + \sigma A$ in the same lattice plane gives rise to an extended and a contracted node as demonstrated by

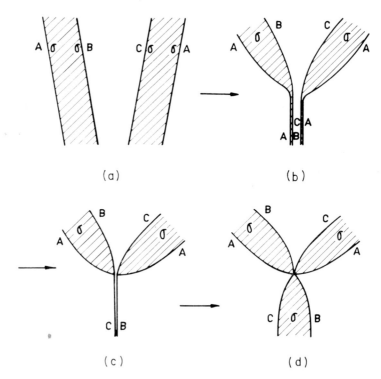

FIG. 162. Formation of an isolated contracted node. (a) The 2 ribbons meet. (b) The ribbons are compressed. (c) The 2 perfect dislocations combine into 1 perfect dislocation CB. (d) The dislocation CB dissociates again.

Fig. 163. The contracted node results because in the region of overlap a fault *a* on *a* forms as can be deduced from the cut *XY* of Fig. 163d. The region containing this fault shrinks to zero and hence the node contracts, because the fault energy is too large.

(3) *The body-centered metals.* Stable Burgers vectors are of two types $(a/2)\,[111]$ and $a\,[100]$. The latter can form from 2 of the first with a gain in energy according to the reaction

$$\frac{a}{2}\,[111] + \frac{a}{2}\,[1\bar{1}\bar{1}] \rightarrow a\,[100] \qquad (10.64)$$

$3/4 + 3/4 > 1$ (relative energies)

which gives rise to *asymmetrical threefold* nodes. This reaction has

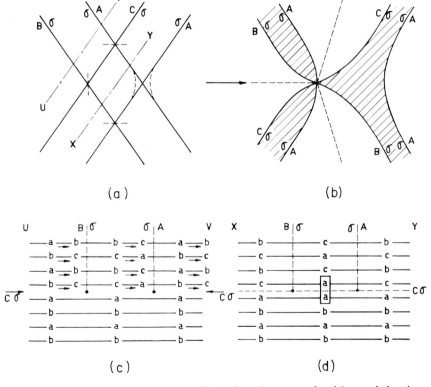

(a)   (b)

(c)   (d)

FIG. 163.   Intersection of 2 ribbons giving rise to 1 contracted and 1 extended node.

been observed in niobium[302] and in iron[303]; it gives rise to hexagonal networks.[251,297] Dissociation of dislocations into ribbons can in principle occur in this structure according to the scheme

$$\frac{a}{2}[111] \rightarrow \frac{a}{3}[111] + \frac{a}{6}[111].$$ (10.65)

Observations are not yet available.

(4) *Hexagonal metals.* The reactions between dislocations with Burgers vectors in the basal plane are not essentially different from those described for the face-centered metals; for undissociated dislocations they are

$$AB + BC \rightarrow AC$$

using the notation introduced in Fig. 124c. Dissociation into partials has not yet been observed in the basal plane of hexagonal metals.

Interesting reactions between loops with Burgers vectors of the type AT and AS, as observed in zinc, are discussed in Section 10d(5)(b).

(5) *Graphite.* The reactions discussed in Section 10b(2)(b) also take place between ribbons in the c plane of graphite. Ribbons in successive lattice planes (separated by $c/2$) dissociate according to a different scheme, e.g., $\sigma A + B\sigma$ against $A\sigma + \sigma B$. This gives rise to typical interactions between such ribbons. We discuss first 2 fusion reactions.

(1) The fusion of 2 ribbons with vectors $A\sigma + \sigma B$ and $\sigma C + A\sigma$ takes place according to the scheme of Fig. 164. It results in the formation of a triple ribbon [see Section 10a(5)]. If the 2 ribbons are pinned somehow, a Y-shaped node results. Such nodes are visible in Fig. 140.

(2) The fusion of 2 ribbons with vectors $A\sigma + \sigma B$ and $\sigma C + B\sigma$ gives rise to the reaction pictured in Fig. 165. On meeting, the 2 inner partials combine according to the scheme $\sigma B + C\sigma \rightarrow A\sigma$. This is only an intermediate stage: $A\sigma$ reacts further with $B\sigma$ to form

[302] A. Berghezan and A. Fourdeux, *4e Coll. Met. Centre Etudes Nucl. Saclay, 1960,* p. 127. Presse Universitaire de France, Paris, 1961.

[303] B. R. Banerjee, J. M. Capenos, J. J. Hanser, and J. P. Hirth, *J. Appl. Phys.* **33**, 556 (1962).

a partial σC. Since $A\sigma$ and σC now repel the final situation of Fig. 165d is stable. We call this node a T node.

Finally we also discuss the intersection of 2 ribbons in successive lattice planes, e.g., $A\sigma + \sigma C$ and $\sigma B + A\sigma$ and which is shown in Fig. 166. After crossing the new segments indicated in Fig. 166a tend to develop. The segments $\pm A\sigma$ will *not* annihilate since they are in different planes; instead a dipole is formed as indicated by Di

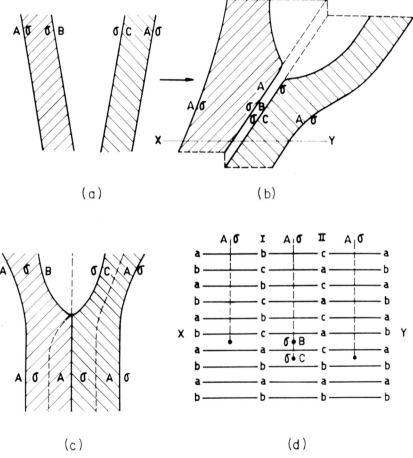

(a)   (b)

(c)   (d)

FIG. 164. Fusion of 2 ribbons containing a different stacking fault resulting in the production of a threefold ribbon (Y-node). The three partials in the threefold ribbon have the same Burgers vector. (a) and (b) Two different stages in the process of meeting. (c) Stable configuration. (d) Cross section through the ribbon.

in Fig. 166b. The area where the 2 ribbons overlap contains a double stacking fault, i.e., 2 layers in rhombohedral arrangement as shown in Fig. 166c. Therefore it will be eliminated by glide of the dipole towards the rest of the dislocations with the formation of constrictions

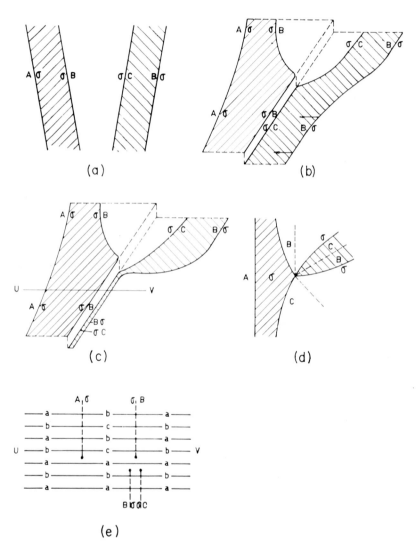

FIG. 165.   Fusion of 2 ribbons containing a different stacking fault resulting in the production of a T-node.

in the 2 ribbons (Fig. 166d). The final situation is described by the lettering pattern of Fig. 166f. The configuration is stable since there is no net repulsion between the dipole and σC, whereas the stacking fault pulls the two together. An observed example is visible in Fig. 140.

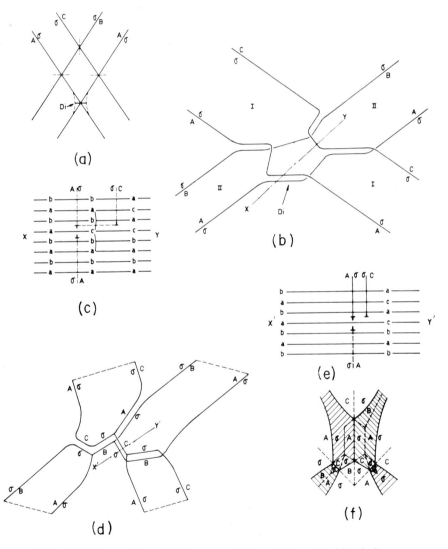

FIG. 166.   Intersection of 2 ribbons containing a different stacking fault.

Many more reactions are described in Amelinckx and Delavig-nette.[272,273]

(6) *Layer structures with close-packed layers.* In this class of structures the same reactions occur as for graphite. Threefold ribbons have been observed in nickel bromide[207] and in tin disulfide[277] (Fig. 141b).

The reactions between multiribbons of the type observed in the chromiumhalides give rise to interesting patterns; we refer to the original paper[304,305a] for a detailed discussion since a large number of illustrations are required for a proper exposition.

(7) *Uranium.*[305b] α-uranium has an orthorhombic structure; how-ever, it can be considered as a deformed hexagonal close-packed structure. The Burgers vector are $[a,\ 0\ 0]$ and $[(a/2)\,(b/2)\,0]$ {as well as $[(a/2)\,(\bar{b}/2)\,0]$}. Three dislocations with such a vector can form a node since

$$[a\,0\,0] + \left[\frac{\bar{a}\ \bar{b}}{2\ 2}\,0\right] + \left[\frac{\bar{a}\ b}{2\ 2}\,0\right] = 0.$$

There is evidence for this reaction.

When lying in the c plane these dislocations can dissociate as in the hexagonal metals, according to the reaction.:

$$[a\,0\,0] \rightarrow \left[\frac{a}{2},\ \alpha b,\ 0\right] + \left[\frac{a}{2},\ -\alpha b,\ 0\right]$$

$$\left[\frac{a\ b}{2\ 2}\,0\right] \rightarrow \left[\frac{a}{2},\ \alpha b,\ 0\right] + \left[0,\ \left(\frac{1}{2} - \alpha\right)b,\ 0\right].$$

where α represents a fraction ($\alpha \simeq \frac{1}{10}$). Networks of extended and contracted nodes, suggesting this dissociation, have been observed.[305b]

c. Dislocation Networks in Particular Structures

In this subsection the geometry of the subboundaries in a number of particular crystal lattices for which observations are available will

[304] P. Delavignette and S. Amelinckx, *Trans. Brit. Ceram. Soc.* **62**, 687 (1963).
[305a] S. Amelinckx and P. Delavignette, *J. Appl. Phys.* **33**, 1458 (1962).
[305b] E. Ruedl and S. Amelinckx, *J. Nucl. Mater.* **9**, 116 (1963).

be discussed. For a detailed derivation of the results summarized here we refer to Amelinckx and Dekeyser[297] and Frank.[306] Decoration methods, as well as electron microscopic observations have permitted a detailed verification of the models predicted by theory.

We will discuss here results obtained by the different methods.

(1) *The primitive lattice.* No crystal has the primitive lattice as its structure. However crystals with the cesium chloride structure have the same configuration of Burgers vectors as in the primitive lattice, i.e., the shortest Burgers vectors are of the type a [100], a [010], and a [001]. With the dislocations having these Burgers vectors one can construct:

(a) Tilt boundaries. (i) Symmetrical tilt boundaries consist of one kind of dislocation and lie in a cube plane.

(ii) Asymmetrical tilt boundaries consist of two kinds of dislocations; for instance, a [100] and a [010]. The contact plane is then of the type ($hk0$), h and k depending on the proportion of the concentrations of the two kinds of dislocations.

All these boundaries consist of sets of parallel lines; they cannot be distinguished from each other on a decoration pattern. By electron microscopy it is possible to distinguish the 2 families of dislocations in asymmetrical boundaries.

(b) Twist boundaries. The 2 dislocation twist boundaries consist of a square grid of mutually perpendicular screw dislocations, for instance, with Burgers vectors a [100] and a [010] and lying in the (001) plane. Threefold nodes are occasionally observed, suggesting that a dislocation with a Burgers vector a [110] forms at the intersecting point. Although there is no gain in elastic energy from this reaction, there may be some gain in core energy that results when 2 dislocations share the same core. Nets lying in other than the cube plane have correspondingly deformed meshes. They correspond to boundaries with some tilt component.

(c) Observations. Dislocations have been decorated in cesium bromide single crystals using the gold diffusion method. Tilt and twist boundaries of the type described have been observed.

(2) *Face-centered cubic lattice.* Most of the work done with decoration as well as with transmission electron microscopy is on crystals

[306] F. C. Frank, *Rept. Conf. Defects in Crystalline Solids, London, 1954* p. 159 (1955).

belonging to this type. A detailed analysis of the networks in face-centered cubic metals is due to Frank[306] and Thompson[307] and for the sodium chloride structure to the present author.[297]

In the face-centered metals the glide elements are (111), $(a/2)$ [1$\bar{1}$0] whereas in the sodium chloride structure they are (110), $(a/2)$ [1$\bar{1}$0]. The configuration of Burgers vectors is the same in both, and therefore the dislocation patterns in annealed crystals are very similar. Of course the same does not apply to subboundaries formed by deformation only.

(a) Tilt boundaries. (i) Symmetrical tilt boundaries. With dislocations having a Burgers vector $(a/2)$ (1$\bar{1}$0] one can form symmetrical pure tilt boundaries in the (110) plane for the NaCl structure and in the (11$\bar{2}$) plane for the face-centered cubic metals.

(ii) Asymmetrical tilt boundaries. For ($hk0$) planes asymmetrical tilt boundaries consisting of dislocations with mutually perpendicular Burgers vectors $(a/2)$ [110] and $(a/2)$ [1$\bar{1}$0] can be formed.

Other asymmetrical tilt boundaries result from the interaction of dislocations with Burgers vector enclosing an angle of 60° (or 120°), e.g., $(a/2)$ [1$\bar{1}$0] and $(a/2)$ [10$\bar{1}$]. In this particular example the boundary plane (hkl) is such that $h + k + l = 0$. Its exact orientation depends again on the proportion of both kinds of dislocations. Many of the observed tilt boundaries probably are of this type.

(b) Twist boundaries. (i) Square grids. If 2 families of dislocations with mutually perpendicular Burgers vectors are available, e.g., $(a/2)$ [110] and $(a/2)$ [1$\bar{1}$0], a square grid of screw dislocations can be formed in the (001) plane, the rotation axis being [001]. If the boundary plane is not the cube plane, the net is such that its projection on the (001) plane is the square grid just described.

Grids of this type have been revealed in the alkali halides by decoration and in platinum,[215] aluminum,[301] etc. by electron microscopy. It was found that at the intersection points small segments of a [100] dislocation develop sometimes according to the reaction

$$\frac{a}{2}[110] + \frac{a}{2}[1\bar{1}0] \rightarrow a[100]. \tag{10.66}$$

This was observed in KCl[299] and in Al.[301] A similar observation was already mentioned for the primitive lattice [see Section 10b(1)].

[307] N. Thompson, *Proc. Phys. Soc.* (*London*) **B66**, 481 (1953).

A common singularity observed in these nets consists of zig-zag lines meandering through the square grids. These are due to stranger dislocations with a third Burgers vector. Examples of this feature are presented in Fig. 37 in KCl and in Fig. 167b in platinum. The analysis in terms of Burgers vectors is shown in Fig. 168. More singularities have been analyzed elsewhere.[226,297]

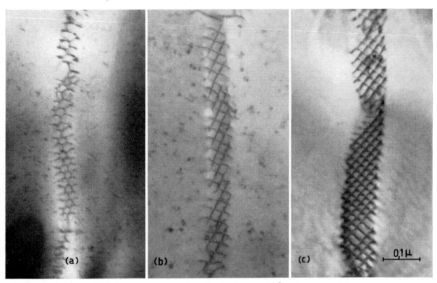

FIG. 167. Networks in platinum. [After E. Ruedl, P. Delavignette, and S. Amelinckx, *J. Nucl. Mater.* **6**, 46 (1962).] (a) Hexagonal network. (b) Square grid with singular line. (c) Square grid.

(ii) Hexagonal grids. With 2 families of dislocations having Burgers vectors at 120° (or at 60°), e.g., $(a/2)$ [$1\bar{1}0$] and $(a/2)$ [$\bar{1}01$], an hexagonal grid of screw dislocations can be constructed. The third family of segments with Burgers vectors $(a/2)$ [$0\bar{1}1$] will be generated by the reaction

$$\frac{a}{2}[1\bar{1}0] + \frac{a}{2}[\bar{1}01] \rightarrow \frac{a}{2}[0\bar{1}1] \tag{10.67}$$

according to the scheme of Fig. 161a,b. This reaction reduces the total length of dislocation line and brings all segments into the screw orientation which has the smallest energy. In this case also the projection rule is valid. If the net is not situated in the (111) plane its projection on that plane is the regular hexagonal grid.

Grids of this type have been observed in AgCl, AgBr[121,308] in the alkali halides,[297,299] and in a number of metals. Figure 40 reproduces a net observed in KCl. The net plane is not far off the cube plane and hence the meshes are elongated. The short segments are pure screws; the other segments have mixed character. Figure 167a shows a similar grid in annealed platinum.

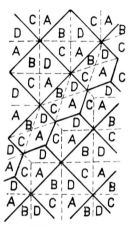

FIG. 168.   Analysis in terms of Burgers vectors of the singularity observed in Fig. 167b.

In the face-centered metals such a grid is glissile, if situated in the (111) plane; i.e., it can glide as a whole in the boundary plane because each dislocation segment has this plane as a glide plane. This is not true for the NaCl structure.

(iii) Dissociation into partials. Until now we have neglected the dissociation into partials, which takes place in metals or face-centered cubic alloys with a small stacking fault energy. As shown by Thompson[307] dissociation in the hexagonal grids will result in the formation of alternatively extended and contracted nodes according to the scheme of Fig. 161c,d. An example as observed in stainless steel is presented in Fig. 130. The area enclosed by the extended node contains a stacking fault, i.e., a lamella in hexagonal closed packing.

In the square grids the glide planes of the dislocations are different and dissociation of the ribbons takes place in different planes. Extended

[308] M. Gardner Miller, Thesis, University of North Carolina, 1961.

and contracted nodes of a somewhat different kind could form,[244] but no observations are as yet available.

(3) *Body-centered metals.* (a) Tilt boundaries. Since there are two types of stable Burgers vectors, $(a/2) \langle 111 \rangle$ and $(a/2) \langle 100 \rangle$, two types of stable symmetrical tilt boundaries in planes perpendicular to each of the vectors may be expected. However, Burgers vector determinations for dislocations in boundaries are not yet available. Berghezan and Fourdeux[302] have observed numerous tilt boundaries in niobium. Figure 169 shows, for instance, a tilt boundary, which evidently contains two kinds of dislocations, as can be judged from the difference in the density of the decoration along them.

(b) Hexagonal grids. Networks result from the interaction between sets of dislocations with $(a/2) [1\bar{1}1]$ and $(a/2) [1\bar{1}1]$ Burgers vectors according to the reaction (10.64), as pointed out in Amelinckx and Dekeyser.[297] The pure twist boundary lies in the (110) plane and the mesh shape is as shown in Fig. 171. The angles are no longer equal

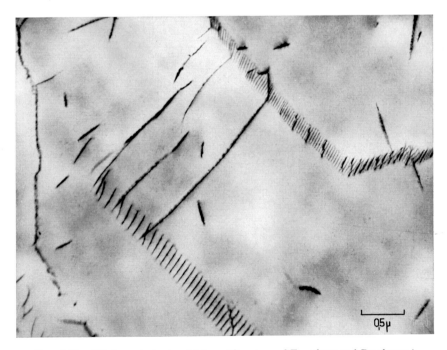

FIG. 169.   Tilt boundaries in niobium. (Courtesy of Fourdeux and Berghezan.)

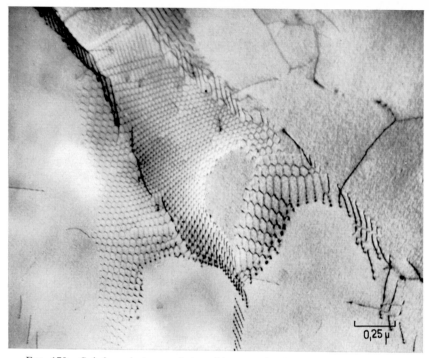

FIG. 170. Sub-boundaries consisting of hexagonal grids in niobium. (Courtesy of Fourdeux and Berghezan.)

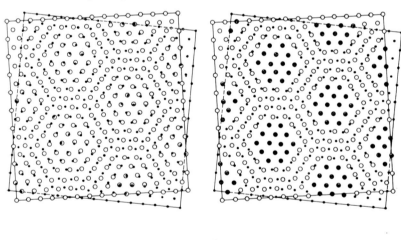

(a)   (b)

FIG. 171. Formation of hexagonal network in the (110) plane of the body-centered lattice. (a) After rotation of the two parts, (b) after relaxation of the atoms.

to 120°, as in the fcc lattice. The superposition of 2 (110) lattice planes with a small orientation difference allows us to visualize this net (Fig. 171a,b). If the net is lying in planes other than the (110) the meshes will be deformed. Nets of this type observed in iron, have been analyzed in detail by Carrington et al.[251] They calculated the mesh shape for different contact planes by minimizing the line energy. Good agreement with the observed nets is found.

The same authors also analyzed singularities resulting from the interaction with stranger dislocations. One of the observed singularities is shown in Fig. 172 and the analysis in terms of the modified Thompson notation is also shown (Fig. 173). The reference system is as shown in Fig. 124b. The striking feature is the formation of a row of rectangles. It results from introducing 1 dislocation with Burgers vector BT parallel to the set with Burgers vector $C'T$. The basic net itself results from the interaction of dislocations with Burgers vector $C'T$ and $A'T$.

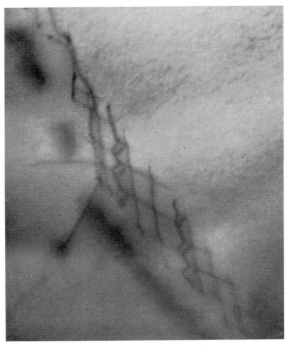

Fig. 172.   Sub-boundary in iron containing a singular line. (Courtesy of Carrington et al.)

Similar boundaries have been observed in niobium by Berghezan and Fourdeux (Fig. 170).

(4) *Hexagonal close-packed metals.* No systematic study of networks in hexagonally close-packed metals is available as yet. Occasional observations of networks in zinc have been described by Berghezan *et al.*[309] The boundaries are extremely mobile when observed with

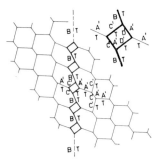

Fig. 173.   Lettering pattern corresponding to the observed net of Fig. 172. (After Carrington *et al.*)

the microscope. Tilt boundaries have been observed to move as a whole, incorporating further dislocations along their path. Hexagonal networks are also very mobile in the *c* plane. The dislocations are not visibly dissociated (Fig. 174).

In α-uranium which has a deformed hexagonally close-packed structure networks of extended and contracted nodes have been observed in the *c* plane.[305b] From these the stacking fault energy was determined to be about 50 ergs/cm².

(5) *Networks in graphite and other layer structures.* In graphite and in most other layer structures all networks are confined to the *c* plane since this is the main glide plane. Extensive nets of contracted and extended nodes are often observed (Fig. 175). Since glide is only possible along the *c* plane, the formation mechanism must be different from that described by Whelan[244] for the face-centered cubic lattice and which involves cross slip. The different stages in the formation of a network are represented in Fig. 176. In (a) the ribbons $B\sigma + \sigma A$ and $C\sigma + \sigma A$ are pushed against one another by the applied shear stress. In (b) σA has combined with $C\sigma$ to form a perfect dislocation

[309] A. Berghezan, A. Fourdeux, and S. Amelinckx, *Acta Met.* **9**, 464 (1960).

CA having a Burgers vector which is perpendicular to $B\sigma$ and which therefore interacts weakly with this partial. The 2 dislocations cross under the influence of the shear stress. In (c) the perfect CA has again dissociated into 2 partials $CA \rightarrow C\sigma + \sigma A$. One more node pair has been formed. This process evidently is "repeatable;" it can go on until all dislocations of the parallel set have been intersected. The different stages shown here have been observed and the Burgers vectors were determined by means of contrast effects.[273]

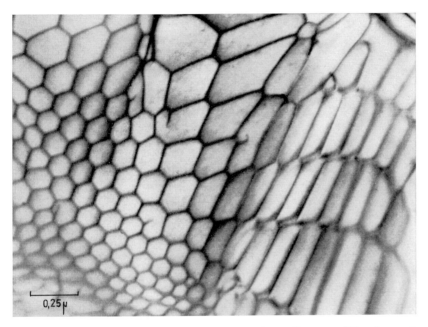

FIG. 174.   Hexagonal network in the basal plane of zinc. (Courtesy of Fourdeux and Berghezan.)

In molybdenum sulfide the stacking fault energy is so small that the extended nodes become uncurved and the network consists of triangular regions alternatively faulted and unfaulted (Fig. 177a). In graphite the interaction between ribbons in adjacent planes gives rise to triple ribbons and other singularities in the nets. For a complete analysis we refer to Delavignette and Amelinckx.[273]

Very well-developed networks of the same type as those shown in Fig. 175 have been observed in nickel bromide[207] which has the cadmium chloride structure and in tin disulfide (Fig. 177b) which has the cadmium iodide structure.

(6) *The diamond structure.* Recently a detailed study of ribbons in silicon was made by Aerts *et al.*[210] Very extended networks are observed in crystals twisted at 1200°C about the ⟨111⟩ axis and slowly cooled. Hexagonal networks lie in the (111) planes perpendicular to the twisting axis; they provided the first example of nets where *all* nodes are dissociated. One family of nodes contains an *intrinsic* fault, the other an extrinsic fault. A schematic view of the

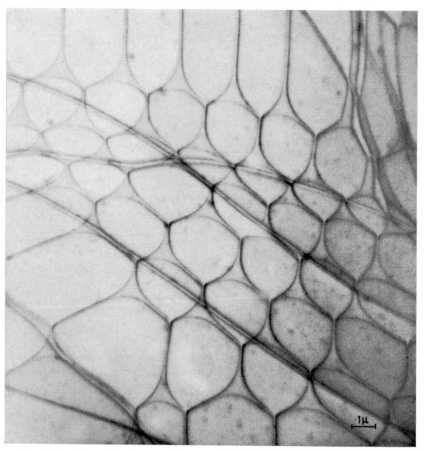

Fig. 175. Network of extended and contracted nodes in graphite. [After S. Amelinckx and P. Delavignette, *J. Appl. Phys.* 31, 2126 (1960.]

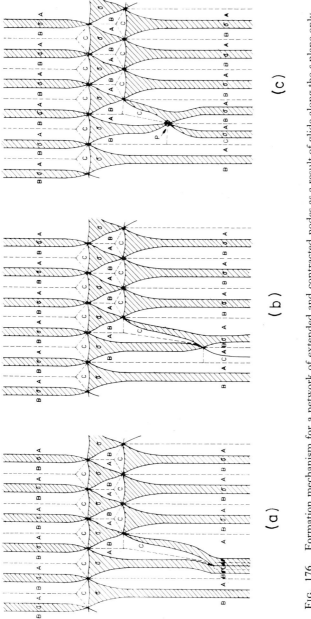

FIG. 176.   Formation mechanism for a network of extended and contracted nodes as a result of glide along the *c* plane only.

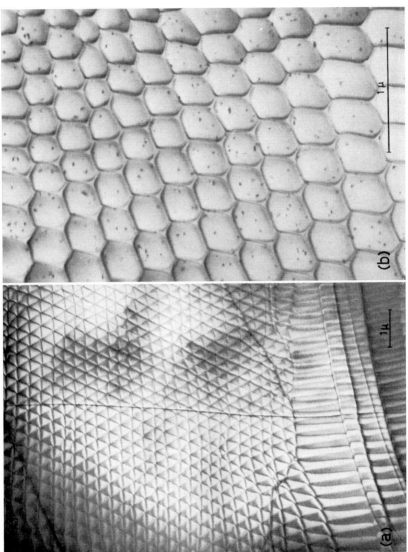

Fig. 177. (a) Network of contracted and extended nodes in molybdenum sulfide. The segments are not curved because the stacking fault energy is very small. [After S. Amelinckx and P. Delavignette, "Direct Observation of Imperfections in Crystals" (J. B. Newkirk and J. H. Wernick, eds.), p. 295. Wiley (Interscience), New York,

geometry as well as the lettering pattern is shown in Fig. 178 while Fig. 179 represents an observed region. Figure 146b,c represent cuts along the lines XY and UV of Fig. 178 showing the structure of intrinsic and extrinsic faults respectively. From the node geometry, and taking anisotropy into account [Section 10a(6)] the stacking fault energies were estimated

$$\gamma_I = 50 \text{ ergs/cm}^2 \qquad \gamma_{II} = 60 \text{ ergs/cm}^2.$$

The values are only first estimates since the image displacement has not yet been taken into account properly.

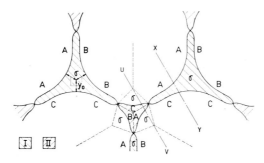

FIG. 178. Lettering pattern of a doubly extended network. Figure 146b is a cut along XY, and Fig. 146c along UV.

Rather similar networks were observed in germanium.[284a] From the extended nodes a value for γ was deduced: $\gamma = 90 \text{ ergs/cm}^2$.

d. Dislocations Resulting from the Agglomeration of Point Defects

(1) *Introduction.* The first evidence for the formation of prismatic loops from point defects was found in alkali halides by the use of decoration techniques.[152] More recently transmission electron microscopy has contributed greatly in extending our understanding of the generation of dislocations from point defects and vice versa.

(2) *Quenching.* (a) Formation of loops. If a solid is heated a certain concentration of point defects, either vacancies or interstitials, is in thermodynamic equilibrium at any given temperature. For example, the concentration of vacancies is given by an expression of the form:

$$C_v = A \exp\left(-U_F/kT\right) \tag{10.68}$$

where U_F is the formation energy of vacancies, A is the entropy factor; it can be considered as temperature independent. C_v becomes appreciable at temperatures not too far below the melting point; with $U_F \simeq 1$ ev there may be about 0.1 atomic percent of vacancies at the melting point. If such a solid is cooled rapidly the vacancies may either disappear at an existing dislocation and induce the latter to

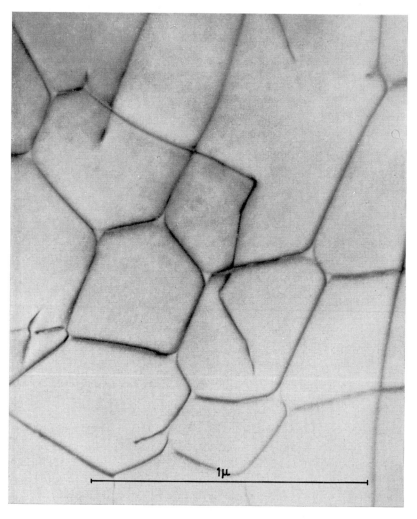

FIG. 179. Observed sequence of nodes in silicon corresponding to Fig. 178.

climb, they may disappear at the surface, or they may form agglomerates which may adopt several shapes. Kuhlmann-Wilsdorf[310] has discussed in detail what happens if the vacancies agglomerate in disk-shaped cavities which subsequently collapse into ring dislocations (Fig. 180). It is easy to see that beyond a certain size it is energetically

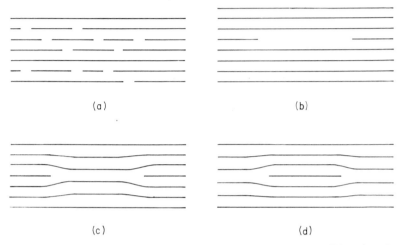

(a)   (b)

(c)   (d)

FIG. 180. Collapsing of a ring-shaped cavity into a prismatic dislocation ring. (a) Vacancies are dispersed. (b) Ring-shaped cavity. (c) Vacancy loop. (d) Interstitial loop.

more favorable to form a dislocation loop with diameter d and Burgers vector b rather than a disk-shaped cavity. The energy of the former is given by

$$E_l = \mu b^2 d + \left(\tfrac{1}{4}\right) \pi d^2 \gamma \tag{10.69}$$

μ is the shear modulus and γ is the specific stacking fault energy. The last term only occurs if the loop contains a fault. The energy of the cavity is roughly

$$E_c = \left(\tfrac{1}{2}\right) \pi d^2 \sigma \tag{10.70}$$

where σ is the surface energy.
It is clear that $E_l < E_c$ provided

$$d > 4\mu b^2 / [\pi(2\sigma - \gamma)]. \tag{10.71}$$

[310] D. Kuhlmann-Wilsdorf, *Phil. Mag.* [8] **3**, 125 (1958).

Presumably the agglomerate is initially more or less spherical; on transforming into a dislocation loop it has to pass through the disk-shaped configuration, which has a higher energy, due to the increased surface area. Some activation energy is required for this process. In substances, with a high surface energy, e.g., platinum, defects may therefore prefer to remain as small spherical cavities up to observable size.

The exact geometry of the prismatic loops, in particular the Burgers vector, depends on the magnitude of the stacking fault energy. Consider, for example, the face-centered cubic structure.

(b) Face-centered cubic structure. (i) Prismatic and sessile loops. Vacancies (as well as interstitials) precipitate in a single sheet on the (111) plane which is the closest packed plane. If the cavities so formed collapse, dislocation loops surrounding a stacking fault result. If the stacking fault energy is large it may become more favorable to nucleate a partial with Burgers vector $(a/6)$ [11$\bar{2}$] at the periphery of the loop, and remove the stacking fault by sweeping it with the partial. The resulting loop is now a perfect dislocation, containing no stacking fault; its Burgers vector is:

$$\frac{a}{3}[111] + \frac{a}{6}[11\bar{2}] \rightarrow \frac{a}{2}[110] \tag{10.72}$$

and it is inclined with respect to the loop plane.

The condition for this process to happen is roughly that the energy of the perfect loop E_2 should be smaller than that of the faulted one, E_1, i.e., $E_2 \leqslant E_1$

$$E_1 = \frac{2}{3}\frac{1}{1-\nu}\mu b^2 R \left(\ln\frac{4R}{r_0} - 2\right) + \pi R^2 \gamma \tag{10.73}$$

where R is the radius of the loops, r_0 is the inner cutoff radius, and the other symbols have their usual meaning. For E_2 one has

$$E_2 = \left[\frac{2}{3}\frac{1}{1-\nu} + \frac{1}{3}\frac{2-\nu}{2(1-\nu)}\right]\mu b^2 R \left(\ln\frac{4R}{r_0} - 2\right). \tag{10.74}$$

The stability condition than becomes:

$$\gamma \geqslant \frac{\mu b^2}{3\pi}\frac{2-\nu}{2(1-\nu)}\frac{1}{R}\left(\ln\frac{4R}{r_0} - 2\right). \tag{10.75}$$

The shear stress exerted by the stacking fault is γ/b; this shear stress is available for the nucleation of the partial. For $\gamma = 200$ ergs/cm^2 and $b = 2 \times 10^{-8}$ cm it is, e.g., 100 kg/mm^2 as compared to the yield stress $\simeq 0.1$ kg/mm^2.

It has been possible to show that this process takes place[311,312] in high-purity aluminum with $\gamma \simeq 200$ ergs/cm^2. This could be deduced from the absence of stacking fault contrast and even more convincingly by observing the glissile character of the loops. They are mobile on a prismatic surface defined by the loop area and the direction of the Burgers vector. Similar dislocation loops have been observed in quenched silver and copper.[313] In quenched aluminum denuded zones are found along the boundaries, proving that the latter act as sinks for vacancies. Figure 181 shows vacancy loops in a quenched aluminum foil.

Loops of this kind tend to leave the original (111) plane, not only by glide but also by climb. In the latter case they tend to rotate around an axis lying in the loop plane. In hyperpure aluminum, or on repeated quenching of the same material, the stacking fault seems to remain in certain loops.[314] If the stacking fault energy is somewhat smaller than the limit set by the inequality (10.75) the loop will keep the stacking fault in its interior. This seems to be the case for aluminum silver alloys.[315]

(ii) Stacking fault tetrahedra. Finally if the stacking fault energy becomes very small, as in gold, it has been shown by Silcox and Hirsch[316] that on quenching from 950°C followed by aging 1 hour at 100-250°C, spatial defects are formed instead of loops. Careful analysis of the contrast effects for different orientations of the foil

[311] P. B. Hirsch, J. Silcox, R. Smallmann, and K. Westmacott, *Phil. Mag.* [8] 3, 897 (1958).

[312] J. Silcox, *Proc. 4th Conf. on Electron Microscopy, Berlin, 1958* 1, 548 (1960).

[313] P. B. Hirsch and J. Silcox, *in* "Growth and Perfection of Crystals" (R. H. Doremus *et al.*, eds.), p. 262. Wiley, New York, 1958. R. E. Smallmann, K. H. Westmacott, and J. A. Coiley, *J. Inst. Metals* 88, 127 (1959); R. E. Smallmann and K. H. Westmacott, *J. Appl. Phys.* 30, 603 (1959); see also, Kin-Pong Chik, A. Seeger, and M. Rühle, *Proc. 5th Intern. Conf. on Electron Microscopy, Philadelphia, 1962* 1, J-11 (1962).

[314] R. M. J. Cotterill, *Phil. Mag.* [8] 6, 1351 (1961); See also R. M. J. Cotterill and R. L. Segall, *Proc. 5th Intern. Conf. on Electron Microscopy, Philadelphia, 1962* 1, J-13 (1962).

[315] R. B. Nicholson and J. Nutting, *Acta Met.* 9, 332 (1961).

[316] J. Silcox, and P. B. Hirsch, *Phil. Mag.* [8] 4, 72 (1959).

revealed that the defect is a tetrahedron bounded by (111) planes. The edges of the tetrahedron are so called *stair rod* dislocations with Burgers vectors of the type $(a/6)$ [110] and the faces are stacking faults. An example is reproduced in Fig. 182 while Fig. 183 is a schematic view.

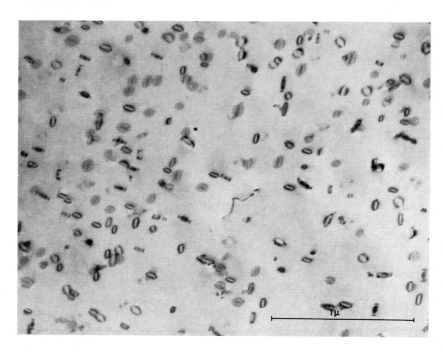

FIG. 181.   Vacancy loops in quenched aluminum. The foil was quenched from 560°C into iced brine. (Courtesy of Hirsch and Cotterill.)

The defects can be generated by the following mechanism. Suppose that the vacancies precipitate initially in a triangular area on a (111) plane limited by $[1\bar{1}0]$, $[10\bar{1}]$, and $[01\bar{1}]$ directions (Fig. 183a). This loop has Burgers vector $(a/3)$ [111] or $A\alpha$ in Thompson's notation [see Section 10a(a)(a)]. The partials limiting the loop can now split in turn on the different (111) planes according to the schemes

$$\alpha A \rightarrow \alpha\beta + \beta A \text{ (in the } b \text{ plane)}$$
$$\alpha A \rightarrow \alpha\gamma + \gamma A \text{ (in the } c \text{ plane)} \qquad \qquad (10.76)$$
$$\alpha A \rightarrow \alpha\delta + \delta A \text{ (in the } d \text{ plane).}$$

The right-hand sides consist each of a stair rod dislocation plus a Shockley partial. The square of the Burgers vector criterion shows that elastic energy is gained. The Shockley partials βA, γA, and δA bow out in their respective (111) planes since they are repelled by the immobile stair rods (Fig. 183b). Along the intersection lines of each pair of (111) planes, 2 Shockley partials meet, attract, and combining

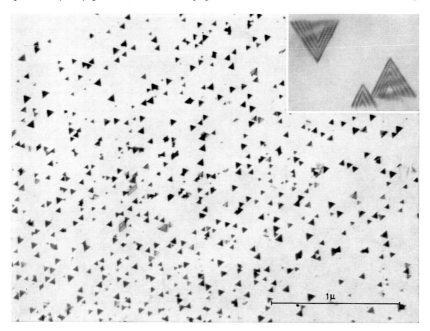

FIG. 182. Stacking fault tetrahedra in gold. The inset shows a magnified view of the defects. The fault fringes are visible. (Courtesy of Hirsch and Cotterill.)

they give rise to new stair rod dislocations according to the reactions

$$\beta A + A\gamma \rightarrow \beta\gamma$$
$$\gamma A + A\delta \rightarrow \gamma\delta \qquad\qquad\Bigg\} \qquad (10.77)$$
$$\delta A + A\beta \rightarrow \delta\beta.$$

The area swept by the Shockley partial clearly contains a stacking fault. The total energy of the tetrahedral defect (E_t) will be smaller than that of a triangular loop E_l containing the same number of defects provided $E_l \geqslant E_t$.

One finds

$$E_l = \frac{\mu a^2 l}{4\pi(1-\nu)} \ln \frac{l}{r_0} + \frac{\sqrt{3}}{4} l^2 \gamma \qquad (10.78)$$

where a is the lattice parameter, l the side of the triangular loop and of the tetrahedron, and r_0 is the core radius. The energy of the cor-

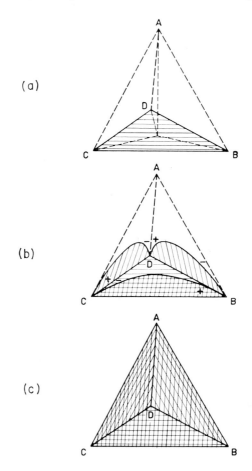

Fig. 183. Formation of a stacking fault tetrahedron. (a) Triangular sessile loop CDB in the (111) plane resulting from the collapse of a disk of vacancies. (b) The partials bordering the loop dissociate and the Shockley partials bow out in the other (111) planes. (c) Final shape of the tetrahedron. The edges are stair rod dislocations; the side faces contain stacking faults.

responding tetrahedron is

$$E_t = \frac{\mu a^2 l}{12\pi(1-\nu)} \ln \frac{l}{r_0} + \sqrt{3}l^2\gamma. \tag{10.79}$$

This puts a limiting size l_0 to the tetrahedron given by $E_l = E_t$ or $l_0/[\ln (l_0/r_0)] = 2\mu a^2/[9\sqrt{3}\pi\gamma(1-\nu)]$. In accordance with the observations, this is about 430 A. Similar tetrahedra have been observed in silver[317] and in nickel-cobalt alloys.[257]

The actual formation mechanism of the tetrahedra presumably proceeds differently; it probably involves nucleation and growth, the nucleus being a tetrahedral arrangement of vacancies.

The mechanism discussed here demonstrates clearly the geometry of the defect, however there is no real evidence that the tetrahedra are generated in this way. It is also difficult to see how the tetrahedra grow. A possible growth mechanism which does not imply the formation of an intermediate loop has been proposed by De Jong and Koehler.[317a]

These authors consider the tetrahedral trivacancy as the nucleus for a tetrahedron. This defect can be considered as one atom in the center of a tetrahedron of four vacancies. The next larger tetrahedral cluster would consist of a tetrahedron of four atoms in the center of a tetrahedron of six vacancies. This is in fact the first defect which can be considered as being a tetrahedron. It would result from the meeting of smaller clusters, e.g., a divacancy and a tetravacancy. The growth of the tetrahedron is now best illustrated by means of the model of Fig. 183[(1)](1–5). Suppose that a vacancy is absorbed in the edge of the tetrahedron as shown in Fig. 183[(1)](2). Edges or corners are the most likely sites where vacancies will be absorbed. The whole row of atoms XY can now move and join the tetrahedron by a displacement over a vector of the type $(a/6)$ [112]; this increases the size of the tetrahedron. It is clear that if the tetrahedron were perfect at the start, a step is now created in one of the side faces as shown in Fig. 183[(1)](6). Absorption of one vacancy along the step will displace it further, and by repeating the operation a whole layer can be deposited. Presumably the growth process may proceed on several faces simultaneously.

[317] R. E. Smallmann, K. H. Westmacott, and J. A. Coiley, *J. Inst. Metals* **88**, 127 (1959); R. E. Smallmann and K. H. Westmacott, *J. Appl. Phys.* **603**, 30 (1959).
[317a] M. De Jong and J. S. Koehler, *Phys. Rev.* **129**, 49 (1963).

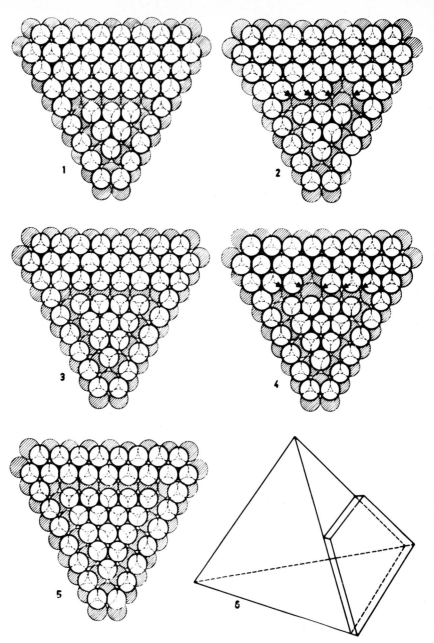

FIG. 183[1]. Growth mechanism for the tetrahedra. The drawing shows the successive atom movements that accompany the addition of vacancies to the tetrahedron. [After M. De Jong and J. S. Koehler, *Phys. Rev.* **129**, 49 (1963).]

It should be pointed out that Fig. 183[(1)](6), which is taken from Ref. 317a, is presumably not correct and should be replaced by Fig. 183[(2)], because the steps propagate away from the corner rather than from the edge. This difficulty was also noticed by Kimura *et al.*[317b]

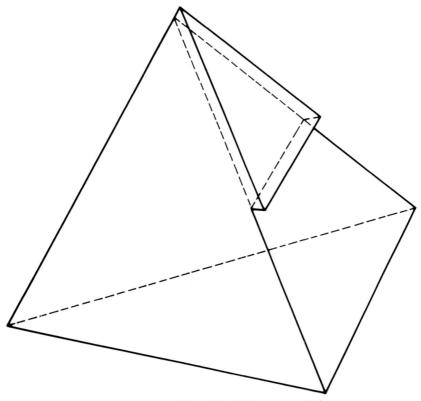

FIG. 183[(2)].   Growth step in side face to tetrahedra.

These authors point out that for steps moving away from the corner the stacking faults form an acute angle and hence have a smaller energy. Furthermore the energy increase resulting from the displacement of the ledge over one atomic distance is now smaller than the formation energy of a vacancy. The large stability of the tetrahedra up to 600°C is now connected with the fact that the growth process cannot be reversed to provide a mechanism for dissolution. No mechanism for dissolution has been proposed as yet.

[317b] H. Kimura, D. Kuhlmann-Wilsdorf, and R. Maddin, *Appl. Phys. Letters* 3, 4 (1963).

(iii) Cavities.   In quenched platinum small cavities were found.[215] It was suggested that the cavities have to go through a high-energy configuration on transforming into disks before collapsing into loops. Because of the large surface energy of platinum, the activation energy required to do this may be too high and cavities may be quenched in.

(iv) Determination of formation energy.   If the assumption is made that all quenched-in vacancies can be agglomerated into observable defects by a mild anneal vacancy concentrations can be measured directly on micrographs. The use of formula (10.68) then allows the determination of the formation energy U_F.

(c) Hexagonal and body-centered metals.   Large loops have been observed in the c plane of quenched magnesium foils.[318] The loops exhibit stacking fault contrast; presumably they have a vector of the type AT. They anneal out with an activation energy of 1.3 ev, which is similar to that for self-diffusion.

So far no quench loops have been observed in body-centered metals; probably they would form on (112) planes.

(d) Graphite.   Loops also have been observed in graphite single crystals quenched from about 3000°C and aged at about 1300°C.[319,320,321] The loops are slightly hexagonal and situated in the c plane. In this particular case it was determined that the loops actually resulted from vacancies. For loops due to vacancies the Burgers vector should be inclined with respect to the c plane, while it should be perpendicular for interstitial loops. This can be shown by reference to Fig. 184. Suppose that part of a layer is removed, as would be the case for vacancies. Should the disk now collapse the 2 layers should come one right on top of the other (i.e., a on a, or b on b). This would create a high-energy stacking fault; in order to remove this, the loop is swept by a partial of the Shockley type. The Burgers vector is now inclined (Fig. 184a). On the contrary interstitials will precipitate in the c-position between a and b and no high-energy stacking fault is formed. The stacking faults that are formed have such a small energy ($\gamma = 0.6$ erg/cm^2) that the equivalent shear stress γ/b is far too small to nucleate a partial. The Burgers vector therefore remains perpendicular (Fig. 184b). The problem is reduced to a Burgers vector determi-

[318] J. S. Lally and P. B. Partridge, *Proc. 5th Intern. Conf. on Electron Microscopy, Philadelphia, 1962* 1, J-12 (1962).

[319] S. Amelinckx and P. Delavignette, *Phys. Rev. Letters* 5, 50 (1960).

[320] G. K. Williamson and C. Baker, *Phil. Mag.* [8] 6, 313 (1961).

[321] C. Baker and A. Kelly, *Nature* 193, 235 (1962).

nation. Using the dark field method it was found that for (11$\bar{2}$0) reflections some loops were plainly visible, while others where practically invisible. The latter are those for which the component of the Burgers vector parallel to c lies in the plane used for contrast. The slight visibility is due to the radial displacement associated with the prismatic loop.[230d] From the fact that other loops were plainly visible it can be concluded that it is an inclined vector. These experiments suggest that in graphite vacancies become mobile at about 1300°C. In Fig. 185 vacancy loops in quenched graphite are visible in stacking fault contrast.

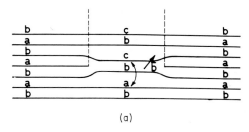

(a)

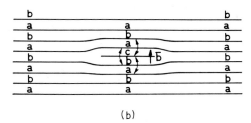

(b)

FIG. 184.   Schematic representation of prismatic loops in graphite or in a hexagonal metal with small stacking fault energy. (a) Vacancy loop; the Burgers vector is inclined with respect to the c plane. (b) Interstitial loop; the Burgers vector is perpendicular to the c plane.

It was found that in graphite quench loops tend to align along dislocation ribbons.[322] This was attributed to an interaction between vacancies and stacking faults, somewhat related to the Suzuki interaction. Figure 137a shows schematically the geometry of a ribbon; from this it is clear that low-energy vacancy loops can form at once in the fault region of a ribbon; no shear parallel to the basal plane is

[322] P. Delavignette and S. Amelinckx, *J. Appl. Phys.* **33**, 554 (1961).

required. Evidently this means that energy is gained if the loop forms
within the ribbon. Such loops are very effective in pinning and they
may well be responsible, at least partly, for quench-hardening.

Williamson[320,323] examined graphite single crystals irradiated at
about 300°C in a reactor. He found large loops which had a Burgers

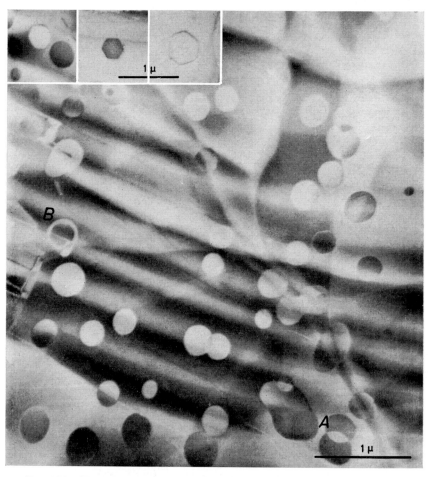

Fig. 185. Vacancy loops in graphite observed in stacking fault contrast. Notice that
the contrast changes where two loops overlap in *A*. Notice also concentric loops in *B*.

[323] G. K. Williamson, "Properties of Reactor Materials and the Effects of Radiation
Damage" (D. J. Littler, ed.), p. 144. Butterworths, London, 1962.

vector perpendicular to the c plane. They were attributed to interstitials formed by fast neutron irradiation.

(e) Alloys. Quenching experiments also were performed on alloy crystals. Mader[257] quenched alloys of different stacking fault energy in the system nickel-cobalt. In the alloys with a small stacking fault energy he found tetrahedra as in gold, while in the alloys with a large stacking fault energy loops were observed. In the intermediate range both loops and tetrahedra were present.

On quenching Al-4% Cu alloys, Thomas and Whelan[324] found that apart from limited loop formation most vacancies were absorbed by existing dislocations. The screw dislocations were induced to climb into a helical shape.[137] Vacancies prefer to condense on existing dislocations, rather than to nucleate loops. In the presence of impurities a coupling apparently takes place between vacancies and impurities and this prevents the formation of a sufficient supersaturation of vacancies to nucleate a loop.

Recently Westmacott et al.[325a] have demonstrated that on quenching from the same temperature and at the same speed, the number and size of the vacancy loops depend very sensitively on the kind of impurity present. The effect demonstrates convincingly that vacancies interact with impurities and presumably facilitate the loop nucleation, since a large number of smaller loops are observed in the aluminium containing the impurity for which the interaction is supposed to be largest. The total concentration of vacancies C_v' present in thermal equilibrium in the impurity containing aluminum is increased with respect to the concentration C_v in pure aluminum:

$$C_v' = C_v \left[1 - 12c + 12c \exp \left(E_b/kT\right)\right] \qquad (10.80)$$

where c is the atomic concentration of solute atoms, and E_b the binding energy between impurity and vacancy.[325b] With a binding energy of 0.2 ev and a value of $c = 5\%$, C_v'/C_v may be as large as 10 at 550°C. Ruedl[326a] observed independently that quenched aluminum containing a small amount of lithium exhibits a larger number of loops than pure aluminum quenched under the same circumstances.

[324] G. Thomas and M. J. Whelan, Phil. Mag. [8] 4, 511 (1959).

[325a] K. H. Westmacott, R. S. Barnes, D. Hull, and R. E. Smallmann, Phil. Mag. 6, 929 (1961).

[325b] M. Lomer, J.I.M. Symp. on Vacancies and Point Defects in Metals and Alloys (1957).

[326a] E. Ruedl, private communication, 1961.

Embury *et al.*[326b] have stressed the importance of deformation during quenching on the distribution of loops. In specimens of pure aluminium deformed about 2% during quenching the loops are inhomogeneously distributed as a consequence of the fact that certain regions have been swept free of vacancies by the moving dislocations. In an aluminium − 7% magnesium alloy quenching results in the formation of rows of loops. This is explained as resulting from the breaking away of a dislocation from its atmosphere of adsorbed vacancies. Two different types of quench behavior are distinguished. In metals like pure aluminum the vacancies are absorbed by existing dislocations, which become tangled as a consequence. Loops are formed in dislocation free regions. In the aluminium-magnesium alloys on the other hand there is an interaction between impurities and vacancies and the vacancies rather precipitate at dislocations which coil up into helices. Breaking away of the dislocations then results in the formation of rows of loops.

(f) Diamond. Evans and Phaal[326c] investigated type I and type II diamonds. They found that type I diamonds contain numerous platelets of nitrogen on cube planes as well as vacancy loops on (111) planes. The loops result from vacancies that are present in thermal equilibrium at the growth temperature of the diamond ($\pm4000 − 4500°$C). Presumably they are associated with the nitrogen impurities. On cooling the nitrogen precipitates and the vacancies are liberated causing a local supersaturation and a consequent formation of vacancy loops. In type II diamond neither nitrogen platelets nor vacancy loops are found; but a larger concentration of dislocations seems to be present.

(3) *Loops formed by irradiation.* (a) Face-centered metals. Loops similar in many respects to those found in quenched metals were first observed in neutron-irradiated copper (dose: 10^{17} nvt to 10^{20} nvt).[327] They have since been observed in other metals,[328,329a] in particular in platinum.[215] In general the loops are too small for determination of their nature by the procedure outlined in Section 9a(5)(6b).

[326b] J. P. Embury, C. M. Sargent, and R. B. Nicholson, *Acta Met.* 10, 1121 (1962).

[326c] T. Evans and C. Phaal, *Proc. 5th Intern. Conf. on Electron Microscopy, Philadelphia, 1962* 1, B-4 (1962); see also T. Evans and C. Phaal, *Proc. Roy. Soc.* A270, 538 (1962).

[327] J. Silcox and P. B. Hirsch, *Phil. Mag.* [8] 4, 1356 (1959).

[328] M. Makin, A. Whapham, and F. Minter, *Phil. Mag.* [8] 7, 285 (1962).

[329a] D. Pashley and A. Presland, *Phil. Mag.* [8] 6, 1003 (1961).

It is worth noting a characteristic difference between interstitial and vacancy loops in the face-centered cubic structure.[329b] A loop resulting from the collapse of 1-layer vacancy disks contains a stacking fault which can be eliminated by the passage of 1 partial; the shear stress available to nucleate such a partial is γ/b. On the other hand, a loop resulting from the agglomeration of 1 layer of interstitials contains a stacking fault which can only be eliminated by the passage of 2 partials on glide planes on each side of the loop plane. The shear stress available for the nucleation of the first partial is only $\Delta\gamma/b$ where $\Delta\gamma$ is the decrease in stacking fault energy after 1 partial has passed. The remaining part of the stacking fault energy can be used to nucleate the second partial. For suitable values of γ and $\Delta\gamma$ it may well be that faults will be eliminated from vacancy loops, but not from interstitial loops. In such a material the distinction between vacancy and interstitial loops would be immediate. It is possible that platinum is such a material.

Mazey and Barnes[329c] have made an elegant attempt to distinguish vacancy and interstitial loops. They bombarded copper with α-particles from a cyclotron. These particles create radiation damage but at the same time helium gas is injected into the copper. On annealing at 350-400°C small cavities ($\sim$ 1000 A in diameter) which are under a large helium pressure are formed. In order to grow, the cavities require vacancies, which are supplied by the vacancy loops. The vacancy loops are therefore suppressed or shrink, while interstitial loops grow. It is found that in the foils as irradiated two kinds of loops are present, small dots and larger resolved loops. The first disappear in the helium containing foils, while the latter grow. Therefore it is concluded that the larger loops are due to *interstitials* and by contrast the small dots should be vacancy clusters.

This conclusion is different from the one reached by Silcox and Hirsch[327] who attributed the large loops to vacancies, which agglomerate in loops by short-range migration in a displacement spike. The small dots would be interstitial clusters. The interstitials would be shot out and leave the displacement cascade by focusing collisions.

Irradiation effects in platinum have been studied in detail by Ruedl et al.[215] Figure 186a represents a foil immediately after irradia-

[329b] E. Ruedl, P. Delavignette, and S. Amelinckx, *Proc. I.A.E.A. Symp. on Radiation Damage in Solids and Reactors Materials, Venice, 1962*, Vol. 1, p. 363. Int. Atomic Energy Agency, Vienna, 1962.

[329c] R. S. Barnes and D. Mazey, *Phil. Mag.* [8] **5**, 1247 (1960).

tion with α-particles, whereas Fig. 186b shows loops after an anneal. Loops similar in appearance to those found in copper have been observed in fission fragment and α-particle irradiated foils. The loops were found to be due to vacancies as well as to interstitials. In the course of this investigation it was found that some boundaries, especially coherent twin boundaries, become decorated with small dots, which disappear on annealing. This phenomenon was attributed to preferential damage formation at these boundaries as a consequence of focusing collisions or to channeling along $\langle 110 \rangle$ directions. The coherent twin would act as an effective barrier for focusons, without being a sink for the defects. Also evidence was obtained for the interaction of point defects and dislocation lines; the latter acquire a ragged shape in neutron-irradiated specimens, even when no visible loops or dots have been formed. Within the grains loops tend to align; this is attributed to the passage of fission fragments causing several primary knock ons, and creating centers for the nucleation of loops.

Pashley and Presland[329a] discovered that evaporated gold films examined in the electron microscope become dotted after some period of irradiation. They could show that the phenomenon is due to the formation of agglomerates of point defects as a consequence of negative ion bombardment in the microscope. The ions originate from the filament; the use of oxide-coated filaments gives rise to enhanced damage formation. The spots appear to anneal out at about 300°C; the damage produced when the foil is maintained above that temperature consists of tetrahedra.

Recently Hirsch et al.[330] reported some interesting experiments. They irradiated quenched gold foils containing tetrahedra by means of α-particles and found that the concentration of tetrahedra decreases. This experiment supports the idea that on irradiation interstitials build up a sufficient concentration to make the tetrahedra collapse. As the tetrahedra anneal out the concentration of black spots increases.

In proton-irradiated copper they find loops in the (111) plane, containing a stacking fault of the extrinsic type. This was shown using the technique described in Section 9h(6) and Appendix A. This observation proves that the loops are due to interstitials. On the other hand it is known that quenching introduces fault-free loops;

[330] P. B. Hirsch, R. M. J. Cotterill, and M. W. Jones, *Proc. 5th Intern. Conf. on Electron Microscopy, Philadelphia, 1962* 1, F-3 (1962).

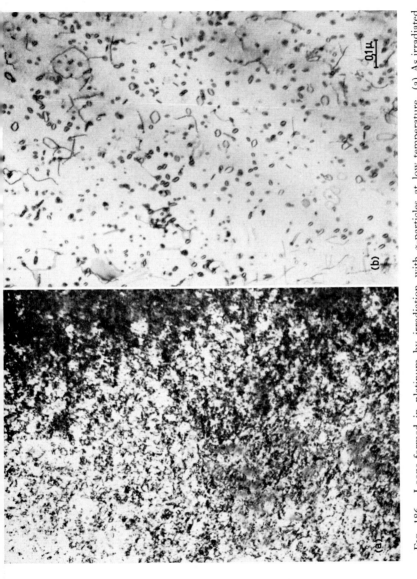

FIG. 186. Loops formed in platinum by irradiation with α-particles at low temperature. (a) As irradiated. (b) After anneal at 500°C during 1 hr. For the larger loops the character was determined; there are both vacancy and interstitial loops.

this experiment can therefore be interpreted as evidence for the idea put forward in the beginning of this paragraph.[329b]

Westmacott *et al.*[331a] have studied the formation of loops in zone refined aluminum under bombardment by fission fragments at an instantaneous flux of 10^{16} fission fragments per cm^2 per sec. It is found that as well interstitial as vacancy loops are formed. The authors suggest that both kinds of loops can grow because each loop attracts only defects of its own kind. Similar to what is observed in platinum[224c] the loops migrate as a whole by the combined effect of conservative climb and glide. Since loops of opposite sign lying on the same glide cylinder attract each other, mutual anihilation by glide only is possible. These observations therefore suggest a new annealing phenomenon: the mutual annihilation of vacancy loops of opposite sign by conservative glide and climb. If the annihilating loops are not of the same size loops having very complicated shapes result; for instance large loops with a number of "punched through" holes in the interior.

The loops were found to lie mainly in the (110) planes perpendicular to their Burgers vector, which is of the type $(a/2)$ [110], as deduced from extinctions.

Mazey, Barnes, and Howie[331b] found that aluminium bombarded at 250°C with 38-Mev α-particles contains *interstitial* loops with an $(a/2)$ [110] Burgers vector normal to their plane. It should be noted that such loops are only seen in foils which contain helium gas, in accord with the earlier assumption of Mazey and Barnes.[329c] The character of the loops was deduced from contrast experiments.

(b) Hexagonal metals. Price[332a] has studied the formation and growth of large loops in the basal plane of perfect cadmium platelets grown by sublimation. The loops are attributed to point defects, presumably vacancies, introduced by the ion bombardment in the microscope. All the loops have an inclined Burgers vector and contain a stacking fault. On growing the loops form networks in the basal plane; the networks are of an interesting kind since all dislocations are partials. The interaction between such loops will be discussed further [Section 10d(5)(b)].

[331a] K. H. Westmacott, A. C. Roberts, and R. S. Barnes, *Phil. Mag.* [8] **7**, 2035 (1962).

[331b] D. J. Mazey, R. S. Barnes, and A. Howie, *Phil. Mag.* [8] **7**, 1861 (1962).

[332a] P. B. Price, *Phys. Rev. Letters* **6**, 615 (1961); "Electron Microscopy and Strength of Crystals" (G. Thomas and J. Washburn, eds.), p. 41. Wiley (Interscience), New York, 1963.

A similar study was made by Crump and Mitchell.[332b] Attention was focussed on the sudden disappearance of the loops after heaving reached a certain size. This phenomenon was noticed earlier in zinc by Berghezan *et al.*[309] However Crump and Mitchell noticed that after the loops disappeared an hexagonal array of linear imperfections was often left behind. The lines are interpreted as narrow ribbons of stacking faults. The stability of the hexagonal arrays of these ribbons remains a problem. Contrary to Price[332a] the authors believe that the vacancy loops do not result from ion bombardment in the microscope but from the agglomeration of vacancies incorporated in the thin crystals during growth from the vapor.

(c) Body-centered metals. Neutron damage in iron, in the form of small black dots, has been observed after integrated fluxes of 1×10^{19} n/cm^2.[332c] The movement of dislocations was found to be hampered by defect clusters already at 5×10^{17} n/cm^2, but clearly resolved clusters are only observed at the higher dose.

(d) Uranium. The "growth" of α-uranium is still not completely understood. If an α-uranium single crystal is irradiated it is found that its dimensions change. It lengthens in the b [010] direction and shortens in the a [100] direction, while the c [001] direction remains unchanged. These changes cannot be recovered by annealing. It is clear that this process implies the transport of material from a (100) planes onto b (010) planes. Since the growth rate is enhanced at low temperature (measurements have been made down to liquid hydrogen temperature)[332d] a diffusion mechanism is excluded. Transport by means of focusing collisions and deposition of defects along the boundaries has been proposed by Gonzer[333a].

Some evidence that such a mechanism is not impossible was obtained in the case of platinum as shown above. More recently it has been proposed that vacancies and interstitials produced on irradiation would precipitate and form loops in different crystallographic planes. Electron microscopic observations by Hudson *et al.*[333b] have revealed the presence of lines of dots in α-uranium specimens

332b J. C. Crump, III, and J. W. Mitchell, *Phil. Mag.* [8] **8**, 59 (1963).

332c B. L. Eyre, *Phil. Mag.* [8] **7**, 2107 (1962).

332d Y. Queré and J. Doulat, *Compt. Rend.* **252**, 1305 (1961); see also S. N. Buckley, UKAEA Report AERE-R3674 (1961).

333a U. Gonzer, private communication, 1961.

333b B. Hudson, K. H. Westmacott, and M. J. Makin, *Phil. Mag.* [8] **7**, 377 (1962).

irradiated with 1×10^{17} n/cm². After 3×10^{17} n/cm² defects become resolvable as loops, not containing stacking faults. The size of the loops appeared to be larger near grain boundaries than in the interior of the grains. The loops are arranged in rows and are often coplanar. (See also Ruedl and Amelinckx.[305b])

Foils were also irradiated at $-195°C$ and at $300°C$. For the specimens irradiated at lower temperature the average loop size is smaller than at 80°, for the same dose. In addition there is no evidence for alignment. The higher temperature irradiation on the other hand resulted in the formation of larger loops aligned in rows; a denuded zone seems to be formed along the grain boundaries. Annealing studies indicate that loops grow at around 300°C.

The loop planes were determined by making the assumption that the loop shape is circular. From the projected shape and from the indices of the foil plane, the loop plane can then be determined. A pronounced preference for (110) and (023) planes is found. The authors have further discussed plausible models for the loops which could lead to the observed macroscopic dimensional changes. It remains however to be shown experimentally that the loops have the Burgers vectors and the character (interstitial or vacancy) required by the model.

(e) Graphite. Radiation damage in graphite has been studied by Bollmann,[334,335a] by Williamson,[323] and by the author.[335b] The most detailed work was performed by the first author who used a combination of dark field transmission electron microscopy and moiré patterns.

He proposed that neutron irradiated graphite contains "holes" and interlamellar clusters of interstitials. The kinematical theory of diffraction is applied to calculate the contrast effects due to both types of defects. It is concluded that holes are observed as black dots in the dark field image formed by the $(10\bar{1}0)$ reflection, while bright dots would represent interstitial clusters. On annealing above the Wigner energy release peak ($\sim 200°C$) the holes would tend to close, occasionally leaving "dislocation dipoles." It should be noted that the dislocation dipoles pictured by Bollmann consist in fact of a short row of missing (or supplementary) atoms in a single graphite plane. They

[334] W. Bollmann, *Phil. Mag.* [8] **5**, 621 (1960).
[335a] J. Bollmann, *J. Appl. Phys.* **32**, 869 (1961).
[335b] Unpublished work, in collaboration with P. Delavignette, 1960.

are the equivalent of a vacancy (or interstitial) loop in a three-dimensional crystal. Like glissile vacancy loops the graphite dipoles can move in a conservative way along certain lattice rows in the sheet. Glide movements of this type would release energy in a continuous fashion in the temperature range 200–1000°C. The annihilation of dipoles by climb would be responsible for the high-temperature annealing peak at 1300°C. The evidence for the existence of the dipoles is based on a comparison between the calculated image for a "dipole" which appears to consist of a pair of black and white dots, and the observed contrast effects. Moiré patterns give a more direct representation of the lattice deformation around the dipoles than does the diffraction contrast.

Williamson[323] found large interstitial loops in graphite irradiated at a temperature above the Wigner release peak. The interpretation of his results does not agree with that of Bollmann.

(f) Magnesium oxide. Groves[335c] has examined damage in magnesium oxide irradiated by means of neutrons up to a dose of 4.3×10^{19} n/cm^2. After annealing at 1100°C the damage anneals partially and resolvable prismatic loops in (110) planes are formed.

(4) *Dislocation loops due to deviations from stoichiometry.* In some cases it appears that deviations from stoichiometry can be accommodated in a crystal by means of dislocation loops. This has been suggested for antimonium telluride (Sb$_2$Te$_3$).[336] This intermetallic compound has a typical layer structure consisting of a cubic close packing of two kinds of atoms. It can be represented by the symbol ... $a\beta c a b c a b\gamma a b\gamma a\beta c$ where the Greek letters represent antimony and the italic letters tellurium: Normally it crystallizes with an excess antimony.[337] On the other hand, electron microscopic observation reveals that slowly cooled crystals invariably contain large prismatic dislocation loops lying in the c plane, which is the layer plane (Fig. 187). The size of these loops is the larger the slower the crystals are cooled; in quenched crystals no loops are observed. Evidently this points to a precipitation phenomenon. Therefore it was proposed that the stoichiometric excess can be taken up in the structure either as interstitial loops of antimony or as vacancy loops of tellurium.

[335c] G. W. Groves, *Proc. 5th Intern. Conf. on Electron Microscopy, Philadelphia, 1962* 1, F-10 (1962).

[336] P. Delavignette and S. Amelinckx, *Phil. Mag.* [8] 6, 601 (1961).

[337] G. Offergeld and J. Van Caeckenberghe, *Phys. Chem. Solids* 11, 310 (1959).

The latter hypothesis is preferred on energetic grounds. It turns out that the loops do not contain a geometrical stacking fault since stacking fault contrast is absent. This implies that a Shockley partial has to be nucleated in order to remove the stacking fault which results from the precipitation of either a layer of antimony inter-

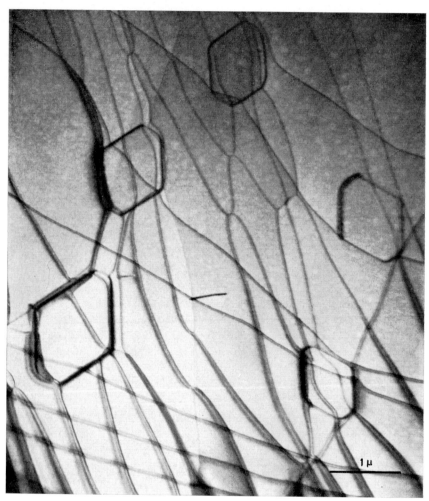

1 μ

FIG. 187.   Large dislocation loops in the c plane of antimonium telluride. The loops are presumably due to tellurium vacancies introduced as a consequence of deviations from exact stoichiometry. [After P. Delavignette and S. Amelinckx, *Phil. Mag.* [8] **6,** 601 (1961).]

stitials or a single layer of tellurium vacancies. The loops still contain a chemical fault, which is however invisible since the diffracting power of antimony and tellurium is practically equal. The contrast effects are in agreement with an inclined Burgers vector. The elongated hexagonal shape of the loops also suggests an inclined Burgers vector, the 2 pure edge segments would tend to be shorter than the mixed segments since this reduces the total energy associated with the loop. In the cubic close-packed structure the geometrical stacking fault in a single layer vacancy loop can be removed by the passage of a single Shockley partial and the full stacking fault energy is available to nucleate this partial. On the other hand the stacking fault in a single layer interstitial loop can only be removed by the passage of 2 partials, and for nucleating the first partial only a fraction of the stacking fault energy is available [Section 10d(3)(a)]. This argument was used to support the hypothesis that the loops were due to the condensation of tellurium vacancies rather than antimony interstitials.

Most loops occur as pairs elongated in directions differing by 60°. This can be understood by observing that the probability for loop nucleation is larger inside an existing loop.[338a] The Burgers vector of a Shockley partial sweeping the second loop will preferentially be at 120° of the Burgers vector of the first one since then the attraction helps the nucleation.

More recent work by Bierly[338b] has shown that the loops occur also in binary alloys of up to 60% Bi_2Te_3 − 40% Sb_2Te_3, but not in the range going from this composition up to pure Bi_2Te_3. Bierly has confirmed the explanation given above.

(5) *The formation of loops during deformation.* It was suggested by Seitz[339] in 1952, that plastic deformation would result in the formation of points defects either interstitials or vacancies. The original idea was inspired by experiments done by Gyulai and Hartley[340] on the increase of the ionic conductivity after plastic deformation and experiments on the increase in residual electrical resistivity of metals after

---

[338a] F. C. Frank, *in* "Deformation and Flow of Solids," p. 73. Colloq., Madrid, 1955 (R. Grammel, ed.). Springer, Berlin, 1956.

[338b] J. N. Bierly, *Proc. 5th Intern. Conf. on Electron Microscopy, Philadelphia, 1962* 1, J-14 (1962).

[339] F. Seitz, *Advances in Phys. (Phil. Mag. Suppl.)* 1, 43 (1953).

[340] Z. Gyulai and D. Hartly, *Z. Physik* 51, 378 (1928).

cold work.[341] The idea has been extremely fruitful in explaining numerous experimental results. By direct observation methods evidence has been accumulated in recent years that indeed "something" is left behind by moving dislocations. This "something" is observed as dislocation loops, but it is not always clear whether these result from the condensation of point defects or are formed as such. Different mechanisms have been proposed[49,145,146,148,342] and it seems from electron optical evidence that some may be operative either separately or simultaneously. We will discuss some of these proposals and summarize the evidence for each of them.

(a) Magnesium oxide.   Crystals deformed at room temperature were subsequently thinned chemically and then studied[145,224a] The most striking feature was the presence of numerous dislocation dipoles which broke up into a number of small loops on heating; these could be made to grow further on continued heating.

The generation of the dislocation dipoles was attributed to the double cross slip of the long screw segments giving rise to large jogs which were left behind. Since such jogs cannot move by glide, long narrow cusps, consisting of "dipoles," are formed. If the screws cross slip back to the original glide plane a long narrow loop is pinched off in the manner shown in Fig. 188b. If the "jog" becomes of macroscopic dimensions the two parts of the cusp which consist of dislocations of opposite sign can eventually pass each other, swing around the jog as a pivot point, and give rise to dislocation multiplication (Fig. 188c).

The two mechanisms described here were suggested earlier under a somewhat different form, to explain Gilman's experiments on LiF.[148] The intersection of screws could eventually produce rather similar features. As pointed out by Cottrell[343] the intersection of screws moving under the same applied shear stress mainly results in the formation of interstitials. The double cross slip mechanism may produce loops of either kind depending only on the sign of the jog. A determination of the nature of the loops could therefore allow differentiation.

[311] J. Molenaar and W. H. Aarts, *Nature* **166**, 690 (1950); M. J. Druyvestein and J. A. Manintveld, *ibid.* **168**, 868 (1951).

[342] J. T. Fourie and H. G. Wilsdorf, *J. Appl. Phys.* **31**, 2219 (1960).

[343] A. H. Cottrell, *in* "Dislocations and Mechanical Properties of Crystals" (J. C. Fisher *et al.*, eds.), p. 509. Wiley, New York, 1957.

Groves and Kelly[224a] succeeded in doing this, using the method described in Section 9e(5)(b). They found only evidence for vacancy loops.[224b]

(b) Zinc. (i) Formation of loops. Price,[146,147] using dislocation iree platelets of zinc, grown by sublimation, and deforming these inside the microscope, found direct evidence for the formation of loops in the wake of a moving screw. Figure 189 reproduces one of his

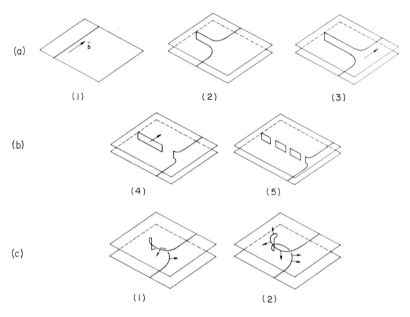

FIG. 188. Formation of defects by moving screws. (a) Long dipoles left in the wake of a jog in a moving screw. (b) The dipole splits up in loops. (c) Dislocation multiplication around a jog as a pivot point.

observations. The loops lie in the basal plane and contain no stacking faults. They are formed by screws with a $[\bar{1}\bar{1}23]$ glide vector and a $(11\bar{2}2)$ glide plane, which double cross slip. According to Price two types of loops, which differ by their contrast effects, are formed in this process. One type would have a $\frac{1}{3}[\bar{1}\bar{1}23]$ Burgers vector, which is the same as that of the generating screw, the other type would have an a [0001] Burgers vector. This second type would result from the

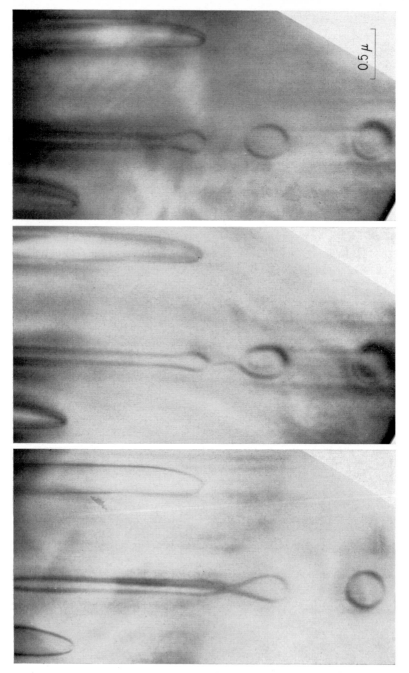

0.5 μ

Fig. 189. Formation of dislocation loops in deformed zinc platelets. (Courtesy of Price.) (a) Formation of loops in deformed zinc platelets by pinching off of dipoles.

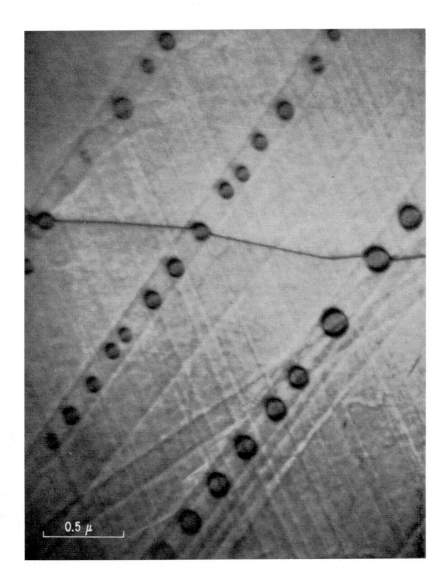

FIG. 189.   (b) Long row of loops left in the wake of moving screws in the basal plane of a zinc platelet. (Courtesy Price.)

first type by the reaction:

$$\frac{1}{3}[\bar{1}\bar{1}23] \rightarrow \frac{1}{3}[\bar{1}\bar{1}20] + [0001]. \qquad (10.81)$$

The first component of the left-hand side is a glissile loop and it disappears by glide along the c plane.

A rough estimate allows us to see that there is a driving force for the splitting of a dislocation dipole into loops; provided the loops exceed a certain dimension.

Loops formed in zinc by cold work (rolling) and fatigue have been studied extensively by Berghezan *et al.*[309] These loops contain a stacking fault. Most of the loops were found to grow in the electron microscope, some shrank. The growth may possibly be due to ion irradiation as in cadmium. These loops, which have Burgers vectors of the type AS, BS, ... (Fig. 191) give rise to typical reactions which we will now discuss with reference to Fig. 190.

(ii) Interaction between loops. The components of the Burgers vectors perpendicular to the basal plane will be omitted in the drawings since they are always the same in magnitude: $c/2$. The segments of loops that impinge in the same or adjacent lattice planes always have opposite components. The sign reversal of the Burgers vector left and right of the loop is indicated by automatic reversing in lettering order.

Four different patterns can result from the reaction between 2 loops.[309,344]

(1) The 2 loops have the same Burgers vector. On meeting the 2 loops fuse and form a single one (Fig. 190a).

(2) If the basal components of the Burgers vectors enclose an angle of 120° the following reaction will occur

$$SA + BS \rightarrow S\sigma + \sigma A + B\sigma + \sigma S \rightarrow B\sigma + \sigma A.$$

The loops are now connected by 2 partials, separated by a stacking fault which has the same energy as the stacking fault within the loops. The partials can therefore become widely dissociated (Fig. 190b).

[344] P. B. Price, in "Electron Microscopy and Strength of Crystals" (G. Thomas and J. Washburn, eds.), p. 41. Wiley (Interscience), New York, 1963.

(3) If the basal components enclose an angle of 60° a single partial results from the reaction (Fig. 190c).

$$SA + TC \rightarrow S\sigma + \sigma A + T\sigma + \sigma C \rightarrow B\sigma.$$

(4) If the basal components are antiparallel the reaction leads to the formation of 3 partials. The reaction is now (Fig. 190d)

$$SA + TA \rightarrow S\sigma + \sigma A + T\sigma + \sigma A \rightarrow B\sigma + \sigma A + C\sigma.$$

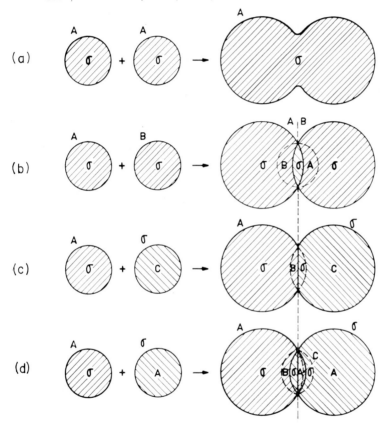

Fig. 190.   Interactions between prismatic loops in the basal plane of zinc. (a) The 2 loops have the same Burgers vector; reaction results in the formation of a large loop. (b) Two partials of the Shockley type $B\sigma$ and σA are formed. (c) A single partial $B\sigma$ separates the 2 regions. (d) The contact consists of 3 repelling partials $B\sigma$, σA, and $C\sigma$. [After A. Berghezan, A. Fourdeux, and S. Amelinckx, *Acta Met.* **9**, 464 (1960); P. B. Price, *in* "Electron Microscopy and Strength of Crystals" (G. Thomas and J. Washburn, eds.), p. 41. Wiley (Interscience), New York, 1963.]

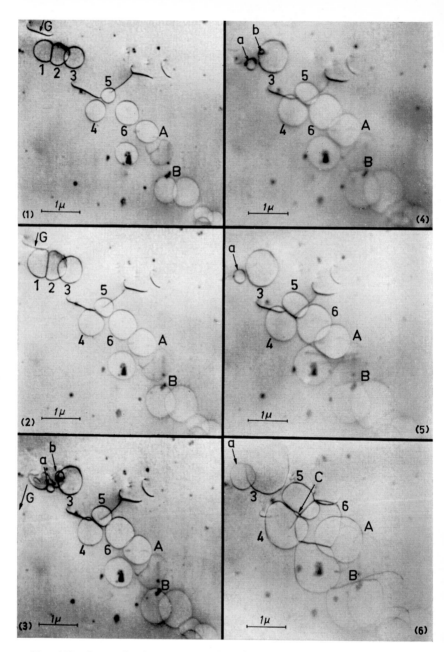

FIG. 191. Interaction between a number of prismatic loops in the basal plane of zinc. [Courtesy of A. Berghezan, A. Fourdeux, and S. Amelinckx, *Acta Met.* **9**, 464 (1960).]

The energy of the dislocations at the left-hand side is proportional to $2a^2$ while the corresponding energy at the right is only $3 \times 1/3\, a^2 = a^2$. More complex patterns result from the interaction between several loops. Numerous examples have been observed; some are reproduced in Fig. 191.

There is also a strong interaction between the sessile loops and glissile basal dislocations. This interaction is probably responsible for a large fraction of the work hardening during basal glide. The possible interactions between loops and glide dislocations are represented in Fig. 192. In (a) the arriving dislocation is repelled by the loop, while in (b) and (c) a reaction takes place which pins the dislocation. In the last example the loop is swept by a partial with vector σB, which changes the stacking inside the loop. Examples of such interactions were also observed in zinc.

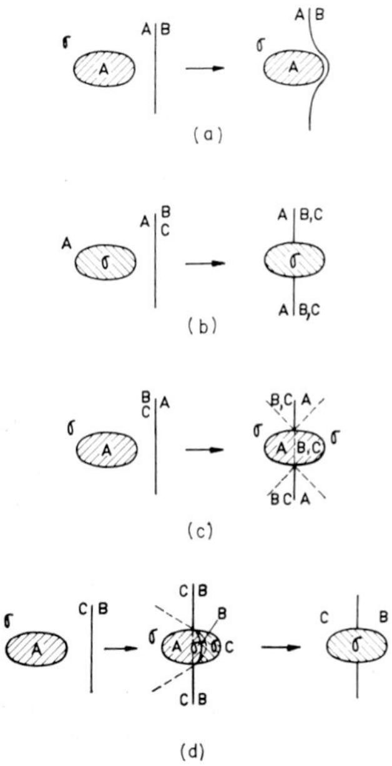

FIG. 192.   Interaction between sessile loops in the basal plane of zinc and glissile dislocations. (a) The arriving glissile dislocation is repelled by the loop. (b) There is attraction between the loop and the glissile dislocation. (c) There is again attraction. (d) The loop is swept by a partial and reaction takes place. [After A. Berghezan, A. Fourdeux, and S. Amelinckx, *Acta Met.* 9, 464 (1960).]

(c) Face-centered and body-centered metals. In all deformed body-centered metals elongated loops are formed.[266,345a] On anneal they break up in smaller rounded loops and they finally disappear by climb. The nature of these loops, either vacancy or interstitial, has not yet been determined.

The formation of dislocation loops in deformed copper has been studied by Ruff.[345b] The concentration of loops seems to increase, as a consequence of annealing at room temperature, above that present immediately after deformation. This observation clearly suggests that isolated point defects are present in high supersaturation immediately after deformation.

(d) Jog formation. We will now discuss a few possible mechanisms that give rise to the jogs required to explain defect formation and the formation of tangles in cold-worked metals.

(1) The intersection of 2 dislocations produces in each dislocation a jog given by the Burgers vector of the other. In particular a moving screw intersecting a stationary screw produces a jog in the moving screw which cannot move conservatively. This mechanism produces only small jogs, unless a large number of screws of the same sign should be intersected. The movement of such a jog would produce isolated point defects, which may of course agglomerate to small loops by diffusion.

(2) Larger jogs may form by the double cross slip mechanism, visualized in Fig. 193a. The jogs of intermediate size are probably responsible for the formation of edge dipoles in the manner shown in Fig. 188a. These dipoles can be pinched off by local diffusion and give rise to loops with a gain in energy. The very large jogs can act as pivot points for spiral sources of dislocations as shown in Fig. 188c. The jog height has to be sufficiently high to allow the 2 spiral arms to pass one another.

Large jogs can also occur in moving edge dislocations in the way sketched in Fig. 193b. In (2) a dipole of screws is left behind around an obstacle. The screw segments cross slip and a prismatic loop is pinched off, while the edge is left with 2 large jogs (4).

[345a] A. Fourdeux and A. Berghezan. Symposium on the Role of Substructure on the Mechanical Behavior of Metals. Air Force Techn. Rept. No. ASD-TDR-63-324 April 1963, p. 437.

[345b] A. W. Ruff, *Proc. 5th Intern. Conf. on Electron Microscopy, Philadelphia, 1962* 1, J-10 (1962).

(3) Jogs of almost any size can result from a combination of crossing of two dislocations and cross glide.[346,347] This mechanism is illustrated in Fig. 194 for the case of ribbons, but it *a fortiori* works for undissociated dislocations. Two ribbons with the same Burgers vector cross each other in neighboring planes. They adopt thereby the antiparallel orientation, and form dipoles. The repulsion between partials is thereby reduced, the stacking fault now pulls the partials together, generating in this way a dipole of perfect dislocations. This dipole can annihilate by cross glide if it adopts the screw orientation. The resulting situation is shown in Fig. 194c,d. The two ribbons contain each a dissociated jog. Examples of extended jogs generated in this way were

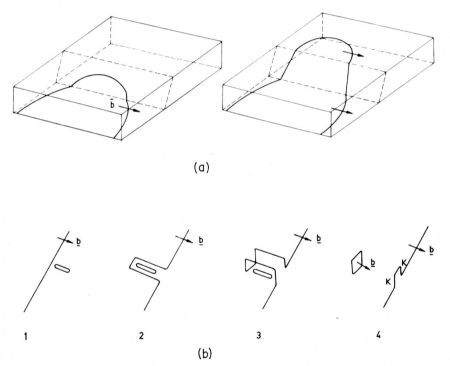

(a)

(b)

FIG. 193. (a) Schematic representation of the double cross slip mechanism. (b) Formation of large jogs in edges as a consequence of avoiding an obstacle and prismatic loops. [After P. B. Price, in "Electron Microscopy and Strength of Crystals" (G. Thomas and J. Washburn, eds.), p. 41. Wiley (Interscience), New York, 1963.

[346] A. S. Tetelman. *Acta Met.* 10, 831 (1962).
[347] R. Siems, P. Delavignette and S. Amelinckx, *Phil. Mag* (in press).

found in tin disulfide single crystal flakes (Fig. 195). From the equilibrium shape of such a jog the specific energy of a jog line could be derived. A jog in a partial in SnS$_2$ can move conservatively but then produces a stacking fault loop on an inclined plane. Evidence has been found for this process.[347]

(4) Recently Seeger and Mader[348] proposed that high jogs can be generated by the action of Frank-Read sources of the conical type. The agglomeration of elementary jogs, created in each successive loop, would give rise to jogs of any size.

(6) *The annealing of prismatic loops.* On heating, the prismatic loops due to quenching, irradiation, or deformation can be annealed out. They shrink as a consequence of climb. A quantitative study of

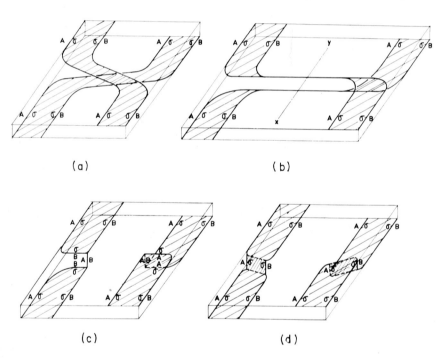

(a)   (b)

(c)   (d)

Fig. 194. Mechanism giving rise to jogs of large size in ribbons. (a) and (b) The two ribbons are in the antiparallel orientation. (c) and (d) They cross slip and large extended jogs result.

[348] A. Seeger and S. Mader, *Phys. Stat. Solidi* 1, 78 (1961).

the kinetics of the process allows a determination of the activation energy.[349]

When applying Friedel's theory of climb,[350] the equation for the rate of shrinking of a dislocation loop under the action of its line tension

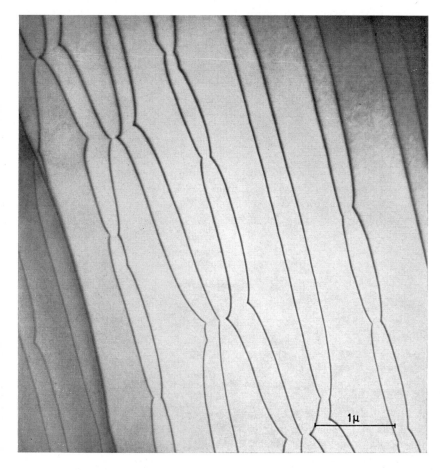

Fig. 195.   Extended jogs in ribbons in tin disulfide probably resulting from the process described in Fig. 194.

[349] J. Silcox and M. J. Whelan, in "Structure and Properties of Thin Films" (C. A. Neugebauer et al., eds.), p. 164. Wiley, New York, 1959.

[350] J. Friedel, "Les dislocations," p. 72. Gauthier-Villars, Paris, 1956.

as the driving force, is given to a good approximation by

$$\frac{dr}{dt} = - \left(\frac{1}{2}\right) z v_a b^2 \frac{\alpha}{r} \exp\left(- U/kT\right) \tag{10.82}$$

where z = atomic coordination factor (~ 11 in fcc crystals)
$\quad v_a$ = atomic vibration frequency ($\sim 10^{13}$)
$\quad b$ = Burgers vector
$\quad U$ = activation energy for self-diffusion.

The constant α is defined by the relation

$$\alpha b/r = \exp\left(F_c b^2/kT\right) - 1 \tag{10.83}$$

where

$$F_c = \frac{\mu b}{4\pi(1 - \nu)} \left[\ln(r/b)/(r/b)\right] \tag{10.84}$$

is the mechanical stress due to line tension and curvature of the loop. In aluminum $\alpha \sim 72$ at $161°C$ and $\alpha \sim 64$ at $200°C$.

In this calculation it is assumed that in the thin foil the surface is a very good sink for point defects, so that no supersaturation of vacancies builds up; in other words the chemical stress is neglected. Further the concentration of jogs was put equal to $1/2$, which is reasonable for a circular loop.

Integration of Eq. (10.82) leads to the relation

$$r = r_0(1 - t/\tau)^{1/2} \tag{10.85}$$

describing the radius as a function of time; (r_0 is the initial radius) and τ is determined by the relations

$$\tau = \tau_0 \exp\left(U/kT\right); \tau_0 = r_0^2/z v_a b^2 \alpha. \tag{10.86}$$

A plot of the radius of a prismatic loop in aluminum as a function of time at $197°C$ is shown in Fig. 196. The curve is in accord with relation (10.85); one finds $U = 1.29$ ev which can be compared with the values ranging from 1.2 to 1.4 ev found by other methods.

Vandervoort and Washburn[351] examined aluminum foils quenched from $600°C$, subsequently heat-treated or cold-worked and finally electrochemically thinned. They find that the quench loops begin to disappear on heating for 10 min to $128°C$; they are completely gone

[351] R. Vandervoort and J. Washburn, *Phil. Mag.* [8] **5**, 24 (1960).

after 10 min at 200°C. Therefore they associate the elimination of the quenched-in loops with the second recovery stage of the electrical resistivity observed by Panseri and Federighi in quenched aluminum.[352] The elimination of loops proceeds in a nonuniform way; the larger loops grow at the expense of the smaller ones. A deformation of 5% by rolling eliminates all of the loops probably because the glide

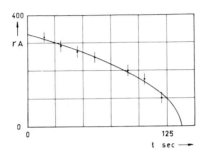

FIG. 196.   Annealing of a vacancy loop in aluminum. The plot shows the loop radius as a function of time at a temperature of 179°C. [After J. Silcox and M. J. Whelan, *in* "Structure and Properties of Thin Films" (C. A. Neugebauer *et al.*, eds.), p. 164. Wiley, New York, 1959.]

dislocations combine readily with the loops since these have the same type of Burgers vector and are glissile. During this process the glide dislocations become kinked and eventually jogged.

The annealing kinetics of loops introduced by deformation in zinc has been studied by Price.[344] He finds $U = 0.93$ ev whereas the activation energy for self-diffusion parallel to the c axis is 0.95 ev. In cadmium the activation energy is also in agreement with the value found by tracer measurements.

Conservative climb of prismatic loops in zinc has been described by Kroupa and Price.[353] The loops are induced to climb sideways under the influence of the stress from a glide dislocation. The process consists in transport of vacancies along the dislocation core from one part of the loop to the other.[354,355a]

(7) *The electrical resistivity due to point defects.* The accuracy of the absolute determination of defect concentrations from electric

[352] M. Panseri and T. Federighi, *Phil. Mag.* [8] **3**, 1223 (1958).
[353] F. Kroupa, and P. B. Price, *Phil. Mag.* [8] **6**, 243 (1961).
[354] C. A. Johnson, *Phil. Mag.* [8] **5**, 1255 (1960).
[355a] F. Kroupa, J. Silcox, and M. J. Whelan, *Phil. Mag.* [8] **6**, 971 (1961).

measurements depends very much on the knowledge of the resistivity increase per atomic percent of defects. Cotterill[314] has made a very direct measurement for the resistivity due to stacking faults in gold. He quenched gold foils from temperatures in the range 850–1000°C, the foils were subsequently aged at 100°C to collect the vacancies in stacking fault tetrahedra. This process was assumed to be complete when the electrical resistivity had reached its asymptotic value. The foils were then polished down and the concentration of defects measured. The foil thickness was determined from slip traces. Foils quenched from below 800°C failed to produce tetrahedra on aging; instead small black spots are produced.

The following value is found for the resistivity due to stacking faults in gold $\rho_{SF} = (1.8 \pm 0.3) \times 10^{-13} \beta\Omega$ cm^2 where β (cm^{-1}) is the total stacking fault density in all planes. The resistivity due to vacancies is: $\rho_V = (2.4 \pm 0.4) \times 10^{-6} \Omega$ cm per atomic per cent. The latter value is a lower limit, the uncertainty is due to the lack of complete knowledge of the annealing mechanism.

(8) *Loops due to prismatic punching.* When discussing the decoration methods we mentioned the elegant experiments of Jones and Mitchell[150] with glass spheres imbedded in silver chloride. Similar phenomena occur on precipitation in metals and have been studied with the electron microscope. Fourdeux and Berghezan[355b] for instance, observed long processions of loops surrounding precipitate particles in magnesium and niobium specimens. The directions of the sequences were parallel to possible Burgers vectors. Lally[318] observed similar prismatic loops and helical dislocations in quenched magnesium, presumably formed around oxide particles. An example due to Fourdeux and Berghezan, observed in niobium, is shown in Fig. 197.

Bailey[355c] found similar effects in zirconium containing hydride particles. In quenched copper Barnes and Mazey[355d] found rows of prismatic loops aligned in $\langle 110 \rangle$ directions, radiating out from small precipitates presumably copper oxide. The loops are presumably of the interstitial type since they are "blown off" to make space for the precipitate particle. The loops form processions similar to pile-ups. A theory relating the spacings with the loop diameter was made by Bullough and Newman.[355e] The observations were compared with the

[355b] A. Fourdeux and A. Berghezan, *Compt. Rend.* **252**, 1462 (1961).

[355c] J. E. Bailey, *Acta Met.* **11**, 267 (1963).

[355d] R. S. Barnes and D. J. Mazey, *Acta Met.* **11**, 281 (1963).

[355e] R. Bullough and R. C. Newman, *Phil. Mag.* [8] **5**, 921 (1960).

predictions of this theory and reasonable agreement was found. The outer loops are usually smaller than the inner ones, this is attributed to quenched in vacancies having diffused to the interstitial loops, decreasing in this way their diameter.

A phenomenon related to prismatic punching was observed by Smallmann[355f] and by Embury and Nicholson[355g] in Al-7%Mg alloy.

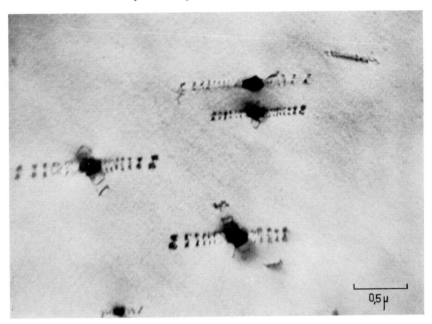

Fig. 197. Prismatic punching observed in niobium. (Courtesy of Fourdeux and Berghezan.)

Foils of this material were quenched and aged. After this treatment small precipitate particles are present at which dislocation sources are nucleated. The configuration consists of prismatic loops, with an $(a/2)$ [110] vector, lying in parallel planes and on a pyramidal surface with the particle at its summit. It is believed that these loops result from the action of a Bardeen-Herring source.[355h] The stress

[355f] R. E. Smallmann, *Metallurgist* 1, 146 (1960).

[355g] J. D. Embury and R. B. Nicholson, *Proc. 5th Intern. Conf. on Electron. Microscopy, Philadelphia, 1962* 1, J-1 (1962).

[355h] J. Bardeen and C. Herring, "Imperfections in Nearly Perfect Crystals", p. 175. Wiley, New York, 1952.

field of the particle and of the newly generated loops pushes the already formed loops further outward by slip, while the supersaturation of vacancies resulting from the quench, causes the simultaneous lateral growth of the loops.

The phenomenon has been studied in greater detail by Westmacott et al.[356a] They find that the prismatic loops are actually formed in cube planes, say in the (001) plane. This formation is attributed, in accord with the previous authors to the operation of a Bardeen-Herring source (Fig. 197[(1)]). The source is an edge segment with a a [001]

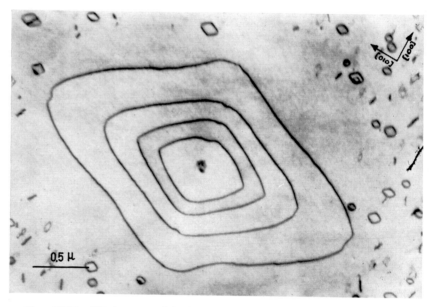

FIG. 197[(1)]. Climb source of the Bardeen-Herring type observed in an aluminium-magnesium alloy. (Courtesy of Westmacott et al.)

Burgers vector ending in two screw segments. The loop dissociates into two concentric loops with vectors of the type $a/2$ [110], according to the reaction:

$$a\,[001] \rightarrow \frac{a}{2}\,[011] + \frac{a}{2}\,[0\bar{1}1]$$

[356a] K. H. Westmacott, R. S. Barnes, and R. E. Smallman, *Phil. Mag.* [8] **7**, 1585 (1962).

This is concluded from the fact that the loops are normally visible in (001) foils, and disappear simultaneously only for a reflection (200). It is further shown by contrast experiments that some loops dissociate over part of their length according to the reaction given above and over the remaining part according to the alternative reaction

$$a\,[001] \rightarrow \frac{a}{2}\,[101] + \frac{a}{2}\,[\bar{1}01].$$

The part dissociated according to the first scheme is invisible with the (200) reflection operating, while the second part is invisible with the (020) reflection operating.

The successive loops originating from one source tend to remain on the same (001) plane during climb, because this minimizes the stresses. The nucleation of the source is attributed to the precipitation of magnesium on the (001) planes of the aluminium matrix. Since magnesium atoms are bigger than aluminium this creates stresses, which can be relieved by the nucleation of a dislocation of the required type.

(9) *Determination of the nature of loops.*   In Section 9e(5)(b) we mentioned already a method for determining the nature of loops.

More recently two more methods were proposed. The first of these methods is based on the change in shape of a loop as the specimen is rotated through a reflecting position[356b]. Suppose that the contrast producing reflection is to the right as in Fig. 197[(2)] and that one rotates the crystal through a reflecting position by turning in the clockwise sense, starting from a position where the image is narrow [Fig. 197[(2)]a]. As one turns the crystal the image becomes wider for geometrical reasons. On approaching the reflecting position s is positive and the image is inside for an interstitial loop and outside for a vacancy loop. On passing through the reflecting position the image side changes, i.e., it goes outside for the interstitial loop and comes inside for a vacancy loop. Hence the interstitial loop increases in width as a consequence of both the geometrical effect and the image side. The vacancy loops on the other hand first expand as a consequence of the geometrical effect but contract afterwards on passing through the reflecting position, due to a change of the image side. Before the next reflection arrives the image has to change side.

[356b] The method is due to G. K. Williamson and B. Edmondson. The description given here is after a presentation by B. Edmondson, *Proc. Joint Conf. on Inorganic and Intermetallic Crystals, Univ. of Birmingham, Engl., 1963.*

However this will in general not result in a decrease in width for the interstitial loop because of the larger increase in width as a consequence of the geometrical factor. One concludes that for the given geometry the interstitial loops continuously increase in size, whereas the vacancy loops first increase but afterwards decrease in width as one passes the reflecting position. The same type of reasoning can be made

Fig. 197[2]. Derivation of the contrast effects associated with a prismatic loop rotated about an axis in the foil plane. The width of an interstitial loop is seen to increase continuously whereas the width of a vacancy loop increases first and then decreases as one passes through the reflecting position. The full line represents the image; the dotted line is the real position of the dislocation loop.

for a rotation in the other sense or for a reflection at the left; the conclusion is the same. The Fig. 197[2] is made for a rotation about an axis perpendicular to the diffraction vector and for purely prismatic loops. In practice however, these conditions are not so strict and the method still works well for a loop which has partly shear character and for a diffraction vector which is not perpendicular to the rotation axis. In order to have an appreciable geometrical effect one should use a loop which is inclined rather steeply.

The second method[356c] is based on so-called "anomalously wide images" observed close to the surface. It was shown by machine

[356c] M. E. Ashby and L. M. Brown, *Proc. 5th Intern. Conf. on Electron Microscopy, Philadelphia, 1962* K-5 (1962).

calculations based on the dynamical theory that the image of a loop seen edge on is asymmetrical. One side consists of a bright crescent, the other side of a black one. In the bright field image this asymmetry is in the opposite sense near top and bottom surface. However, in the dark field image the asymmetry is the same at top and bottom. This is a consequence of the anomalous absorption. The asymmetry in the dark field image is such that for a vacancy loop the white crescent is in the positive sense of the diffraction vector. The opposite is true for an interstitial loop.

Using the method outlined here the loops in neutron irradiated magnesium oxide were found to be due to interstitial platelets on the (110) plane[356d-f] whereas the loops in deformed magnesium oxide are of the vacancy type.[356d]

e. Deformation in the Microscope

(1) *Face-centered cubic metals.* Several investigators have made observations on thin foils undergoing deformation. Usually aluminum was used in these studies because of its large transparency. These observations have illustrated in a striking way the role played by dislocations in plastic deformation. In the initial stages of deformation most dislocations originate from sub-boundaries, and they often disappear again at sub-boundaries.

On movement the dislocations leave traces located at upper and lower surfaces. The origin of these traces is not very well understood. According to Hirsch *et al.*[356g] they may be due to a dislocation trapped between the surface and the oxide film. Sometimes the whole glide plane exhibits contrast. The traces record the path followed by the glide dislocations: in this way it is found that cross slip is very prominent in aluminum (Fig. 198). As pointed out by Wilsdorf[357] this is due to the particular deformation conditions. In a thin foil, that glide

[356d] G. W. Groves and A. Kelly, *Proc. Joint Conf. on Inorganic and Intermetallic Crystals, Univ. of Birmingham, Engl.,* 1963.

[356e] D. H. Bowen, *Proc. Joint Conf. on Inorganic and Intermetallic Crystals, Univ. of Birmingham, Engl.,* 1963.

[356f] B. Edmondson, *Proc. Joint Conf. on Inorganic and Intermetallic Crystals, Univ. of Birmingham, Engl.,* 1963.

[356g] P. B. Hirsch, R. W. Horne and M. J. Whelan, *in* "Dislocations and Mechanical Properties of Crystals" (J. C. Fisher *et al.*, eds.), p. 92. Wiley, New York, 1957.

[357] H. Wilsdorf, *Proc. Symp. Advances in Electron Metallog.,* Special Technical Publ. No. 245 (ASTM 1958), p. 43.

system is preferred which gives the smallest thickness of material as measured along the direction of the Burgers vector. Further since a dislocation will tend to shorten as much as possible it has a tendency to adopt an orientation perpendicular to the foil and hence parallel to the Burgers vector of the preferred glide system. Moreover cross slip is very easy in aluminum as a consequence of the large stacking

FIG. 198. Slip traces left in the wake of moving dislocations in aluminum. It is evident that some dislocations have cross slipped. (Courtesy of Fourdeux and Berghezan.)

fault energy. Therefore pile-ups are not observed since dislocations can avoid obstacles by cross slipping around them. A limited amount of slip in non-octahedral planes is sometimes observed.

It is found that slip steps are formed at the edge where a number of dislocations have left the foil, which proves that slip steps result from dislocation movement.[358] This was shown in a somewhat diffe-

[358] A. Berghezan and A. Fourdeux, J. *Appl. Phys.* **30**, 1913 (1959).

rent way by Hirsch *et al.*[201] They coated one side of a deformed stainless steel foil with a protecting varnish, while thinning from the other side. Correspondence was found between slip steps and the internal arrangement of dislocations in the slip plane. In stainless steel pileups are often found to correspond to slip steps.

In very pure aluminum the cooperative movement of dislocations on parallel glide planes was observed; tilt boundaries move as a whole through the specimen. The elastic interaction between like dislocations in parallel glide planes is large enough to keep the walls together, especially in very pure material where the lattice friction is small.[358,359] Whelan *et al.*[243] found that dislocations in stainless steel nucleate at foil edges or leave boundaries. In particular they suggested that sources may be active in twin boundaries. The deformation of stainless steel foils was studied in detail by Wilsdorf.[360] The author stresses the importance of the presence of vacancies in the crystal; they produce a friction on the dislocations and are responsible for their jerky motion. If a dislocation has been stationary for some time, after having collected a significant number of vacancies, it is found that it leaves behind small surface pits, when moving away. These pits are due to the agglomeration of vacancies that diffused along the dislocation core. The same author studied the different dislocation sources active in thin foils; the fact that grain boundaries and twin boundaries act as sources was confirmed but moreover it was found that polygon walls emit dislocations. Precipitates are preferred nucleation centers. Frank-Read sources of the double-ended type as well as of the spiral type are occasionally found.

Static observations of Frank-Read sources are due to Westmacott *et al.*[361a] These authors observed concentric glide loops originating from a quench loop in Al-4% Cu alloys quenched from 525°C into acetone. The glide loops apparently were subject to further climb and partly developed helices.

Pashley and Presland[361b] have convincingly demonstrated that the movement of dislocations on examination in the electron microscope is due to the stresses resulting from the formation of an organic poly-

---

[359] R. Wilson and P. Forsyth, *Phil. Mag.* [8] 6, 453 (1961).

[360] H. G. F. Wilsdorf, *in* "Structure and Properties of Thin Films" (C. A. Neugebauer *et al.*, eds.), p. 151. Wiley, New York, 1959.

[361a] K. H. Westmacott, D. Hull, R. S. Barnes, and R. E. Smallman, *Phil. Mag.* [8] 4, 1089 (1959).

[361b] D. W. Pashley and A. E. B. Presland, *Phil. Mag.* [8] 7, 1407 (1962).

mer film on the metal foil. When allowing oxygen to enter the specimen chamber, the movement could be suppressed completely.

(2) *Hexagonal metals.* Price has studied in detail the deformation in the microscope at different temperatures ($+25$ to $-150°C$) of perfect crystal platelets of zinc and cadmium grown by sublimation in helium.[362] As in the face-centered metals dislocations are generated at low-angle boundaries, high-angle boundaries, at twin tips, and at crack tips. Similar results were found by Berghezan *et al.*[249] In their study of electrothinned zinc. Even in the perfect crystals dislocation motion was not uniform; unseen obstacles seem to be always present.

(a) Basal glide. If the geometry is favorable most glide is on the *c* plane. Under a shear stress, for which the resolved stresses on the 2 partials are opposite in sense, the basal dislocations can be separated into their partials.

(b) Nonbasal glide. Nonbasal glide was observed by Price[362] as well as by Berghezan *et al.*[249] Different glide systems were identified in perfect platelets.

(i) First-order pyramidal glide $(10\bar{1}1)$ $\langle 1\bar{2}10 \rangle$. This type of glide is observed in cadmium platelets developed along (0001) and strained with the tensile axis parallel to that plane. It is also found in magnesium but not in zinc. The dislocations do not cross glide from the basal plane onto the $(10\bar{1}1)$ plane, but they are nucleated immediately in the $(10\bar{1}1)$ plane. No glide is observed in the $(10\bar{1}0)$ plane.

(ii) Pyramidal glide $\langle 11\bar{2}3 \rangle$. Platelets of zinc and cadmium strained parallel to the basal plane deform by twinning at high strain rates[362] but at small strain rates by glide in the $\langle 11\bar{2}3 \rangle$ directions and on pyramidal planes, mainly $(11\bar{2}2)$. Frequent cross glide takes place in this system over the whole temperature range studied. Zinc and cadmium behave similarly at temperatures corresponding to the same fraction of the absolute melting temperature. On motion these dislocations become jogged. Jogged screws produce long edge dipoles, which are pinched off by cross slip or climb. The elongated loop breaks up further into smaller loops in the basal plane as described in Section 10d(5)(b).

Howie *et al.*[363a] have made a detailed study of dislocation movement

---

[362] P. B. Price, *J. Appl. Phys.* **32**, 1746 (1961).
[363a] L. M. Howie, J. L. Whitton, and J. F. McGurn, *Acta Met.* **9**, 773 (1962).

and dislocation interactions in hexagonal zirconium. They confirmed the main glide system as being $(10\bar{1}0)$ $[\bar{1}2\bar{1}0]$, but they observed glide on other slip planes as well. They also found evidence for cross slip induced by obstacles. The most striking feature of their observations was the formation of very pronounced slip traces which were shown to consist of dislocations held up close to the upper and the lower surface of the foil by a strongly adherent zirconium oxide film. Interactions between such surface dislocations give rise to the formation of networks located close to upper and lower surface. Zirconium hydride inclusions were found to give rise to dislocation sources.

(3) *Motion of dislocations in irradiated metals.* The motion of dislocations in irradiated stainless steel was studied by Wilsdorf,[363b] in aluminium by Thomas and Whitton,[363c] and in iron by Eyre.[332c] In all cases the mobility of the dislocations is found to be reduced, even though no visible damage is present. The motion becomes slow and jerky and often the dislocations develop cusps. In aluminium no complete recovery of mobility is found even after heat treatment at 400°C during 16 hr.[363c]

f. The Arrangement of Dislocations in Deformed Metals.

A large number of papers have been devoted to this subject. The use of thin foil methods for this type of study has been questioned because it is difficult to be sure that no rearrangement takes place during the thinning process. Another objection is that one observes only a very thin section through a three-dimensional pattern which extends over distances large compared to the foil thickness. This last objection can be avoided partly by the use of oriented monocrystalline foils with the foil plane parallel to the glide plane.[363d] This is in fact the situation in layer structures.[202] With this geometry one can hope to reveal the dislocation geometry after deformation.

(1) *Face-centered cubic metals.*[364a,b,c] An important aspect of the work hardening theory is the relation between the strain rate depend-

[363b] H. G. F. Wilsdorf, in "Structure and Properties of Thin Films" (C. A. Neugebauer *et al.*, eds.), p. 151. Wiley, New York, 1959.

[363c] W. R. Thomas and J. L. Whitton, *Acta Met.* **9**, 1075 (1961).

[363d] S. Mader, in "Electron Microscopy and Strength of Crystals" (G. Thomas and J. Washburn, eds.), p. 183. Wiley (Interscience), New York, 1963.

[364a] R. Segall and P. Partridge, *Phil. Mag.* **4**, 912 (1959).

[364b] See also M. J. Whelan, *Mémoires Scientifiques Rev. Métallurg.* **56**, 153 (1959).

[364c] See also P. B. Hirsch, P. Partridge, and H. Tomlinson, *Proc. 4th Intern. Conf. on Electron Microscopy, Berlin, 1958* **1**, 536 (1960).

ence of the beginning of stage III of the stress strain curve and the stacking fault energy. The importance of the stacking fault energy in determining the characteristics of deformation of face-centered cubic metals seems to be confirmed by observations at small deformation levels. Dislocations in materials with a small stacking fault energy (e.g., stainless steel, α-brass) remain in piled-up groups, against boundaries (Fig. 199). Obviously this reflects the difficulty for cross slip to occur. In metals with a large stacking fault energy, like aluminium, cross slip is very frequent and as a result the dislocations tend to form well-defined sub-boundaries and a pronounced cell structure develops. In metals with an intermediate stacking fault energy the cells are less well-defined.

At higher deformation levels and for metals with a small stacking fault energy networks are formed in the slip planes. These networks result from the interaction between dislocations on intersecting slip planes. Some of the reactions have been discussed in another paragraph For very large deformations wide stacking faults and possibly twins are formed. In other metals very complicated "tangles" are formed. Tangle formation is probably the result of a large number of different processes of which some will be disccused further.

Lomer-Cottrell barriers are frequently observed in the metals or alloys with small stacking fault energy, usually their length is small and it is doubtful whether they are of importance as barriers against which pileups can form.

Aluminum deformed in unaxial stress and in fatigue has been studied by Segall and Partridge.[364a]

In specimens deformed in unaxial stress the distribution of dislocations is irregular up to stresses of 3 kg/mm^2 and the resistance to the deformation is provided therefore not by a uniform internal stress field, but by the regions of high dislocation density. The strain results in local bending, i.e., in a given region there is a predominance of dislocations of a given sign, which favors their rearrangement into well-defined boundaries, the interior of the cells being relatively free from dislocations.

The situation is rather similar for specimens fatigued at high stress levels; the subgrain size is larger and the number of dislocations within the cells is greater than in specimens with the same dislocation density but deformed unidirectionally. Specimens fatigued at low stress levels show different features. No polygonization is apparent although the dislocation density may be as high as $6.10^9/cm^2$. This is due to

FIG. 199. Pileup in Cu-7% Al alloy, which has a small stacking fault energy. (Courtesy of Art.)

the absence of plastic curvature. Small prismatic dislocation loops are observed in large concentrations (5.10^{14} cm^3 after about 10^6 cycles). The dislocations have become very irregularly shaped; many have become helices. From these observations it is concluded that during deformation large numbers of points defects are generated which agglomerate into disks, which in turn collapse into vacancy loops. The observations give no indication as to the mechanism that produced the vacancies.

In stainless steel specimens deformed in unaxial stress, and polished from one side only so as to preserve the surface structure at the other side, correspondence has been found between the slip steps and the dislocation arrangement in the slip plane. Fatigued specimens show evidence for the formation of intrusions and extrusions in regions where there is a high concentration of active slip planes.[201]

The deformation substructure of copper was studied by Warrington[365a] and that of silver by Bailey.[365b] The results are rather similar. In particular there is again no evidence for pileups; a cell structure forms at all deformation temperatures. In copper the cell size, after a standard 20% deformation in tension, increases with deformation temperature.

The yield stress for copper crystals, identical to those examined in the electron microscope, was also determined; it was found to be proportional to the inverse square root of the grain size.

The arrangement of dislocations in nickel and in nickel cobalt alloys with small stacking fault energy has been studied by Mader and Thieringer.[366a] The authors prepared foils parallel to the primary glide planes of deformed single crystals. They come to the conclusion that dislocations leave the foil during the thinning operation. They find that the observed configurations of dislocations lend support to theories of work hardening based on long-range stresses rather than to those based on the intersection of forest dislocations.

Fourie and Murphy[366b] have made an extensive study of deformed copper single crystals in an attempt to correlate the electron microscopic picture with the different stages in the stress-elongation curve.

[365a] D. H. Warrington, *Proc. European Reg. Conf. on Electron Microscopy, Delft, 1960* 1, p. 354 (1961).

[365b] J. Bailey, *Phil. Mag.* [8] **5**, 833 (1960).

[366a] S. Mader and H. M. Thieringer, *Proc. 5th Intern. Conf. on Electron Microscopy, Philadelphia, 1962* 1, J-3 (1962).

[366b] J. T. Fourie and R. J. Murphy, *Phil. Mag.* [8] **7**, 1630 (1962).

In particular attention is paid to the formation of elongated loops from large jogs. In stage I of the deformation screw dislocations of the primary slip system seem to interact with sub-boundaries resulting in the formation of elongated dislocation loops. These loops act as new nucleation sites for more loops, resulting in an increased work-hardening rate. At the point where the density of loops becomes sufficient to prevent the formation of elogated loops by bypassing existing loops, stage II of the deformation sets in.

(2) *Hexagonal metals: zinc.* Large dislocation loops were found in the basal plane of zinc monocrystals fatigued at low temperature. These loops were also attributed to the condensation of point defects.[249] Substantial recovery takes place in zinc and cadmium at room temperature.

In zirconium Bailey[355c] has found a pronounced influence of the presence of impurities on the dislocation arrangement after deformation. In pure zirconium irregular tangles are formed, while in commercial zirconium long and straight dislocations are found. This is apparently due to the operation of fewer glide planes or glide systems in the impure material.

(3) *Body-centered metals.* A very comprehensive survey of the deformation substructure in body-centered cubic metals is due to Keh and Weissmann.[266] The resulting substructure after deformation seems to be rather similar in all metals investigated: iron, molyb-denum, tungsten, tantalum, and niobium. The early stages of deformation are characterized by the occurrence of nonuniformly distributed kinked dislocations forming so-called "tangles." No pileups are found, probably as a consequence of the ease of cross glide. After increased strain a cell structure develops.

The influence of the temperature of the deformation characteristics of α-iron was investigated. After room temperature deformation the dislocations are jogged and inhomogeneously distributed. The slip traces are wavy as a consequence of frequent cross slip. As the temperature is lowered the dislocations become straighter and the tendency to form cells diminishes.

There is a decreasing linear relationship between the cell size at a given strain and the temperature of deformation between room temperature and lower. At higher temperatures the cell size increases

[367] J. R. Low, and A. M. Turkalo, *Acta Met.* 10, 215 (1962).

more rapidly due to the recovery during deformation. In all deformed metals elongated loops and high jogs are found. In particular in silicon-iron, Low and Turkalo[367] found that the elongated loops are frequently oriented along $\langle 112 \rangle$ in accord with the hypothesis that they should be edge dipoles formed from jogs in screws (see also Sestak and Libovicky.[368])

The dislocation density at a given plastic strain does.not depend on the temperature of deformation provided it is higher than the temperature of recovery. However it does depend on the grain size, since the smaller the grain size the larger is the dislocation density. In the recovered material and also in creep specimens, well-developed, regular hexagonal networks were often observed. The geometry of these networks has been described in Section 10c(3).

Walter and Koch[369] have analyzed the dislocation structures in cold rolled high purity iron crystals after different annealing treatments. A very similar work was done independently by Hsun Hu.[370] In particular the geometry of the lattice rotations that accompany heavy reduction of single crystals by cold rolling in the (001) [100] orientations are studied. The results are not easily summarized and we refer therefore to the original papers for details.

(4) *Germanium.* The rearrangement of dislocations during thinning of specimens can be avoided in substances where the mobility of dislocations is negligible at room temperature. This is for instance the case in germanium. With this idea in mind, Alexander and Mader[371a] have studied the dislocation arrangement in germanium deformed in tension at 600°C under reducing atmosphere. The specimens are cut along the (111) primary glide plane. The most striking feature was the appearance of numerous elongated loops, like those seen in metals, formed by dipoles of edges. No cross slip is observed in accord with the fact that the dislocations are dissociated in partials.[371b] The dislocation density increases suddenly at the lower yield point and becomes $10^9/cm^2$ in the easy glide region (stage I) and $10^{10}/cm^2$ in stage II.

[368] B. Sestak and S. Libovicky, *Czech. J. Phys.* **12**, 131 (1962).

[369] J. L. Walter and E. F. Koch, *Acta Met.* **10**, 1059 (1962).

[370] Hsun Hu, *Acta Met.* **10**, 1116 (1962).

[371a] H. Alexander and S. Mader, *Acta Met.* **10**, 887 (1962).

[371b] A. Art, E. Aerts, P. Delavignette, and S. Amelinckx, *Appl. Phys. Letters* **2**, 40 (1963).

g. Mechanical Twinning

(1) *Tin.* Fourie *et al.*[372] made a ciné film of the growth of deformation twins in tetragonal tin. They studied both the nucleation and the growth of the twins. Nucleation is found to take place mainly at the foil edges, but also at etch pits, at other twin boundaries, and at defects in general. No evidence is found for the necessity of plastic deformation prior to twinning. In particular no pileups are found which could give rise to twinning. The observations are consistent with the nucleation mechanism proposed by Bell and Cahn[373] and Orowan[374] according to which homogeneous twinning takes place at the shear stress required for twinning. In particular there is no evidence for a pole mechanism.

The propagation of the twin takes place at stress levels below that required for the nucleation. The observation can be explained on the assumption that the advancing noncoherent interface consists of an array of twinning dislocations. The propagation speeds, as deduced from the film, are of the order of 10^{-6} cm/sec, at the stress resulting from the irradiation with the electron beam.

(2) *Face-centered cubic metals.* The deformation at 77°K of a number of face-centered cubic metals and alloys gives rise to mechanical twinning. An important parameter in determining the twin behavior seems to be the stacking fault energy. Venables[375,376] has examined mechanical twinning in a number of Cu-Al alloys with varying stacking fault energy. It is found that the propagation of twin dislocations is considerably impeded by stacking faults. In the alloys with small stacking fault energy twins cannot propagate very far.

(3) *Hexagonal metals.* Price studied mechanical twinning in zinc and cadmium.[344] With the tensile axis parallel to the basal plane twinning occurs frequently in deformed zinc and cadmium: the twin elements are $(10\bar{1}2) \langle 10\bar{1}1 \rangle$. In agreement with the previous work on tin, it is found that twins are homogeneously nucleated at crystal edges, at reentrant corners, or at corrosion pits. Twinning is preferred

[372] J. T. Fourie, F. Weinberg, and F. W. C. Boswell, *Acta Met.* **8**, 851 (1960).

[373] R. L. Bell and R. W. Cahn, *Proc. Roy. Soc.* **A239**, 494 (1957).

[374] E. Orowan, "Dislocations in Metals," p. 116. A.I.M.E., New York, 1954.

[375] J. Venables, *Proc. European Reg. Conf. on Electron Microscopy, Delft, 1960,* 1, p. 443 (1961).

[376] J. Venables, *Phil. Mag.* [8] **6**, 379 (1961).

over nonbasal glide at sufficiently large strain rates. Sometimes twinning takes place without any plastic deformation. The twins are always wedge-shaped, the wedge angle being between 1° and 6°. The growth of the twins was found to be impeded by the presence of dislocations.

As for tin no evidence was found that a pole mechanism should be responsible for the propagation of the twin interface; on the contrary the observations suggest that, at least in thin foils, twin dislocations are nucleated continuously. The twin dislocations cannot be observed since even for a 1° wedge angle they are too closely spaced.

(4) *Body-centered metals.* Votava and Sleeswijk[377] studied deformation twinning in a 15% rhenium-molybdenum alloy. Evidence was found for the occurrence of so-called "emissary" slip dislocations in the prolongation of twin tips. These slip dislocations with an $(a/2)$ [111] Burgers vector result from the dissociation under applied stress of $(a/6)$ [111] twin dislocations, present in the noncoherent boundary at the twin tip. The reaction is

$$\frac{a}{6}[111] \rightarrow \frac{a}{2}[111] - \frac{a}{3}[111]. \tag{10.87}$$

In this way a low-energy noncoherent twin boundary composed of 1 $(a/3)$ [111] dislocation for every 2 $(a/6)$ [111] dislocations is left behind by the emissary dislocations.

Evidence for emissary dislocations had previously been obtained from optical surface studies by Sleeswijk[378] but the interpretation was not unambiguous.

h. Recrystallization

We will not review the subject in detail but summarize only a few representative papers.

(1) *Mechanism of recrystallization.* The recrystallization of nickel was studied by Bollmann.[379] In the heavily deformed nickel dense dislocation clouds are found with very little dislocation free areas. The first result of heating is a contraction of the clouds and a gradual growth of the dislocation-free areas with formation of sharp sub-

[377] E. Votava and A. Sleeswijk, *Acta Met.* 10, 965 (1962).
[378] A. Sleeswijk, Thesis, University of Amsterdam, 1961.
[379] W. Bollmann, *J. Inst. Metals* 87, 439 (1958, 1959).

boundaries. At 200°C recrystallization nuclei become visible. They have the following characteristics: (1) they are larger than the surrounding subgrains; (2) they have a rounded shape and smooth outlines; (3) they are often twinned. The observation can be interpreted in terms of Cahn's theory according to which certain subgrains serve as nuclei and grow preferentially, the driving force being the difference in dislocation density between the deformed and the recrystallized grains. This implies that polygonization is a preliminary stage for the recrystallization.

Recrystallization in silver has been studied by Bailey.[365b] As deformed the specimens have a cell structure. The misorientation between cells increases with increasing deformation, and some cells develop subcells. Some specimens were annealed in bulk and thinned subsequently; others were annealed in the microscope. Significant differences were noticed between the two kinds of specimens. The grain size for an equivalent heat treatment is smaller in the thin foils; and it turns out to be an increasing function of the foil thickness. This is due to a decrease in mobility of the grain boundaries with decreasing foil thickness. The recrystallization process takes place in a very inhomogeneous way; the same specimen shows large recrystallized areas, while in other areas no obvious changes have taken place. The observations seem to rule out a process based on the random nucleation of new grains; but it supports, in agreement with Bollmann's observations in nickel, the theory due to Cahn. In very heavily deformed foils (95% reduction) recrystallization seems to occur by the migration of high-angle boundaries, although it is difficult to distinguish high- and low-angle boundaries.

The recrystallization of deformed single crystal foils of aluminum has been studied by Granzer and Haase[380]; these authors also come to the conclusion that Cahn's theory explains best the observed features.

The recrystallization and polygonization of copper has been studied by Votava.[381]

(2) *Recrystallization twins; growth twins.* Recrystallization twins in copper have been studied by Votava and Hatwell.[382] Evidence is presented that twinning dislocations are present in the noncoherent

[380] F. Granzer and G. Haase, *Z. Physik* **162**, 468 (1961).
[381] E. Votava, *Acta Met.* **9**, 870 (1961).
[382] E. Votava and H. Hatwell, *Acta Met.* **8**, 874 (1960).

interfaces of recrystallization twins in accord with earlier optical observations[20,21] by the same authors.

Recrystallization twins are common in platinum, which is heat-treated as a foil.[215] They were also found in niobium.[265]

Reimer et al.[383] have found numerous microtwins and dislocations in nickel films made by electrodeposition on a copper substrate. The films grow epitaxially on the corresponding crystal planes of copper. Twin formation is found if deposition takes place on (100) and (111) planes but not on (110) planes. The dislocation density is a sensitive function of the substrate; the largest dislocation density $(4 \times 10^9 - 1.3 \times 10^{10}/cm^2)$ is found in (110) foils. The observed features are very similar to stacking faults, but electron diffraction allows to conclude to the presence of twins rather than stacking faults.

i. Phase Transformation of Cobalt and Zinc Sulfide

The phase transformation of cobalt from hexagonal close-packed below 417°C to face-centered cubic above that temperature has been studied by Votava.[261,262] The foils thinned at room temperature contain numerous stacking faults which represent in fact the hexagonal phase. They disappear on heating at 450°C by coalsecence of pairs of partials into perfect dislocations, which on further heating are eliminated partly at grain boundaries. By cooling down the reverse process takes place, the face-centered cubic phase transforms back into the hexagonal phase by the splitting of perfect dislocations into partials connected by wide stacking faults. In thin foils the transformation is considerably hindered and large undercooling is observed. It is not possible to decide on the basis of thin foil observations whether a pole mechanism as proposed by Bilby[384] and Seeger[385] is responsible for the transformation or not. It is now established, however, that the transformation is produced by the movement of partials.

Zinc sulfide crystals grown at high temperature are of the wurtzite (hexagonal) modification. Below about 1100°C the sphalerite (cubic) phase is the stable one. The wurtzite phase, containing stacking faults, is usually frozen in. If dislocations are introduced in such

---

[383] L. Reimer, J. Ficker, and T. Pieper, Z. Metallk. **52**, 735 (1961).

[384] B. A. Bilby, Phil. Mag. [7] **44**, 782 (1953).

[385] A. Seeger, Z. Metallk. **44**, 247 (1953); **47**, 653 (1956).

crystals, the two partials split up into a stacking fault extending over the whole crystal. In this way one layer of sphalerite is formed. The freezing-in of the high temperature phase is apparently due to the difficulty of nucleating dislocations that perform the shear transformation.

j. Interaction of Impurities and Dislocations; Helical Dislocations

(1) *Helical dislocations in fcc metals and alloys.* As discussed previously the quenching of the Al-4% Cu alloy results in the formation of helical dislocations which seem to play an important role as nucleation sites for precipitates. Several papers are devoted to this subject.[386-389] Figure 200 shows an example due to Thomas and Whelan.[387]

Nicholson et al.[386,390] examined two systems in detail, aluminum-copper and aluminum-silver. In both, decoration of helices takes place on quenching, followed by aging, but the crystallography and also the interaction mechanisms are different.

In the aluminum-copper system the ϑ' precipitates form in 2 families of cube planes, but not in the cube face containing the axis of the helix. This suggests that nucleation of the precipitates does not take place on the pure screw parts, but only in the edge parts. Nicholson considered as a driving force the relief of the long-range strains associated with the nucleation of the ϑ' particles.

In the aluminum-silver system the γ' precipitates form on all {111} planes without exception, also in the one containing the axis of the helix. This different behavior is explained as follows. It is first pointed out that on quenching the alloy, Frank sessile loops containing a stacking fault are formed; this proves that the effect of the silver is to lower the stacking fault energy of aluminum. Further since there is a coupling between vacancies and silver atoms[388,389] the vacancies will transport silver towards the helix and increase locally the silver concentration, thereby lowering the stacking fault energy. The climbing helix will thereby dissociate into partials. The silver will now

---

[386] R. B. Nicholson, G. Thomas, and J. Nutting, *J. Inst. Metals.* **87**, 429 (1958, 1959).

[387] G. Thomas and M. G. Whelan, *Phil. Mag.* [8] **6**, 1103 (1961).

[388] G. Thomas, *J. Inst. Metals* **90**, 57 (1961, 1962).

[389] A. Eikum and G. Thomas. To be published.

[390] R. B. Nicholson, *Proc. European Reg. Conf. on Electron Microscopy, Delft, 1960,* **1**, p. 375 (1961).

be adsorbed by the fault ribbons as a consequence of Suzuki inter-action. Since this process does not differentiate between edge and screw dislocations precipitates will form on all (111) planes.

The aluminum-copper 4% system also was studied by Thomas and Whelan.[387] The authors point out that a different structure is formed depending on whether the aging takes place in a thin foil or in the

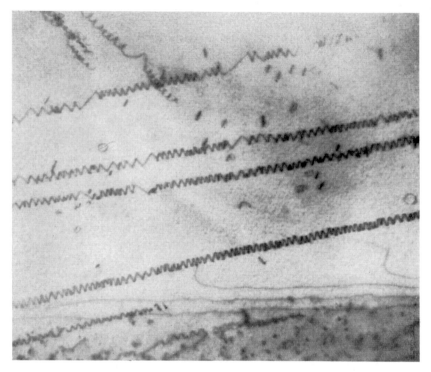

FIG. 200. Helical dislocations in quenched Al-4% Cu alloy. The helices are parallel to [110] in agreement with the screw character of the initial dislocation. (Courtesy of Thomas and Whelan.)

bulk. In thin foils after aging in the microscope, surface nucleation is predominant. In bulk specimens the decoration of helices takes place, as described by Nicholson.

Frank et al.[391] have examined further the aluminum-20 weight% silver alloy. In accord with Nicholson they find that on quenching

[391] G. Frank, D. L. Robinson, and G. Thomas, J. Appl. Phys. 32, 1763 (1961).

Frank sessile loops with a Burgers vector $(a/3)$ [111] containing a stacking fault are formed. Further the authors measured the extinction distance from the fault fringes observed in the loops. They find a value which is almost equal to the one expected in silver, and much smaller than the one to be expected for aluminum. From this the authors conclude that segregation of silver takes place, resulting in a lowering of the stacking fault energy, in accord with Nicholson.[390] The sessile loops and also the quench helices are found to be preferred nucleation sites for precipitation. Helices, which are connected to boundaries are found to straighten on continued aging, and at the same time large precipitates form at the points where the helices join the boundary. This is interpreted as evidence for the diffusion of silver along these dislocations.

Helical dislocations were also studied in the Al-5% Mg system.[389] In particular attention was focused on the transformation of helices into loops by a mechanism described earlier in the case of fluorite.[133,134] Segments of the helix intersect and loops are pinched off. This process does not take place in other alloy systems presumably because they become pinned by precipitates. It is suggested that helices do not form in pure metals because they transform readily into loops, since they are not stabilized by impurity pinning (Fig. 201).

(2) *Decoration of dislocations.* "Decoration" of dislocations by small precipitates also is observed with the electron microscope. Such dislocations are then visible by absorption contrast. In some cases the extraction of the precipitates, by disolving the surrounding matrix, leaves a skeleton which is some kind of internal replica of the dislocation network.[391a]

Campbell and Muldawer[392] found decoration in a gold-cobalt alloy: the precipitate particles are cobalt. Gaunt and Silcox[393] have examined the same system. They find that the cobalt precipitates consist of small particles clustered in sheets on (100) planes.

McLean[394] investigated iron + 0.03% carbon foils quenched, aged at 100°C, subsequently strained to various amounts, and finally thinned electrolytically. The precipitates are dendritic and they decorate isolated dislocations and dislocations in sub-boundaries.

[391a] J. Plateau, private communication, 1961.

[392] R. Campbell and L. Muldawer, *Phil. Mag.* [8] **6**, 531 (1961).

[393] P. Gaunt and J. Silcox, *Phil. Mag.* [8] **6**, 1343 (1961).

[394] D. McLean, *Proc. European Reg. Conf. on Electron Microscopy, Delft, 1960,* 1 p. 425 (1961).

The dendrites grow rapidly during the first few hours at 100°C, afterwards the growth slows down rapidly. On straining the dislocations seem to loop around the precipitates; the hardening is attributed to a "barrier-effect."

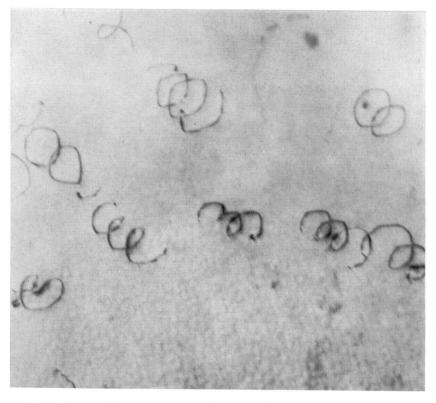

FIG. 201. Slightly polygonal helices in quenched Al-5% Mg alloy. (Courtesy of G. Thomas.)

Berghezan and Fourdeux[302] found that dislocations in niobium are decorated by oxides or nitrides if the heat treatment is not done in ultra high vacuum. As in the case of silver decoration in potassium chloride, impurities seem to be attracted less to the pure screw segments than to the edge and mixed segments.

Hatwell and Votava[395] described the preferential precipitation of

[395] H. Hatwell and E. Votava, *Acta Met.* **9**, 945 (1961).

chromium carbide on twinning dislocations in austenite Fe-Cr-Ni-C sheets.

(3) *Precipitates and moving dislocations.* The thin foils technique permits direct observation of both the dislocations and the precipitates, and it is therefore possible to study their interaction. Nicholson *et al.*[386] have found that a dislocation can pass through zones and through coherent and partly incoherent intermediate precipitates, but avoids noncoherent precipitates. In particular the ϑ' precipitates in Al-4% Cu are sheared by the passing dislocation. When a dislocation passes through partially coherent precipitates jogs are probably formed.

(4) *Segregation at stacking faults.* Direct evidence for the segregation of niobium carbide at stacking faults in a stainless steel containing 18% of chromium has been presented by Van Aswegen and Honeycombe.[396] This is interpreted as a result of the Suzuki interaction of niobium atoms with dislocations, resulting in a lowering of the stacking fault energy and consequently in a wide dissociation of the dislocations.

k. Thin Plate Effects

(1) *Introduction.* The question to what extent the finite specimen size influences the geometry of dislocations in thin foils is often raised. A number of effects have been observed which unambiguously show that the influence of the surface may be important. Because the surface can give way more easily than the bulk of the crystal the stress due to a dislocation is altered in its vicinity. The energy of the dislocation will diminish, since the effective outer cutoff radius will decrease.

The influence of the decreased stress field reflects itself for instance in the decreased width of ribbons. Ribbons are in fact particularly useful "probes" to explore changes in the stress field. The 2 partials are kept together by a constant force γ, numerically equal to the stacking fault energy, provided corresponding units are used. They repel each other by a force which depends on the distance to the surface. A study of the ribbon width as a function of the depth below the surface is therefore the most direct way of studying the influence of the surface on the stress field.

Expressions for the stress fields of edge and screw dislocations in isotropic semi-infinite solids and in plates have been derived by

[396] J. S. T. Van Aswegen and R. W. K. Honeycombe, *Acta Met.* **10**, 262 (1962).

Dietze and Leibfried.[397a] The anisotropic hexagonal semi-infinite solid has been treated by Spence[397b] and Siems *et al.*[252a] The first author considered the screw dislocation only while the second also treated the case of the edge dislocation in the basal plane. This is required for the adequate treatment of an arbitrary ribbon. Spence also calculated the stress field of a screw dislocation in an anisotropic plate. One should be aware of these surface effects when using the ribbon widths to deduce stacking fault energies.

(2) *Change in ribbon width.* We will treat only the simplest case to demonstrate the order of magnitude of the effect. For a more complete treatment we refer to the paper by Siems *et al.*[252a] and by Spence.[397b] We consider a 30° ribbon in a semi-finite isotropic solid. If one of the partials is a pure screw and the other a 60° dislocation the total Burgers vector forms an angle of 30° with the ribbon. This case is particularly simple since one only needs to consider the interaction between the screw components b and $b/2$. Then only the stress for a screw is required.

The stress field for the screw in a semi-infinite crystal can be obtained by superposing on the stress field for a screw in an infinite medium that due to a mirror screw dislocation at a symmetrical position with respect to the surface. In this way all stresses vanish at the surface and the boundary conditions which require that the surface should be stress-free are satisfied. This simple procedure does not work for an edge dislocation. The only stress of interest for our problem is σ_{xz}. For the geometry shown in Fig. 202a it becomes

$$\sigma_{xz} = -\frac{\mu b}{2\pi}\left(\frac{y}{y^2 + (x-d)^2} - \frac{y}{y^2 + (x+d)^2}\right). \qquad (10.88)$$

It is easy to verify that at the surface (i.e., for $x = 0$)σ_{xz} vanishes. We are particularly interested in the stress at the point like P at the same depth below the surface as the dislocation, i.e., for $x = d$.

$$\sigma_{xz}(x = d) = -\frac{\mu b}{2\pi}\frac{1}{y}\left(1 - \frac{1}{1 + (4d^2/y^2)}\right). \qquad (10.89)$$

The factor between brackets is obviously the surface correction; it

[397a] H. Dietze and G. Leibfried, Diplomarbeit, Göttingen, 1949.

[397b] G. B. Spence, *Proc. 5th Conf. on Carbon, Pennsylvania State Univ., 1961; J. Appl. Phys.* **33**, 729 (1962).

reduces to 1 for $d \to \infty$. The equilibrium distance y_d between partials in a 30° ribbon at a depth d below the surface is determined by:

$$-\frac{b}{2}\sigma_{xz}(x = d) = \frac{\mu b^2}{4\pi}\frac{1}{y_d}\left(1 - \frac{1}{1 + (4d^2/y_d^2)}\right) = \gamma. \quad (10.90)$$

If we introduce as a natural unit for the distances y_∞, the width of the ribbons in the infinite material, we obtain:

$$\frac{d}{y_\infty} = \frac{1}{2}\left(\frac{y_d}{y_\infty}\right)^{3/2}\left(1 - \frac{y_d}{y_\infty}\right)^{-1/2} \quad (10.91)$$

where

$$y_\infty = \mu b^2/4\pi\gamma. \quad (10.92)$$

A graph of d/y_∞ versus y_d/y_∞ is shown in Fig. 202b. From this it is clear that the ribbon width is reduced to one-half its value in the infinite solid for a depth $d = 1/4\, y_\infty$. Figure 203 shows an example of an observation. The surface exhibits growth steps; at every crossing of a growth step the ribbon width decreases. Figure 203a shows the

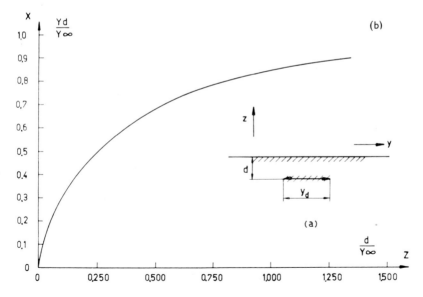

FIG. 202. (a) To illustrate the notation used in calculating the influence of the surface on the ribbon width. (b) Variation of relative ribbon width y_d/y_∞ as a function of distance to the surface d/y_∞.

gradual closing of a ribbon on approaching the surface in a wedge-shaped part of the crystal. The effect of anisotropy has been taken into account in the calculations of Spence[397b] and Siems *et al.*[252a]

(3) *The energy of a dislocation close to a free surface—the refraction effect.* One can account for the presence of the surface by taking for the outer cutoff radius in the expression for the energy the value $2d$, which is, incidentally, the harmonic average of the distance to the closest surface, on the one hand, and the distance to the other surface, which is supposed to be infinitely far, on the other hand. We then obtain in the isotropic case:

$$E = \frac{\mu b^2}{4\pi K} \ln \left(\frac{2d}{r_0} \right) \tag{10.93}$$

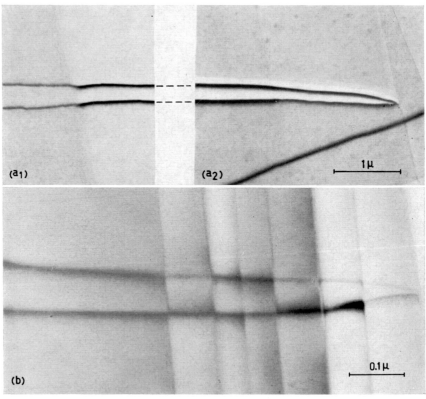

Fig. 203.   (a) Every time the ribbon is crossed by a surface step its width changes. (b) Gradual narrowing of a ribbon on approaching the surface in SnS_2.

where $K = 1$ for a screw and $K = 1 - \nu$ for an edge.

The expression for a screw can be obtained, for instance, by calculating the work done in creating the screw by shearing along the glide plane $x = d$. This energy is given by [see Eq. (10.89)]

$$E = \left(\frac{1}{2}\right) b \int_{r_0}^{R} \sigma_{xz} \, dy = \frac{\mu b^2}{4\pi} \int_{r_0}^{R} \left(\frac{1}{y} - \frac{y}{4d^2 + y^2}\right) dy \qquad (10.94)$$

where r_0 is respectively the outer and the inner cutoff radius

$$E = \frac{\mu b^2}{4\pi} \left[\frac{1}{2} \ln \frac{y^2}{y^2 + 4d^2}\right]_{r_0}^{\infty} . \qquad (10.95)$$

Taking into account that $R \gg 4d^2$ and $r_0^2 \ll 4d^2$ leads immediately to formula (10.93). From this expression it is obvious that the energy becomes a function of d. In the anisotropic case the expressions become more complicated (see Siems et al.[397c]), but unless the anisotropy is very large, as in graphite, the energy differs only by about 10% from that given by (10.93) if the proper values for μ and ν are used.

Consider now the situation of Fig. 204 where a dislocation is crossed by a surface step. The line energy of the dislocation in the 2 halves is now different, since d is different. As a consequence the shape of minimum energy is no longer the straight line AB. The dislocation is "refracted" and adopts some shape like ACB. It is easy to see that

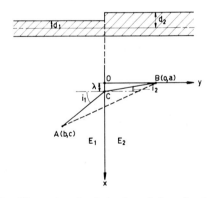

FIG. 204.   Illustrating the derivation of the refraction effect.

[397c] R. Siems, P. Delavignette, and S. Amelinckx, *Phys. Stat. Solidi* **2**, 421 (1962).

if the orientation dependence of the energy is neglected, Snell's law for the refraction of light applies:

$$\sin i_1/\sin i_2 = E_2/E_1. \qquad (10.96)$$

This can easily be shown by minimizing the total energy:

$$E = E_1 \sqrt{a^2 + \lambda^2} + E_2 \sqrt{(b - \lambda)^2 + c^2} \qquad (10.97)$$

with respect to the parameter λ

$$\frac{\partial E}{\partial \lambda} = \frac{\lambda E_1}{\sqrt{a^2 + \lambda^2}} - \frac{(b - \lambda) E_2}{\sqrt{(b - \lambda)^2 + c^2}} = 0. \qquad (10.98)$$

This last relation is obviously equivalent to Eq. (10.96) since $\lambda(a^2 + \lambda^2)^{-1/2} = \sin i_1$ and $(b - \lambda) [(b - \lambda)^2 + c^2]^{-1/2} = \sin i_2$. The "index of refraction" is just the ratio of the energies. From Eq. (10.93) it is clear that the energy only depends on d by a logarithmic factor. As a consequence of this, the index of refraction will not be very different from 1, unless the dislocations are at distances from the surface which differ by a factor which is appreciably larger than 1. Nevertheless an observable refraction effect still occurs if the angle of incidence is large. Figure 205 shows, for instance, the refraction of a ribbon. It is quite clearly apparent that the ribbon changes its width simultaneously. In the region where the ribbon is widest it is further from the surface and hence its energy is larger than in the region where it is narrowest. The behavior is therefore also in accord with the prediction of the previous paragraph. In a more exact treatment one has also to take into account the orientation dependence of the energy.

(4) *Edge dislocations in a thin plate.* As shown by Siems et al.[397c] and by Kroupa,[397d] an edge dislocation parallel to the surface of a thin isotropic plate causes a lattice bending

$$\theta = \frac{6b\xi}{t^3} (\xi - t) \qquad (10.99)$$

(ξ = distance of the dislocation from the surface). Intuitively it can be seen immediately that the bending angle for a foil of thickness t

[397d] F. Kroupa, *Czech. J. Phys.* **9**, 488 (1959).

should be of the order of b/t in analogy with the relation valid for a symmetrical tilt boundary.

The effect results again from the image forces with respect to the surfaces. For plates with a thickness of the order of 1000 A the angle becomes of the order of minutes of arc. It is possible to detect this angular difference between regions both sides of the dislocation as a striking difference in shade in transmission electron microscopy. This can be understood by reference to Fig. 106 which represents the diffracted intensity as a function of s the interference error. From this

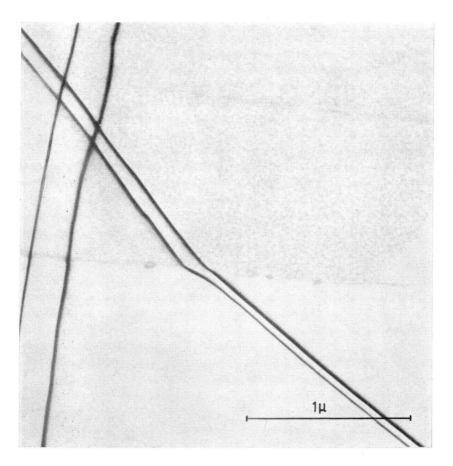

FIG. 205. Refraction by a growth step of a ribbon in SnS_2. [After R. Siems, P. Delavignette, and S. Amelinckx, *Phys. Stat. Solidi* **2**, 636 (1962).]

graph it is clear that if for one region the diffracted intensity is a maximum ($s \simeq 0$) a change in s of the order of $1/2\ t_e$ is the extinction depth, is sufficient to bring the diffracted intensity to zero, i.e., to cause maximum contrast. In order to cause the required change in s an angular difference of 10^{-3} rad is sufficient. In thick foils the effect is due to the asymmetry in s of the transmitted intensity. This can be checked by observing the shade difference in a dark field; the dark field image is *not* complementary to the bright field image.

It is possible to obtain a quantitative estimate of the angle by measuring the displacement of the appropriate Kikuchi line on crossing the dislocation. Such measurements have been made for SnS$_2$ crystals and values in agreement with formula (10.99) were found.[397c]

A complete verification of formula (10.99) requires knowledge of the foil thickness t, of the distance dislocation-surface ξ, and of the magnitude and direction of the Burgers vector. The foil thickness can be measured in the way described in Section 9g(2). As yet there is no simple and straight forward way for measuring the distance of a dislocation to the surface (Section 9e(4)(g) and Appendix C). However one can look for the largest observed ϑ and assimilate this with a dislocation in the middle of the foil. Formula (10.99) then gives a means for measuring *the magnitude* of the Burgers vector. On the other hand, an accurate measurement of the bending angle and of the foil thickness can be used to determine the approximate distance of the dislocation from the surface.

l. Miscellaneous

(1) *Dislocations due to epitaxy—Interfacial dislocations.* Frank and Van der Merwe[398] postulated in 1949 that epitaxial growth would give rise to dislocations at the interface if there is a misfit between the substrate and the oriented overgrowth. The interfacial dislocations are generated in order to relieve the stress set up as a consequence of the misfit. The fit between substrate and overgrowth can be assured by elastic deformation only within small islands separated by dislocations. The density of dislocations is evidently the smaller the better the fit. For very small misfits one can hope to observe the individual dislocations directly by electron microscopy. Just recently the first

[398] F. C. Frank and A. Van der Merwe, *Proc. Roy. Soc.* **A198**, 205, 216 (1949); **200**, 125 (1949); **A201**, 261 (1950).

direct evidence for such interfacial dislocations was presented independently by Delavignette *et al.*[399] and by Matthews.[400]

The first authors observed cross grids of dislocations in the contact plane between a chromium bromide substrate and an overgrowth consisting of small crystals of an unknown nature. The network of dislocations was confined to the periphery of the small crystals. The interfacial dislocations consist of single lines, while the dislocations in chromium bromide are known to dissociate in at least 4 partials (Fig. 206). Matthews[400] observed interfacial dislocations between evaporated sandwich layers of PbS and PbSe. He verified that the dislocations have edge character in accord with the theoretical picture. The actual elastic misfit can be deduced by measuring the spacing of the moiré fringes, which are superposed on the dislocations. This spacing is systematically somewhat larger than the one calculated on the basis of the lattice parameters by the use of formula (12.13). From this it can be concluded that the elastic misfit is only about 4%; the remaining 96% of the misfit is accomodated by the interfacial dislocations.

Excellent examples of interfacial dislocations were observed in uranium carbide (UC) containing platelike precipitates of uranium dicarbide (UC_2). The lattice spacings of these two compounds differ by somewhat less than 1% (UC: 3.54 A; UC_2: 3.50 A). The misfit at the interface is taken up by regularly spaced dislocations[401a].

Other nice examples of interfacial dislocations have been found in a Ni — Cr — Ti alloy at the interface of the η lamellae with the matrix.[401b]

(2) *The measurement of dislocation densities.* For an objective statistical estimate of the dislocation density in thin foils certain precautions are required. Different methods for measuring dislocation densities have been proposed. Bailey and Hirsch[401c] measure the total projected length R_p of dislocation lines in a given area A of a representative photograph. Assuming the dislocations to be oriented at random

[399] P. Delavignette, J. Tournier, and S. Amelinckx, *Phil. Mag.* [8] **6**, 1419 (1961).

[400] J. W. Matthews, *Phil. Mag.* [8] **6**, 1347 (1961).

[401a] J. L. Whitton, *Proc. Joint Conf. on Inorganic and Intermetallic Crystals, Univ. of Birmingham, Engl. 1963.*

[401b] H. F. Merrick and R. B. Nicholson, *Proc. 5th Intern. Conf. on Electron Microscopy, Philadelphia, 1962* K-8 (1962).

[401c] J. Bailey and P. B. Hirsch, *Phil. Mag* [8] **5**, 485 (1960).

Fig. 206.   Network of interfacial dislocations between an oriented overgrowth and the chromium bromide substrate. The nature of the overgrowth is not known. [After P. Delavignette, J. Tournier, and S. Amelinckx, *Phil. Mag.* [8] **6**, 1419 (1961).]

with respect to the foil plane the dislocation density is given by:

$$\rho = 4R_p/\pi At \qquad (10.100)$$

where t is the foil thickness.

A somewhat faster method has been proposed by Ham.[401d] He uses a set of random lines of total length L drawn on the micrograph and counts the number of intersections N of dislocations and lines within the area. One finds that $R_p = \pi NA/2L$ and hence

$$\rho = 2N/Lt. \qquad (10.101)$$

A very similar method has been proposed by Keh.[266,401e] On the enlarged micrographs a cross grid of lines with lengths L_1 and L_2 is drawn and the average numbers $\bar{n}_1$ and $\bar{n}_2$ of intersections of dislocation lines with the two sets of grid lines is determined. The average dislocation density ρ is given approximately by

$$\rho = \left(\frac{\bar{n}_1}{L_1} + \frac{\bar{n}_2}{L_2}\right)\frac{1}{t} \qquad (10.102)$$

where t is the foil thickness.

Care should be taken because some dislocations may be out of contrast; tilting the specimen is therefore required.

Ham[401f] has demonstrated convincingly that around 60% of all dislocations are lost during polishing down of foils of an aluminium-silver solid solution. One should be aware of this difficulty when estimating dislocation densities because similar conclusions probably hold also for pure metals and other alloys.

(3) *Chemical decomposition under electron irradiation.* A. J. Forty[208,402,403a] has made a detailed study of the decomposition of lead iodide under electron irradiation in the microscope. He finds that in the initial stage of the decomposition excessive climb takes place, presumably as a consequence of vacancies created in the hot irradiated part, and precipitation on dislocations in the cooler parts.

[401d] R. K. Ham, *Phil. Mag.* [8] **6**, 1183 (1961).

[401e] A. S. Keh, *J. Appl. Phys.* **31**, 1501 (1960); E. C. Bain Laboratory Research Rept. No. 908.

[401f] R. K. Ham, *Phil Mag.* [8] **7**, 1177 (1962).

[402] A. J. Forty, *Phil. Mag.* [8] **6**, 895 (1961).

[403a] A. J. Forty, *Phil. Mag.* [8] **6**, 587 (1961).

This climb process reveals itself by the formation of flattened helical dislocations. Climb seems to proceed at a much larger rate parallel to the basal plane than perpendicular to it, causing the extremely flat shape of the helices. Intersection of the helix with itself results in the formation of dislocation loops.[134] In a later stage of the decomposition polyhedral cavities are formed and metallic lead precipitates. Similar studies have been made on sodium chloride[403b] and potassium chloride.[403c] In the initial stage of irradiation small dislocation loops and lenticular shaped cavities are formed. At higher irradiation doses cubic cavities are present. The cubic cavities do not seem to migrate as opposed to the "bright patches" seen in lead iodide.

(4) *Surface effect associated with ribbons.* Where a ribbon emerges in a surface step on a cleaved surface, steps equal in height to $h_1 = \bar{\nu} \cdot \bar{b}_1$ and $h_2 = \bar{\nu} \cdot \bar{b}_2$ are formed on the lateral surface of the step; $\bar{b}_1$ and $\bar{b}_2$ are the Burgers vectors of the partials of the ribbon; $\bar{\nu}$ is the unit normal on the surface step. The sign of the step depends on the sign of the Burgers vectors and on the exact way in which cleavage took place. Different configurations are possible; they are represented schematically in Fig. 207.

Lenghtening of these steps results in a change of the free surface area. A lateral displacement of the dislocation over dx results in an increase in surface energy of $\Delta E = h\sigma dx$ (σ is specific surface energy). This means that effectively a force $F = h\sigma$ is acting on the emergence point of the dislocations, tending to shorten the step.

The surface energy increase that accompanies a lateral displacement over a unit distance is of the order of 10^3 ergs/cm^2 $\times$ 10^{-8} cm $=$ 10^{-5} erg/cm. The change in self-energy of the partial due to lengthening resulting from this displacement is of the order $\frac{1}{4}\mu b^2 \simeq 10^{11} \times 10^{-16}$ erg/cm, i.e., of the same order of magnitude. As a consequence of the forces acting on their emergence points the 2 partials are pulled apart if the configuration of steps is that of Fig. 207d. This last configuration is the easiest to observe and Fig. 208 is an example.

The equilibrium shape has been studied quantitatively by Gevers et al.[403d] The total energy of the configuration, i.e., (1) the interaction energy, (2) the self energy, (3) the stacking fault energy, and (4) the surface energy, is minimized by varying the shape of the partials.

[403b] T. Hibi and K. Yada, *J. Electromicroscopy* (*Tokyo*) **9**, 101 (1960).

[403c] A. J. Forty, *Phil. Mag.* [8] **7**, 709 (1962).

[403d] R. Gevers, S. Amelinckx, and P. Delavignette, *Phil. Mag.* [8] **6**, 1515 (1961).

From this equilibrium configuration it is possible to deduce the specific surface energy σ if the other material constants are known.

(5) *Anti-phase boundaries and dislocations.* Many solid solution alloy systems show a tendency for ordering when heat-treated below a critical temperature. In the copper-zinc system, for instance, the alloy of composition CuZn will adopt the cesium chloride structure. The copper-gold system is of particular interest in this respect because of the various ordered structures that can occur. We discuss in particular the alloy CuAu.[404,405] When heat treated just below 380°C it orders adopting a face-centered tetragonal layer structure, so called CuAu I. Layers perpendicular to the c axis consist alternatively of copper and gold. Since the ordering starts from a number of nuclei

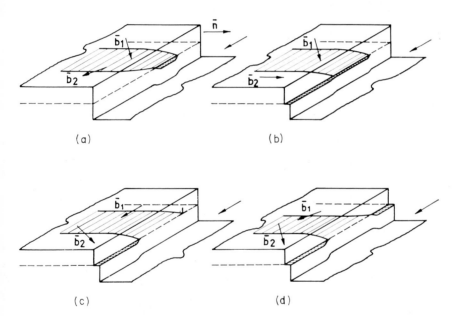

FIG. 207.   Surface topography for dislocation ribbons emerging in a cleavage step. (a) The 2 steps are in the opposite sense. A ledge connects the 2 emergence points. (b) The 2 steps are in the same sense and at the same side of the ribbon. (c) A step is attached to only one of the dislocations. (d) Steps on both sides of the ribbons.

[404] A. Glossop and D. Pashley, *Proc. Roy. Soc.* **A250**, 132 (1959).

[405] S. Ogawa, D. Watanabe, H. Watanabe, and T. Komoda, *Acta Cryst.* **11**, 872 (1958).

within a single crystal it is clear that where domains meet there is a large probability that a gold layer will join a copper layer and build a so-called antiphase (A.P.) boundary. In a thin foil where the A.P. boundaries extend from top to bottom such boundaries are either

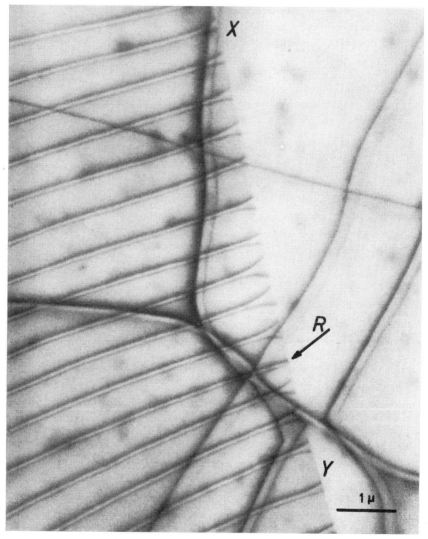

FIG. 208.   The effect of Fig. 207d as observed in graphite. A sequence of ribbons emerges in a cleavage step.

closed or end at dislocations.[404] Conversely the passage of a dislocation of which the Burgers vector has a component parallel to the c axis will generate an antiphase boundary. To illustrate this last point we refer to Fig. 209. The Burgers vectors in this structure, which is tetragonal and based on a slightly deformed face-centered cubic lattice ($c/a = 0.92$), are of the type $(a/2) \langle 110 \rangle$. Such vectors are either perpendicular to the c axis (which is along a cube edge of the fcc lattice) or have a component $c/2$ parallel to the c axis and therefore give just the required shift to make a gold layer join a copper layer.

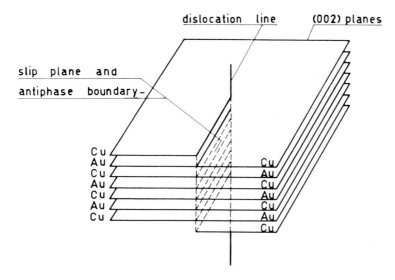

FIG. 209. Illustrating the relation between dislocations and antiphase domain boundaries in the alloy CuAu I.

When CuAu I is heated at a temperature between 380°C and 410°C, which is the disordering temperature, a structure develops called CuAu II. It consists in fact of a periodic array of antiphase boundaries, and was the first such structure to be observed directly in the electron microscope[404-406] (Fig. 210). The period, i.e., the distance between antiphase boundaries, is about 20 A in accord with the structure deduced from X-ray diffraction[407] and of which 1 unit cell is shown

[406] D. Pashley and P. Presland, *J. Inst. Metals* **87**, 419 (1958, 1959).

[407] C. H. Johansson and J. O. Linde, *Ann. Physik* [5] **25**, 1 (1936).

in Fig. 211. It consists in fact in a regular succession of domains containing 5 unit cells, of the CuAu I structure. Hence the new unit cell is 10 unit cells long. The domain boundaries are always along cube planes, and the majority of them are perpendicular to the foil plane.

Fig. 210.   (a) Periodic antiphase boundaries in CuAu II. (Courtesy of Glossop and Pashley.)

The origin of the contrast at these perpendicular boundaries is somewhat obscure. According to Glossop and Pashley[404] the contrast is due to interference between the directly transmitted beam and the doubly diffracted beam which has suffered a phase change. On the other hand, Ogawa et al.[405] have explained the line structure as being due to a periodic change in lattice parameter. In fact it is not unreasonable to assume that in the antiphase boundary the distance between like atoms will be different from the distance between unlike atoms within the cells. If this is true contrast can arise in the same way as for an edge dislocation seen end-on. The s value along a column in the boundary will be different from one within the cell and hence the transmitted and scattered intensities will be different.

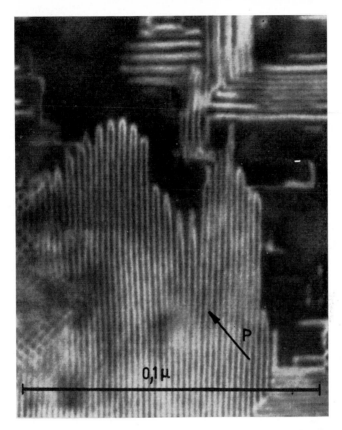

FIG. 210. (b) The effect of a dislocation in P on the pattern of antiphase boundaries. (Courtesy of Glossop and Pashley.)

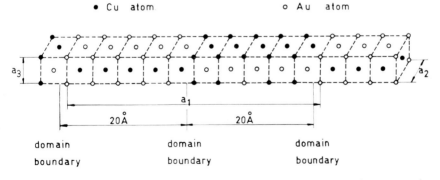

FIG. 211. One unit cell of the ordered alloy CuAu II. (After Johannson and Linde.)

Perio and Tournarie[408a] have made still another proposal. They assume that the composition of the alloy is "modulated" with a period equal to the distance between antiphase boundaries. The boundaries would be visible because of a deficiency of copper in them. This would mean that the visibility is due at least partly to absorption contrast, and it would explain the noncomplementary nature of the bright and dark field image. Perio and Tournarie's assumption explains the diffraction pattern in a natural way. It also explains why contrast can still be observed in a nonperiodic array.

More recently Hashimoto et al.[408b] have given a detailed dynamical theory. In this theory the image results from interference between the direct beam and the beam diffracted by the superstructure due to the periodic array of antiphase boundaries. However this theory does not explain the contrast at an isolated perpendicular antiphase boundary.

For antiphase boundaries either parallel or inclined to the foil plane the origin and form of the contrast is of the same nature as for stacking faults. A discussion has been given in Section 9h(7).

The transition between CuAu I and CuAu II has been studied in the microscope by Pashley and Presland.[406] Contrary to what one might expect dislocations do not seem to play a role in the transformation. The new domains with a width of 20 Å are nucleated within the larger domains of the CuAu I structure. The interaction between a family of newly formed and growing A.P. boundaries of the CuAu II structure and a pre-existing boundary PQ of the CuAu II structure, ending in dislocations, is shown in Fig. 212. An observed example is reproduced in Fig. 210b.

Similar periodic antiphase structures occur in Cu_3Pd,[409,410] Au_3Mn,[411] Ag_3Mg,[412] Cu_3Pt, Au_3Zn, and Au_4Zn[405]; these were deduced from electron diffraction work. Nonperiodic antiphase boundaries occur in most ordered alloys and they have been observed in a number of them, especially in relation with dislocations in Fe_3Al,[294] $AuCu_3$.[292]

[408a] P. Perio and M. Tournarie, Acta Cryst. 12, 1044 (1959).

[408b] H. Hashimoto, M. Mannami, and T. Naiki, Phil. Trans. Roy. Soc. A253, 459 (1961).

[409] D. Watanabe, M. Hirabayashi, and S. Ogawa, Acta Cryst. 8, 510 (1955).

[410] D. Watanabe and S. Ogawa, J. Phys. Soc. Japan, 11, 226 (1956).

[411] D. Watanabe, J. Phys. Soc. Japan 13, 535 (1958).

[412] K. Fujiwara, M. Hirabayashi, D. Watanabe, and S. Ogawa, J. Phys. Soc. Japan, 13, 167 (1956).

(6) *The observation of domain structures.* (a) Introduction. Although the title of this book does not cover this subject explicitly we shall nevertheless discuss very briefly the possibility of observing ferromagnetic, antiferromagnetic, and ferroelectric domain walls in the electron microscope. The justification for this degression is that one expects possible interactions between domain walls on the one hand and dislocations and other imperfections on the other hand.

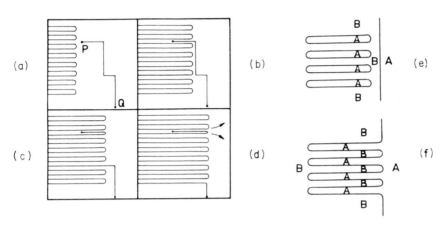

FIG. 212.   (a), (b), (c), and (d) Interaction between a set of growing antiphase boundary loops and a boundary ending in dislocations *P* and *Q*. (e) and (f) Mechanism generating the zig-zag patterns which are frequently observed. Compare with Fig. 210b. (After Glossop and Pashley.)

(b) Ferromagnetic domain walls.   Fuller and Hale[413] have been the first to demonstrate how ferromagnetic domains can be revealed in the electron microscope. As a consequence of the changing magnetic field in a Bloch or Néel wall the Lorentz force acting on the electrons and hence the deflection angle of the electron beam changes gradually as one passes from one domain to the adjacent one. At the exit face of the thin foil, i.e., in the well-focused image this does not give rise to appreciable contrast. However, at some distance behind (or in front of) the true focal plane one observes a black or a white line, depending on the sense of the magnetization both sides of the wall. When using a small aperture the unsharpness resulting from defocusing can be tolerated and for instance dislocations can still

[413] H. W. Fuller and M. E. Hale, *J. Appl. Phys.* **31**, 238 (1960).

be observed. Figure 212[1] shows a sequence of domain walls in a cobalt foil; a stacking fault is still faintly visible. The formation of the black and white images is shown schematically in Fig. 212[2] (a) for the case where the microscope is focused on a plane behind the foil plane. The reverse contrast would be observed if the local plane is brought in front of the specimen; this is made clear by Fig. 212[2] (b). Knowing the location of the focal plane with respect to the exit face of the foil it is possible to decide whether a wall causes divergency or convergency of the electron beam and hence the sense of the magnetization both sides of the wall can be determined.

In thin foils the magnetic field is often parallel to the foil surface and to the domain wall. For a 180° wall between two domains for which these last conditions are satisfied and which moreover is perpendicular to the foil the wall thickness can be measured. The average of the width of parallel bright and dark traces measured on the same

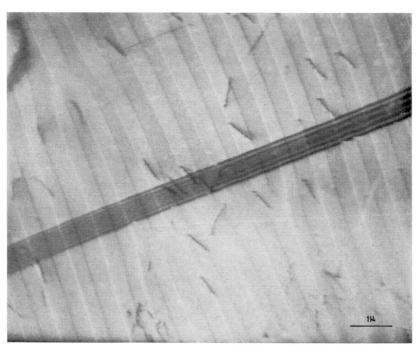

Fig. 212[1]. Sequence of ferromagnetic walls in cobalt. Stacking faults and dislocations are still visible. (Courtesy J. Silcox.) The domain walls are lying along the "easy" axis [0001].

photograph is a measure for the wall thickness. For a $180°$ wall in iron one finds for instance about 2000 A.[413,414]

Within the domains a fine structure consisting of fine striations is sometimes observed. This is for instance the case in Permalloy films. These striations are due to fluctuations of the local magnetization vector with respect to the mean direction. The deviations create

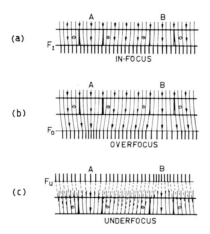

FIG. 212[(2)]. Origin of black and white lines at magnetic domain walls (a) In focus: no contrast. (b) Over focus: A is marked by a black line, B by a white line. (c) Under focus: A is marked by a white line, B by a black line.

small stray fields which are revealed as small fluctuations of electron intensity. Fuller and Hale (*loc. cit.*) have shown that the magnetization vector is *perpendicular* to the directions of these fine striations. The sense of this vector is such that the nearest bright wall is at the left and consequently the nearest dark wall is at the right of it if the focal plane is behind the foil.[413,415] These remarks show that the electron microscope allows a detailed determination of the domain characteristics.

(c) Antiferromagnetic domains. It is not clear at present whether antiferromagnetic boundaries can be revealed by the electron microscope in all cases. It has however been shown that in nickel oxide, so called T-domain boundaries can be revealed. Nickel oxide is cubic and it has the sodium chloride structure above the Néel point

[414] J. T. Mitchalak and R. C. Glenn, *J. Appl. Phys.* **32**, 1261 (1961).
[415] E. Fuchs, *Z. Angew. Physik* **13**, 157 (1961).

(525°K). Below this temperature it is antiferromagnetic.[416] The spin structure is such that one family of (111) planes has a ferromagnetic arrangement of spins, the spin direction being also in these (111) planes. Successive (111) sheets have oppositely oriented spins, making the three dimensional arrangement antiferromagnetic. As a consequence of the magnetic ordering the lattice becomes slightly deformed: it contracts along the $\langle 111 \rangle$ direction which is perpendicular to the (111) sheets. The symmetry is lowered, it becomes approximately rhombohedral ($\alpha = 90°4'$) (in fact orthorhombic). On cooling through the Néel temperature domains are formed as a result of contractions along different $\langle 111 \rangle$ directions. A fraction of the walls between such domains are twin planes,[416] so called T-domain walls. The spin direction is in this case the same in both domains, but the ferromagnetic (111) sheets are in twin relationship. Low energy domain walls of this type adopt (110) and (100) orientations. Such domain walls have been observed in nickel oxide.[417] Figure 212[(3)](a)

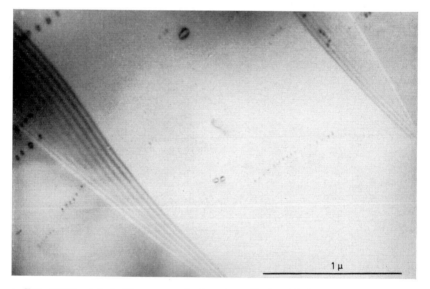

Fig. 212[(3)]. (a) Antiferromagnetic domain walls in nickel oxide. The specimen plane is a cube plane; the walls are inclined under 45° (courtesy P. Delavignette).

[416] G. A. Slack, *J. Appl. Phys.* **31**, 1571 (1960); W. L. Roth and G. A. Slack, *ibid.* p. 325S; W. L. Roth, *ibid.* p. 2000.

[417] P. Delavignette and S. Amelinckx, *Appl. Phys. Letters* **2**, 236 (1963).

shows an example. The contrast is due to the difference in lattice above and below the domain wall. A fringe system somewhat similar to stacking fault fringes or wedge fringes results [Fig. 212$^{(3)}$(a)]. The fringe system has been studied in detail by Gevers et al.[417a]
The main properties are:

(1) The bright field image is asymmetrical, i.e., the first and last fringes are opposite in nature.

(2) The nature of the first and last fringes only depends on the sign of Δs (i.e., the s difference for the lattices both sides of the boundary).

(3) The fringe spacing close to the entrance face may be different from that close to the exit face.

(4) In the dark field image the first and the last fringes have the same nature.

(5) Bright and dark field images are similar at the entrance face, but pseudocomplementary at the exit face.

From the nature of the extreme fringes in bright and dark field images one can deduce the spin orientation in the different domains in NiO, or the direction of polarization of the ferroelectric domains in BaTiO$_3$ (see Section 10.1. 6d). The domain walls are found to move frequently in a direction perpendicular to themselves; they interact visibly with dislocations. The orientation difference between domains both sides of a wall can be deduced from the doubling of Kikuchi lines on a selected area diffraction pattern taken accross a wall.

(d) Ferroelectric domains.   Ferroelectric domains have so far only been observed by electron microscopy in barium titanate. Several authors have independently produced pictures of them.[418-420] Barium titanate is cubic above the transition temperature of 120°C; between 0°C and 120°C it is tetragonal. On passing through the transition temperature, domains result because the spontaneous polarization is different in different parts of the crystal. This polarization is such that the tetragonal c axis of the domains coincides with one of the three $\langle 001 \rangle$ directions of the cubic phase. As a consequence 180° and 90° domain walls are possible. The walls tend to adopt {110} orientations.

[417a] R. Gevers, P. Delavignette, H. Blank, and S. Amelinckx, *Phys. Stat. Solidi,* **4**, 383 (1964).

[418] H. Ffisterer, E. Fuchs, and W. Liesk, *Naturwissenschaften* **49**, 178 (1962).

[419] M. Tanaka, N. Kitamura, and G. Honjo, *J. Phys. Soc. Japan* **17**, 1197 (1962).

[420] H. Blank and S. Amelinckx, *Appl. Phys. Letters* **2**, 140 (1963).

They are therefore either perpendicular to, or inclined under, 45° with respect to the plane of the foil, which is in view of the crystal habit a cube plane. The perpendicular walls are observed as lines, separating regions of different brightness [Fig. 212$^{(3)}$(b)]; whereas the inclined

Fig. 212$^{(3)}$.   (b) Ferroelectric domain walls in barium titanate (courtesy H. Blank).

walls are seen as systems of fringes similar to stacking fault fringes. The origin of the contrast is similar to that due to the antiferromagnetic walls in nickel oxide.

11. DIRECT RESOLUTION OF THE CRYSTAL LATTICE

a. Introduction

With present high-resolution microscopes it is possible to resolve the individual lattice planes in suitably chosen thin crystals. This was first achieved by Menter[421] using sublimed flakes of platinum and copper phthalocyanine. Platinum phthalocyanine is a large ring molecule with a planar structure of fourfold symmetry, the platinum atoms being located in the center. The platinum atoms can be regarded as lying in $(20\bar{1})$ planes of the crystal structure with a spacing of about 12 A and practically normal to the (001) habit plane. Therefore the circumstances are particularly favorable since the strongly scattering planes of platinum atoms are in fact imbedded in a matrix of much smaller scattering power. It was found that a fringe pattern with a spacing of 12 A due to the $(20\bar{1})$ planes could indeed be observed as shown in Fig. 213.

b. Image Formation

The image of the lines is formed by the interference between the directly transmitted beam and the beams diffracted by the $(20\bar{1})$ and $(\bar{2}01)$ lattice planes with a spacing of 11.94 A and the $(\bar{4}02)$ planes with a spacing of 5.97 A. This can be proved by using apertures of different size so that fewer and fewer beams contribute to the image. The situation is similar to the image formation by a periodic structure in the optical microscope according to the Abbe theory. A detailed theory of the image formation has been given by Hashimoto et al.[408b]

It is possible to obtain their result for a plane parallel plate of thickness t immediately by the use of formula (9.95). We shall calculate the intensity distribution in the well-focused image (which is equivalent to the intensity distribution at the surface of the foil), as a function of $\bar{r}$, the position vector of a point on the back surface. We choose the reference system in such a way that the x axis is along

[421] J. Menter, *Proc. Roy. Soc.* **A236**, 119 (1956).

the reciprocal lattice vector $\bar{n}$, and the y axis perpendicular to it; one has then $\bar{n} \cdot \bar{r} = nx$. One has a constant value for the wave function, i.e., for the intensity at the back surface, along lines $\bar{n} \cdot r =$ constant, i.e., the image consists of straight lines perpendicular to the diffraction vector. Further it is clear since x only enters in the factor $\exp(2\pi i n x)$ that the system of parallel lines will be periodic with period $1/|\bar{n}| = d_{hkl}$.

FIG. 213. Lattice planes $(20\bar{1})$ in platinum phthalocyanine directly observed in the microscope. The upper part of the pattern contains one more fringe than the lower part, as a consequence of the presence of a dislocation. (Courtesy of J. W. Menter.)

The expression for the intensity is [from Eqs. (9.96), (9.97), (9.123) and (9.124)]

$$I = TT^* + SS^* + TS^* e^{-2\pi i n x} + T^*S e^{+2\pi i n x} \qquad (11.1)$$

which can be written as

$$I = I_t + I_s + 2\sqrt{I_s I_t}\sin(2\pi n x + \varphi), \qquad (11.2)$$

where

$$\sin \varphi = TS^* + T^*S = -\frac{2s}{\sigma^2 t_e} \sin^2 \pi \sigma z$$

and

$$\cos \varphi = i(T^*S - S^*T) = -\frac{2}{\sigma t_e} \sin \pi \sigma z \cos \pi \sigma z,$$

and consequently

$$\text{tg } \varphi = \frac{s}{\sigma} \text{ tg } \pi \sigma t. \tag{11.3}$$

I_s and I_t are defined by Eqs. (9.127) and (9.129); they have to be taken for $z = t$.

Since absorption has been neglected so far $I_t = 1 - I_s$ and formula (11.2) reduces to

$$I = 1 + 2\sqrt{I_s(1 - I_s)} \sin (2\pi n x + \varphi). \tag{11.4}$$

From this expression it follows that the image consists of a uniform background equal to the incident intensity 1, modulated by a sinusoidal ripple with has maximum amplitude if $I_s = 1/2$. The amplitude of the ripple is then unity (the contrast which is twice the amplitude is 2).

For $s = 0$ one has $T = T^*$ and $S = -S^*$ and from Eq. (11.3) follows that $\varphi = 0$; the intensity is now simply

$$I = 1 + \sin 2\pi \frac{t}{t_e} \sin 2\pi n x. \tag{11.5}$$

For s different from zero, $\varphi \neq 0$ and the image will be shifted. It is therefore clear that in general the observed image will not translate directly variations in "mass thickness." The period of the image is the same as that of the period of the variations in mass thickness; the phase depends, however, on the thickness and on the interference errors.

c. Discussion of the Image Profiles

In practice the crystal usually is not an ideal plate; it has a wedge shape or it may be bent. Both deviations from the ideal plate shape are reflected in the fringe pattern. We will briefly discuss these effects. A detailed discussion is given by Hashimoto et al.[408b]

27

(1) *Wedge-shaped crystal.* (1) We will first assume $s = 0$. Let both the wedge edge and the lattice planes be parallel to the y axis. The thickness may then be represented as $t = px$ where p is the tangent of the wedge angle. The intensity distribution in the fringe system becomes

$$I = 1 + \sin \frac{2\pi p x}{t_e} \sin 2\pi n x. \qquad (11.6)$$

When considering the order of magnitude of the parameters involved, i.e., $t_e \simeq 5 \times 10^{-6}$ cm; $1/|\bar{n}| = 2 \times 10^{-7}$ cm; $p < 10^{-1}$ it becomes clear that the period of the first factor is $t_e/p \simeq 5 \times 10^{-5}$ cm and it is large compared to the period of the second factor which is 2×10^{-7}. Therefore the pattern is a modulated sine wave ripple on a constant background. Along lines $x = kt_e/2p$; k = integer [i.e., where the crystal has a thickness $t = k(t_e/2)$] the contrast disappears and the fringes shift phase by half an interspacing (Fig. 214a,b) i.e., contrast reversal takes place. Evidently this phenomenon will cause spacing anomalies which are purely electron optical.

(2) If the fringes are perpendicular to the edge of the wedge-shaped crystal we can put $t = py$ and the intensity is now described by

$$I = 1 + \sin \frac{2\pi p y}{t_e} \sin 2\pi n x. \qquad (11.7)$$

Along the lines $y = k(t_e/2p)$, i.e., for which $t = k(t_e/2)$, the fringes again disappear. At both sides of the line, they gradually increase in contrast; one set is shifted with respect to the set on the other side of the line over half the interfringe spacing [Fig. 214c)] as a consequence of contrast reversal.

In the more general case where the fringes have an arbitrary direction with respect to the direction of the wedge edge, contrast reversal again takes place along the lines of equal thickness $t = kt_e/2$.

If $s \neq 0$ it is clear from Eq. (11.3) that the phase angle φ will also change with thickness. Apart from the spacing anomalies just mentioned, there is also a continuous change in spacing if the wedge edge is parallel to the fringes. If the lattice rows are perpendicular to the wedge edge, the fringes are no longer perpendicular to this edge since there will be a continuous lateral shift of the fringes due to the change of φ with thickness (Fig. 214d).

(2) *Bent crystal plate.* (i) Crystal bent about the y axis. There is now a continuous change in s when going in the x direction. The fringes are still parallel to the y axis. Choosing the origin at the point where $s = 0$, we can write $s \simeq |\bar{n}|\, x/R$ for $x \ll R$, where R

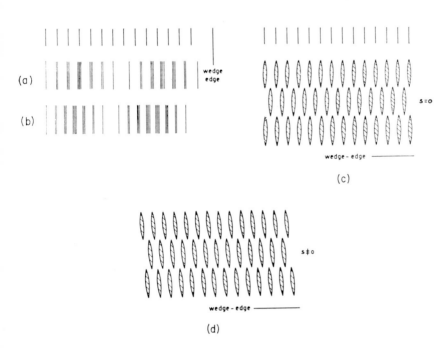

(a)

(b)

(c)

(d)

FIG. 214.   Direct resolution of the lattice for a wedge-shaped crystal. (1) *The wedge edge is parallel to the fringes.* (a) Along the lines $t = kt_e/2$ the fringes shift over one-half the interspacing distance giving rise to spacing anomalies. The position of the lattice rows is indicated by the small bars. The picture is valid for $s = 0$ only. (b) If $s \neq 0$ the phase angle changes gradually with the thickness giving rise to a fringe spacing which differs from the lattice spacing.
    (2) *The wedge edge is perpendicular to the fringes.* (c) Along the lines $t = kt_e/2$ the fringes shift again over one-half an interfringe spacing in a direction perpendicular to the fringes. For $s = 0$ the fringes remain parallel to the lattice rows. (d) If $s \neq 0$ the phase angle changes continuously with thickness. The fringes form a small angle with the lattice rows.

is the radius of curvature of the specimen. The intensity distribution is now given by Eq. (11.4) where I_s and φ now depend on x through s (or σ).

The contrast will disappear if $I_s = 0$, i.e., for $\sigma t = k$ (integer different from zero) or for

$$s = \pm \sqrt{\frac{k^2}{t^2} - \frac{1}{t_e^2}} \tag{11.8}$$

for these values of σt, tg $\varphi = 0$.

The contrast will be optimum for $I_s = 1/2$, i.e., for $\sin \pi \sigma t = \pm (1/\sqrt{2})\sigma t_e$. This condition will be satisfied for $s = 0$ if $t = \frac{1}{4}t_e \pm kt_e/2$ (k: integer); tg φ for these values of t is zero. The fringe system is now for $k = 0$ as represented in Fig. 215.

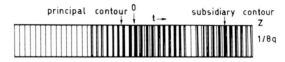

FIG. 215. Lattice fringes for a bent crystal of thickness $t = 1/4t_e$; the contrast is now optimum for $s = 0$, i.e., in the center of the main contour.

(ii) Crystal bent about the x axis. The quantity s is now changing along the y axis, i.e., along the lattice rows, and the contrast will then vary along the fringes. Along the lines for which $I_s = 0$ or 1 the contrast again disappears whereas it will be optimum for $I_s = 1/2$. As s varies the phase angle φ varies also. Therefore the fringes will form a small angle with the lattice rows. Moreover there is a contrast reversal at each line where the contrast disappears.

(3) *Arbitrary deformation.* For a wedge-shaped crystal, which is bent at the same time, a combination of spacing anomalies occurs. This can give rise to the formation of ending fringes. One has to be aware of this possibility when studying dislocations by this method.

d. Graphical Construction of Fringe Profiles

One can devise a simple complex plane construction for the fringe profile which may be useful. The construction is somewhat similar to that given for stacking fault and antiphase boundary fringes. The first part of the construction is the same (Fig. 216). The angle $\angle A_1OA_2 = \pi\sigma t$, $|OA_1| = \cos \pi\sigma t$,

$$|A_1A_2| = \sin \pi\sigma t, \ |A_1B| = s/\sigma \sin \pi\sigma t = \gamma \sin \pi\sigma t, \ |BP| = \frac{1}{\sigma t_e} \sin \pi\sigma t$$

$$= \sqrt{1 - \gamma^2} \sin \pi st = |S|, \ OB = |T|.$$

As x varies the point P describes the dotted circle. The point corresponding to $x = 0$ is P_0. The background amplitude is 1; it is given by OQ. The amplitude for an arbitrary x is given by OP where $\psi = 2\pi\, x/d_{hkl}$. The phase angle can easily be deduced from the construction, too; it is simply $\angle\, A_1OB$ since (see Eq. 11.3):

$$| \operatorname{tg} \varphi | = + \frac{\gamma \sin \pi\sigma t}{\cos \pi\sigma t} = + \gamma \operatorname{tg} \pi\sigma t. \qquad (11.9)$$

From this construction can, e.g., be deduced that the dotted circle shrinks to a point if $\sin \pi\sigma t = 0$ or $1 - \gamma^2 = 0$. It is then clear that the contrast $C = 2BP = 2\sqrt{1 - \gamma^2} \sin \pi\sigma t$ disappears. In the kinematical region where $\gamma \simeq 1$ the fringes will be practically invisible.

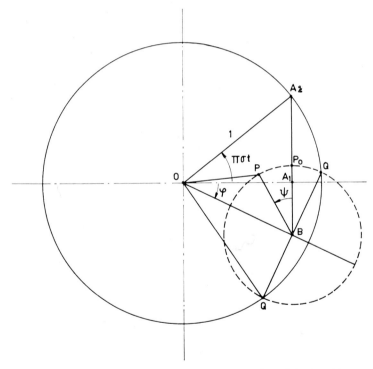

Fig. 216. Complex plane construction for the profile of lattice fringes. The angle $A_1OA_2 = \pi\sigma t$; $OA_1 = \cos \pi\sigma t$; $A_1A_2 = \sin \pi\sigma t$; $A_1B = -(s/\sigma) \sin \pi\sigma t$; $BP = \sqrt{1 - \gamma^2} \sin \pi\sigma t = | S |$; $OB = | T |$. As x varies P describes the dotted circle and OP gives the amplitude. The phase angle $\varphi = < A_1OB$.

Experimentally it is found that fringes are best visible in the contours, i.e., for small s (or small γ), and sin $\pi \sigma t \neq 0$.

The amplitude goes through zero if the dotted circle passes by 0. The condition for this to occur is $OB = BP$ or $\cos^2 \pi \sigma t + \gamma^2 \sin^2 \pi \sigma t = (1 - \gamma^2) \sin^2 \pi \sigma t$ or

$$\text{tg } \pi \sigma t = \pm (1 - 2\gamma^2)^{-1/2}. \tag{11.10}$$

If we require furthermore that C be as large as possible we must take $\gamma = 0$; one then has $\pi \sigma t = (\pi/4) \pm k(\pi/2)$ or $t = \frac{1}{4} t_e \pm k t_e/2$ in accord with the condition derived previously. This particular situation is shown in Fig. 217b. As P describes the dotted circle, OP fluctuates between 0 and $\sqrt{2}$, i.e., $C = \sqrt{2}$, corresponding to a contrast of 2 for the intensities.

We now show how the graphical construction can be used to derive the fringe pattern for a wedge crystal. We consider only the case $s = 0$ ($\gamma = 0$). As the thickness increases A_2 describes the circle with unit radius (Fig. 217a). If A_2 is in the point P_1 the radius of the dotted circle is zero, and $C = 0$; the phase φ is zero also. If A_2 comes in P_2 ($t = 1/4\ t_e$) the situation of Fig. 217b is created and the contrast is 2. If A_2 comes in P_3 the dotted circle coincides with the full line circle and OP is constant and equal to 1, again $C = 0$ (Fig. 217c). As P passes P_3 the phase changes by π and the contrast reverses, as mentioned already above (Fig. 217d).

e. Applications

The possibility of observing the Bravais lattice rows opens the way to a direct observation of dislocations. Figure 213 shows a terminating lattice row which is the configuration characteristic for a dislocation. Unfortunately the method is at present only applicable to crystals with a relatively large spacing. The smallest lattice spacing which could be resolved at present is 6.9 A, the spacing of the (020) planes in molybdenum trioxide.[422] A method allowing to represent lattice rows after some "geometrical" magnification, will be discussed further (Section 12).

Platinum phthalocyanide crystals, which have been subjected to fission fragment damage, have been studied by Bowden and Chadder-

[422] G. A. Bassett and J. W. Menter, *Phil. Mag.* [8] **2**, 1482 (1957).
[423] F. Bowden and L. Chadderton, *Nature* **192**, 31 (1961).

ton.[423] The damage appears as a local destruction of the lattice planes without any "recrystallization" in most cases. Some well-defined features consisting of "dislocation dipoles" are also observed. Such dipoles have been found earlier by means of moiré fringes in radiation-damaged graphite [see Section 10d(3)(d)].

Bowden and Chadderton also discovered that the electron beam causes a disarrangement of the phthalocyanide lattice. The fringe pattern due to the (20$\bar{1}$) planes does not disappear, but the diffraction

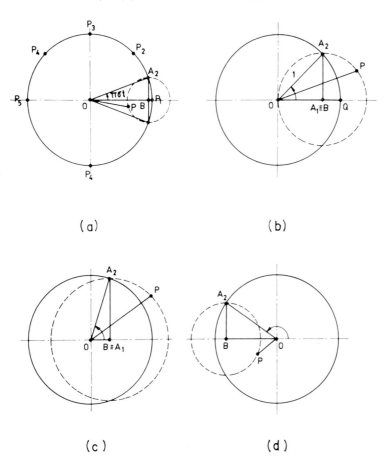

(a)

(b)

(c)

(d)

Fig. 217.   Graphical derivation of the fringe profile in the case $s = 0$ for a wedge-shaped crystal. The 3 graphs correspond to increasing thickness of the foil. (a) $t < 1/4t_e$. (b) $t = 1/4t_e$. (c) $1/4t_e < t < 1/2t_e$. (d) $1/2t_e < t < 3/4t_e$.

pattern shows that practically all reflections except the (20$\bar{1}$) and ($\bar{2}$01) fade gradually on irradiation. It appears as if the platinum atoms remain in place amidst a disordered matrix.

Komoda et al.[424] have observed twinning in copper phthalocyanide. The spacing of the lines in the image changes abruptly across a line from that corresponding to the (20$\bar{1}$) planes to that corresponding to the (001) planes.

Another substance for which the lattice could be resolved directly is Faujasite, a cubic sodium aluminosilicate with a unit cell of 24.8; the fringe spacing is 14.4 A.[425] After the early work of Menter on the phthalocyanides a number of workers have described their observations on other substances, indanthrene scarlet with spacings of 15.4, 19.3, and 28.1 A, nickel phthalocyanide with a spacing of 9.8 A,[426] and molybdenum trioxide with a spacing of only 6.93 A.[422] Observations were also made on the superlattice of antigorite (spacing 90 A)[427,428] and of the copper-gold II ordered alloy.[404,405]

Recently Mannami[238b] has calculated the effect of the presence of a dislocation on the fringe pattern.

12. Moiré Patterns[231b,408b,429–431]

a. Introduction

When 2 thin crystals of the same or similar crystal structure are superposed with a small orientation difference one observes in the electron microscope a fringe system resulting from interference between the directly transmitted beam and the doubly diffracted beam. If the orientation difference is small the doubly diffracted beam forms a very small angle with the incident beam and it enters the object aperture causing the observed interference effect. An example of this is shown in Fig. 218.

[424] T. Komoda, E. Suito, N. Uyeda, and H. Watanabe, *Nature* **181**, 332 (1958).

[425] J. W. Menter, *Proc. Stockholm Conf. on Electron Microscopy, 1956* p. 88 (1957).

[426] R. Neider, *Proc. Stockholm Conf. on Electron Microscopy, 1956* p. 93 (1957).

[427] G. W. Brindley, J. J. Comer, R. Uyeda, and J. Zussmann, *Acta Cryst.* **11**, 99 (1958).

[428] J. A. Chapman and J. Zussmann, *Acta Cryst.* **12**, 550 (1959).

[429] D. W. Pashley, J. W. Menter, and G. A. Bassett, *Nature* **179**, 752 (1957).

[430] H. Hashimoto and R. Uyeda, *Acta Cryst.* **10**, 143 (1957).

[431] G. A. Bassett, J. W. Menter, and D. W. Pashley, *Proc. Roy. Soc.* **A246**, 345 (1958).

b. Intuitive Theory

One can consider the crystal as a line grating, the lines being parallel to the lattice planes that are in the Bragg condition. When superposing two such line gratings with a small orientation difference ϑ a fringe system perpendicular to the considered lattice planes and with

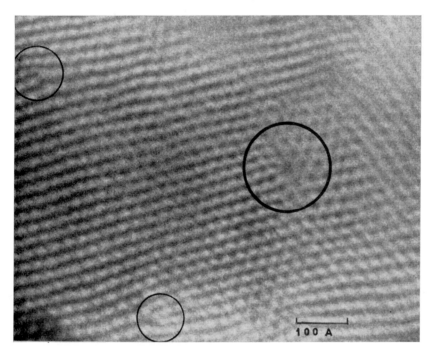

FIG. 218.   Moiré pattern from a sandwich of palladium-gold. The encircled areas contain dislocations. In this case 2 different operating reflections produce a two-dimensional dot pattern. (Courtesy of Bassett, Menter, and Pashley.)

a larger spacing results as visualized in Fig. 219a by means of an optical analog. The spacing is given to a good approximation by

$$D = d/\vartheta \tag{12.1}$$

where d is the spacing of the line gratings. If ϑ is small enough D becomes resolvable in the electron microscope although d is of atomic dimensions. Hence it is possible to produce some kind of "geometrical magnification" which makes it possible to observe lattice rows directly. In this case "the observed lattice rows" are perpendicular to the lattice rows used for the diffraction.

Superposing 2 line gratings with slightly different spacings d_1 and d_2 in the parallel orientation also produces fringes with a larger spacing D, given by

$$D = \frac{d_1 d_2}{d_1 - d_2}. \tag{12.2}$$

Figure 219b is the optical analog for the situation. The fringes are now parallel to the lines of the original grating. By the use of gratings which differ only slightly in parameter the distance D can again be resolved in the microscope.

A combination of a difference in spacing and in orientation leads to fringes of spacings

$$D = d_1 [(d_1 - d_2)^2 / d_2^2 + \vartheta^2]^{-1/2} \tag{12.3}$$

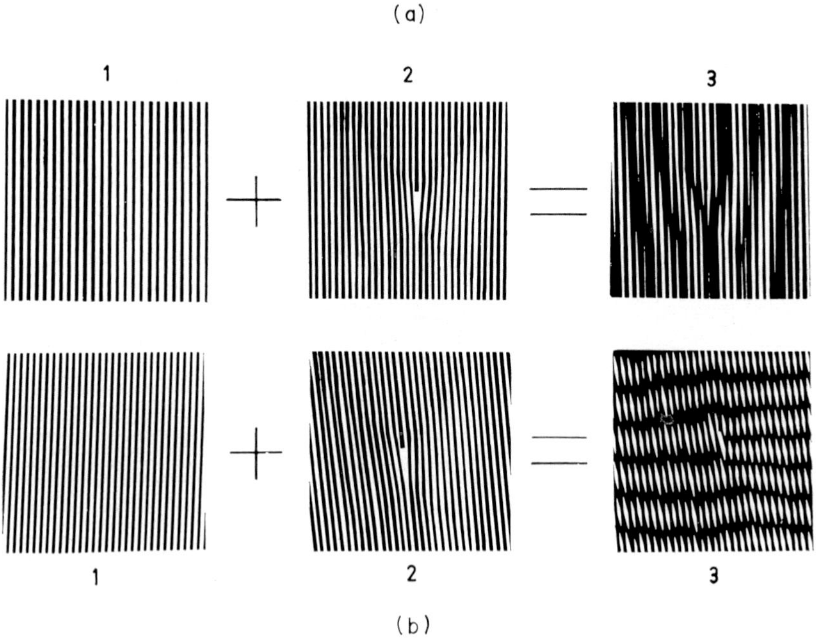

(a)

(b)

Fig. 219. Optical analog illustrating the formation of moiré fringes by superposition of 2 lattices. (a) 1, 2, 3: rotation moiré fringes. (b) 1, 2, 3: parallel moiré fringes. (After J. W. Menter.)

and having a direction which encloses an angle

$$\text{arctg } [(d_1 - d_2)/d_2 \vartheta] \qquad (12.4)$$

with the diffraction vector.

Experimentally one can use sandwiches which are formed by chance in specimens of layer structures, e.g., graphite, talc.[430] Some methods of specimen preparation favor the occurence of such sandwiches, e.g., shearing graphite between 2 object glasses.[432]

A more reproducible method for the production of suitable metal specimens was developed by Bassett et al.[431] Use is made of evaporated metal layers with only slightly different lattice parameters. The first layer is evaporated on a crystal face that causes epitaxial deposition, e.g., mica or sodium chloride. The second layer is evaporated on top of the first and grows again epitaxially. The sandwich can be freed from the substrate by dissolving the latter. A number of suitable sandwiches are listed in Table III.

TABLE III

Metal	Crystal spacing		Moiré spacing (A) when combined with parallel gold		Moiré magnification,
	$(02\bar{2})$	$(42\bar{2})$	$(02\bar{2})$	$(42\bar{2})$	M
Nickel	1.24	0.719	9.2	5.3	7.4
Cobalt	1.26	0.725	9.8	5.7	7.8
Copper	1.28	0.737	11.3	6.5	8.8
Rhodium	1.34	0.773	19.7	11.4	14.7
Palladium	1.37	0.792	29	17	21
Platinum	1.38	0.798	35	20	25

An example of a moiré pattern from a sandwich palladium-gold is shown in Fig. 218. The dot pattern results from the presence of 2 (or more) active reflections.

c. Geometry of Moiré Fringes: Exact Theory

The geometry of a moiré patterns will now be formulated in a more general way, which is due to Gevers.[231b] The sandwich crystal

---

[432] I. M. Dawson and E.A.C. Follett, *Proc. Roy. Soc.* **A253**, 390 (1959).

producing the moiré pattern is considered to be derived from a single crystal by displacing the top half (I) in a nonuniform way with respect to part II of the same column. The displacement is described by a function

$$\bar{\varDelta r} = \bar{u}(r).$$ (12.5)

The resulting phase shift between waves diffracted by the two parts of the crystal is

$$\alpha = 2\pi\bar{n} \cdot \bar{u}$$ (12.6)

if $\bar{s} \cdot \bar{u}$ is neglected. This expression is the same, of course, as for a stacking fault, except that $\bar{u}$ is now a function of $\bar{r}$. The intensities of the transmitted and scattered beam can be calculated from the set of Eq. (9.182). From these equations it is clear that the intensities only depend on the position as a consequence of the term $e^{i\alpha}$. The loci of points of equal intensity will thus be given by $\alpha =$ constant.

The expression for α is now transformed. Since $\bar{r}$ is a lattice vector and $\bar{n}$ a reciprocal lattice vector $\bar{n} \cdot \bar{r}$ is either an integer or zero. For small differences one has

$$- (\varDelta\bar{n}) \cdot \bar{r} = + (\varDelta\bar{r}) \cdot \bar{n} = \alpha'.$$ (12.7)

If the displacement $\varDelta\bar{r}(= \bar{u})$ is such that $\varDelta\bar{n}$ is independant of $\bar{r}$ the loci of equal intensity are straight lines perpendicular to $\varDelta\bar{n}$. In view of the fact that $i\alpha$ enters as the argument of an exponential function this line system will be periodic.

We now choose the reference system in the contact plane and such that $\bar{e}_x$ (unit vector along the x axis) is perpendicular to the fringes and $\bar{e}_y$ is parallel to them. We can then write

$$-\varDelta\bar{n} \cdot \bar{r} = Kx = x/\varLambda$$ (12.8)

where $\bar{K}$ is a vector of length $1/\varLambda$ and perpendicular to the fringes; $\varLambda$ is the period of the fringe system. $\bar{K}$ could be called the "moiré wave vector." One simply has from the definition and from the fact that $\varDelta\bar{n}$ is independent of $\bar{r}$:

$$\bar{K} = -\varDelta\bar{n}.$$ (12.9)

The vector $\bar{K}$ defines completely the geometry of the moiré pattern. The conditions under which a deformation $\bar{u}(\bar{r})$ gives rise to a moiré pattern consisting of straight fringes reduces in fact to the requirement that it must be possible to define $\bar{K}$ by relation (12.9).

We now consider the specific cases.

(i) Rotation moiré. We have

$$| \Delta\bar{n} | = 2n \sin \frac{\alpha_r}{2} \simeq n\alpha_r. \tag{12.10}$$

The vector $\Delta\bar{n}$ forms an angle of $90° - (\alpha_r/2)$ with $\bar{n}$. From this we conclude that for small α_r the fringes are parallel to $\bar{n}$ and that the spacing is

$$\Lambda_r = \frac{1}{n\alpha_r} \simeq d_n/\alpha_r \tag{12.11}$$

where d_n is the interplanar spacing corresponding to the reflection $\bar{n}$.

(ii) Parallel moiré. One has

$$\Delta\bar{n} = \alpha_p \bar{n}$$

where

$$\alpha_p = (d_n - d_{n'})/d_{n'}. \tag{12.12}$$

The fringe direction is now evidently perpendicular to $\bar{n}$; the spacing is still given by Eq. (12.11) provided one takes for α the expression (12.12); hence

$$\Lambda_p = d_n d_{n'}/(d_n - d_{n'}). \tag{12.13}$$

(iii) Mixed moiré. We can always decompose $\Delta\bar{n}$ into a rotation component $(\Delta\bar{n})_r$ and a parallel component $(\Delta\bar{n})_p$. The fringes being perpendicular to $\Delta\bar{n}$ they form an angle $-\beta$ with n; β is given by (Fig. 220):

$$\operatorname{tg} \beta = (\Delta n)_p/(\Delta n)_r = \alpha_p/\alpha_r. \tag{12.14}$$

Further one has (Fig. 220) that $| \Delta\bar{n} |^2 \simeq | (\Delta n)_r |^2 + | (\Delta n)_p |^2$

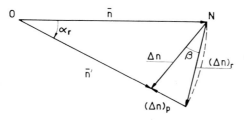

FIG. 220. Illustrating the derivation of the spacing for a mixed moiré pattern.

and hence

$$\frac{1}{\Lambda} = \sqrt{\frac{1}{\Lambda_r^2} + \frac{1}{\Lambda_p^2}}. \tag{12.15}$$

d. Intensity Distribution in the Fringes: Kinematical Theory

A simple kinematical theory of moiré fringes has been given by Rang.[433] We will use here a somewhat different approach similar to the one used for the kinematical theory of contrast due to a stacking fault parallel to the foil plane. It is sufficient, in fact, to use in Eq. (9.23) the adequate expressions for

$$\alpha = -2\pi\Delta\bar{n} \cdot \bar{r} = 2\pi\bar{K} \cdot \bar{r} = 2\pi x/\Lambda. \tag{12.16}$$

For the stacking fault α was a constant; this is no longer true.

We have from formula (9.23) for the amplitude A of the wave diffracted by the sandwich

$$A_s = A^{(1)} + A^{(2)} e^{i\alpha} e^{2\pi i s t_1} \tag{12.17}$$

where

$$\pi s A(z) = e^{\pi i s z} \sin \pi s z. \tag{12.18}$$

The upper indices refer to the first and second component of the sandwich, which are respectively t_1 and t_2 thick.

Relation (12.17) expresses, in fact, the interference between the 2 waves giving rise to the moiré fringes:

(1) the wave diffracted in crystal I, $A^{(1)}$ and transmitted by crystal II, (2) the wave transmitted through crystal I and diffracted by crystal II with amplitude $A^{(2)}$.

The exponential factor takes care of the correct phase relationship and of the phase shift due to the moiré displacement. On interpreting this formula one has to keep in mind that according to the kinematical approximation there is *no* loss in intensity on transmission, although some intensity is diffracted away.

The intensity of the diffracted beam becomes

$$I_s = AA_s^* = I_s^{(1)} + I_s^{(2)} + 2\sqrt{I_s^{(1)}I_s^{(2)}} \cos\left[\pi s(t_1 + t_2) + \alpha\right] \tag{12.19}$$

[433] O. Rang, *Proc. 4th Intern. Conf. on Electron Microscopy, Berlin, 1958* p. 371 (1960).

where $I_s^{(1)}$ and $I_s^{(2)}$ are the intensities diffracted by the separate components

$$I_s^{(i)} = A^{(i)}A^{(i)*} = \sin^2 \pi st/\pi^2 s^2. \tag{12.20}$$

Explicitly one can write

$$(\pi s)^2 I_s = 4 \sin^2 \left[\frac{1}{2}\pi s(t_1 + t_2)\right] \cos^2 \left[\frac{1}{2}\pi s(t_1 - t_2)\right]$$

$$- 4 \sin \pi st_1 \sin \pi st_2 \sin^2 \frac{\pi}{\Lambda}(x - x_0) \tag{12.21}$$

where $x_0/\Lambda = (s/2)(t_1 + t_2)$. This formula has been derived by Gevers[231b] as a limiting case of the dynamical theory to be discussed below.

In Rang's theory $t_1 = t_2 = t$ and $s = 0$ Eq. (12.19) then reduces to:

$$I_s = 2I_s^{(1)}(1 + \cos \alpha) = 4I_s^{(1)} \cos^2 \frac{\pi x}{\Lambda} \tag{12.22}$$

which is Rang's formula.[433] It is now clear that the latter theory is kinematical but with $s = 0$. It is a consistent description of the optical analog it should however only be applied if $t/t_e \ll 1$.

The contrast C of the fringes, as defined by the difference between the maximum and minimum intensity in the fringe system, is given by [from Eq. (12.19)]:

$$\left.\begin{array}{l} C = (4 \sin \pi st_1 \sin \pi st_2)/(\pi s)^2 \\ = \pm 4(I_s^{(1)}I_s^{(1)})^{1/2}. \end{array}\right\} \tag{12.23}$$

From this it is clear that for certain values of st_1 and st_2 the contrast disappears.

e. Intensity Distribution in the Fringes: Dynamical Theory

It is possible to obtain the amplitude and the intensity distribution in the moiré fringes by directly solving the Schrödinger equation and the use of dispersion surfaces with several branches. This method was followed by Hashimoto et al.[408b] However, it gives rise to long and tedious calculations and one obtains complicated expressions, which are not easily interpreted. Equivalent results can be obtained by means of the much shorter and more elegant method used by Gevers. We will follow this last procedure.

The expression for the amplitude can in fact be deduced by a simple adaption of formulas (9.189) and (9.190) derived in the paragraph on stacking fault contrast. These formulas are based on the column approximation. It is sufficient to replace α by its suitable value

$$\alpha = -2\pi \overline{\Delta n} \cdot \bar{r}. \tag{12.24}$$

Whereas α was a constant for the stacking fault, it now becomes a function of the position coordinate $\bar{r}$. With the choice of the reference system adopted in Section 12c an amplitude profile along the direction perpendicular to the fringes will be obtained if we put

$$-\overline{\Delta n} \cdot \bar{r} = x/\Lambda. \tag{12.25}$$

This simple procedure is subject to the following limitations:

(1) Differences in s values for the two parts of the sandwich have been neglected.

(2) The extinction distances are assumed to be the same in both crystals.

The more general treatment by Gevers[231b] is not subject to these limitations and it will be discussed now. Let Fig. 221 represent the $(hk0)$ level of the reciprocal lattice. C is the projection of the center of the reflecting sphere on this plane. The vectors $\bar{n}_1 = O\bar{N}_1$ and $\bar{n}_2 = O\bar{N}_2$ are the diffraction vectors operating, respectively, in crystal

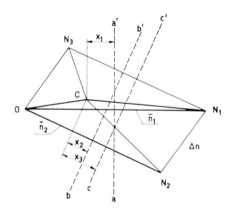

Fig. 221. The $(hk0)$ level of the reciprocal lattice. C is the projection of the center of the reflecting sphere on this plane, $\bar{n}_1$ and $\bar{n}_2$ are the reciprocal lattice vectors operating in the 2 crystals; aa' and bb' are the corresponding Brillouin zones.

I and II; aa' and bb' are the corresponding Brillouin zone boundaries. As shown experimentally by Bassett et al.[431] the moiré fringes result from the interference between the directly transmitted and the twice diffracted beam. We will consider this double diffraction process. The beam scattered by crystal I (vector $\bar{n}_1$) is subsequently scattered by crystal II. For this second scattering process N_1 is the origin of the reciprocal lattice and since the scattering is in the opposite sense to that for the first scattering process in crystal II (vector $\bar{n}_2$) the active reciprocal lattice vector is $\overline{N_1 N_3} = -\overline{ON_2}$. The corresponding Brillouin zone boundary is cc'. The interference errors are determined by the x_i (see Fig. 221), i.e., the distances from C to the respective Brillouin zone boundaries; they are related to the s_i by the formulas

$$s_i = n_i \lambda x_i \ (i = 1, 2, 3). \tag{12.26}$$

The twice transmitted wave resulting from an incident wave of unit amplitude and wave vector k_0 is given by [see Section 9h(2)]

$$T(s_1, t_1, t_{e_1}) \, T(s_2, t_2, t_{e_2}) \, e^{2\pi i \bar{k}_0 \cdot \bar{r}} \, e^{\pi i (s_1 t_1 + s_2 t_2)} \tag{12.27}$$

where t_{e_1} and t_{e_2} are the extinction distances in the 2 crystals.

The component of $\bar{k}_0$ parallel to the plane of the drawing is $\overline{CO}$. On the other hand the wave vector for the twice diffracted wave has a parallel component $\overline{CN_3} = \overline{CO} + \overline{ON_3} = \overline{CO} - \overline{\Delta n}$, i.e., $\bar{k}_0' = \bar{k}_0 - \overline{\Delta n}$. Therefore the twice diffracted beam is

$$S(s_1, t_1, t_{e,1}) \, S(-s_3, t_2, t_{e_2}) \, e^{-2\pi i \overline{\Delta n} \cdot \bar{r}} \, e^{2\pi i \bar{k}_0 \cdot r} \, e^{\pi i (s_1 t_1 + s_3 t_2)} \tag{12.28}$$

and the total amplitude of the transmitted wave, allowing for interference between these 2 beams, is

$$T_m' = T(s_1, t_1, t_{e_1}) \, T(s_2, t_2, t_{e_2}) \, e^{\pi i (s_1 t_1 + s_2 t_2)}$$

$$+ S(s_1, t_1, t_{e_1}) \, S(-s_3, t_2, t_{e_2}) \, e^{\pi i (s_1 t_1 + s_3 t_2)} \, e^{-2\pi i \overline{\Delta n} \cdot \bar{r}}. \tag{12.29}$$

Similarly the scattered wave has the amplitude

$$S_m' = S_1(s_1, t_1, t_{e_1}) \, T(-s_3, t_2, t_{e_2}) \, e^{\pi i (s_1 t_1 + s_3 t_2)}$$

$$+ T(s_1, t_1, t_{e_1}) \, S(s_2, t_2, t_{e_2}) \, e^{\pi i (s_1 t_1 + s_2 t_2)} \, e^{2\pi i \overline{\Delta n} \cdot \bar{r}}. \tag{12.30}$$

The first term results from scattering by the first crystal and transmission through the second, the second term from transmission through I and scattering by II. These expressions are clearly generalizations of formulas (9.192) and (9.194) to which they reduce for $\alpha = 2\pi \bar{K} \cdot \bar{r}$ and $t_{e_1} = t_{e_2} = t_e$, $s_1 = s_2 = s_3 = s$. Gevers has shown that in most cases the last two simplifications are justified, but that putting $t_{e_1} = t_{e_2}$ may introduce somewhat larger quantitative errors. The qualitative conclusions will not be invalidated however by taking $t_{e_1} = t_{e_2}$. In what follows we limit ourselves to this simple case and the notation can then be simplified

$$T_m = T_1^{(+)}T_2^{(+)} + S_1^{(+)}S_2^{(-)} e^{2\pi i \bar{K}\cdot\bar{r}} \tag{12.31}$$

$$S_m = S_1^{(+)}T_2^{(-)} + T_1^{(+)}S_2^{(+)} e^{-2\pi i \bar{K}\cdot\bar{r}} \tag{12.32}$$

with

$$\bar{K} = -\bar{\Delta n}; \; \bar{K} \cdot \bar{r} = x/\Lambda. \tag{12.33}$$

The subindices denote the crystal part; the upper sign denotes the sign of s.

From Eqs. (12.31) and (12.33) it is easy to deduce the intensity distribution; one finds:

$$I_T = TT^* = I_T^{(1)}I_T^{(2)} + I_S^{(1)}I_S^{(2)} + 2(I_S^{(1)}I_T^{(1)}I_S^{(2)}I_T^{(2)})^{1/2} \cos \frac{2\pi}{\Lambda}(x - x_0) \tag{12.34}$$

where x_0 is determined by the relations:

$$\cos 2\pi x_0/\Lambda = \pm (I_T^{(1)}I_T^{(2)})^{-1/2}$$

$$\times \left[\cos \pi\sigma(t_1 + t_2) + \frac{1}{(\sigma t_e)^2} \sin \pi\sigma t_1 \sin \pi\sigma t_2 \right], \tag{12.35}$$

$$\sin 2\pi x_0/\Lambda = \pm (I_T^{(1)}I_T^{(2)})^{-1/2} \left[-\frac{s}{\sigma} \sin \pi\sigma(t_1 + t_2) \right]. \tag{12.36}$$

The sign to be chosen is opposite to that of the product

$$\sin \pi\sigma t_1 \sin \pi\sigma t_2.$$

The $I_{S,T}^{(i)}$ are intensities scattered (S) and transmitted (T) by the separate components of the sandwich as indicated by the upper indices (1) and (2). As was the case for the amplitude it turns out that it is possible to express the total transmitted intensity in terms of the intensities transmitted and scattered by two parts separately. However, the position of the fringes, as determined by x_0, involves the amplitudes. The explicit expressions for I_T and I_S have been given before [Eqs. (9.138) and (9.139].

If absorption is neglected it is clear that

$$I_S = 1 - I_T.$$

f. Discussion of the Fringe Profile

The two main features which are accessible to observation are the fringe geometry and the contrast.

(1) *Fringe position.* The intensity minima and maxima are given (from Eq. 12.34), respectively, by

$$x = x_0 + (r + \tfrac{1}{2})\Lambda \tag{12.37}$$

and

$$x = x_0 + r\Lambda \tag{12.38}$$

where r is an integer or zero.

(2) *Contrast.* A convenient measure for the contrast C is the difference between the maximum and the minimum value of I_T. From Eq. (12.34) follows

$$C = 4(I_S^{(1)}I_S^{(2)}I_T^{(1)}I_T^{(2)})^{1/2}. \tag{12.39}$$

(One makes use of the relation $[(x + y)^2 - (x - y)^2 = 4\,xy]$). This formula can be compared with (12.23) of the kinematical theory; it shows immediately that the contrast is optimum if $I_S^{(1)} = I_S^{(2)} = I_T^{(1)} = I_T^{(2)} = 1/2$.

A complete discussion of both features is long in view of the large number of parameters involved and it is beyond the scope of this review. We refer to Gevers[231b] for this; we shall only summarize the conclusions.

(3) *Effect of a Surface Step.* A surface step equal to an integral number of distances $t_e'' = (1/\sigma)$ will not change the fringe pattern since all expressions involved in Eqs. (12.34) and (12.35) are periodic with this distance as a period.

In all other cases more or less drastic changes of the fringe pattern can occur. In the simple case $s = 0$, for instance, the fringes may stop at a step, the contrast may invert, the positions may remain the same, but the contrast and the intensities may change, etc. If $s \neq 0$ the situation is still more complex, since then arbitrary shifts are possible.

(4) *Effect of tilting.* On tilting the crystal, the fringes change continuously in intensity, contrast, and position.

An important conclusion emerges from this discussion. Most moiré patterns are probably observed under conditions where the dynamical theory should be applied.

(5) *Anomalous absorption.* Gevers[231b] further considered the influence of anomalous absorption which is taken into account phenomenologically, as described above, by the substitution $(1/t_e) \rightarrow (1/t_e) + (i/\tau_e)$. It is shown that the interfringe spacing and the fringe direction are not influenced. On the other hand the position, i.e., x_0, the profile and especially the contrast are influenced, in particular in the thick crystals. The most striking feature due to absorption is the asymmetry in s; for plus and minus the same s value the pattern should be different.

g. Multiple Moiré Patterns

If 2 or more reflections are active in producing the moiré pattern a fringe system in 2 or more directions will arise and the moiré pattern may become an array of dots formed at the intersection points of the two sets. An example is visible in Fig. 218. Particularly attractive patterns result when all reflections of the $(10\bar{1}0)$ type of an hexagonal crystal collaborate in producing the contrast. The hexagonal array of Fig. 222 was observed in such circumstances in a graphite crystal.

The dynamical theory for moiré patterns due to more than one reflection has been formulated by Gevers[434] using his column approximation.

h. The Effect of Dislocations on the Moiré Patterns

(1) *Optical analog.* Consider a perfect crystal foil and one containing an edge dislocation emerging in the foil, as represented in Fig. 219a1, 2. On superposing the two crystals with a slight misorientation the image shown in Fig. 219a3 results. It is evident that the

[434] R. Gevers, *Phys. Stat. Solidi* **3**, 2289 (1963).

moiré pattern exhibits a dislocation also. This shows that a dislocation in one of the crystals will be imaged as a dislocation in the moiré pattern. Since the moiré spacing can be resolved in the microscope, the moiré dislocation can be observed directly as for instance in Figs. 218 and 223a. The supplementary half-fringe is perpendicular to the supplementary half-plane of the edge dislocation, if the diffraction vector is taken perpendicular to the supplementary half-plane.

A similar reasoning applies to the parallel moiré; the optical analog is demonstrated in Fig. 219b. The supplementary half-fringe is now parallel to the dislocation half-plane.

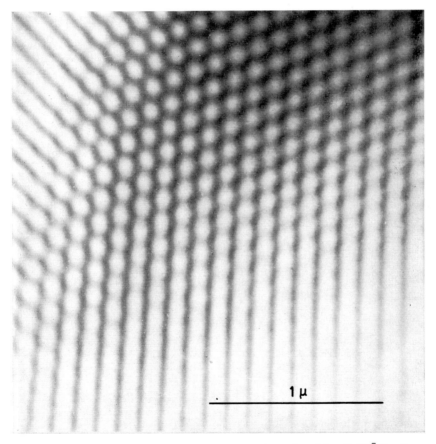

FIG. 222. Moiré patterns in graphite; contrast is produced by all {10$\bar{1}$0} reflections.

(2) *Relation between dislocation arrangements and moiré fringes.*
(a) Rotation moiré. The rotation of 2 crystallites about an axis
normal to their contact plane can be considered as resulting from
the presence of a twist boundary in the contact plane. In fact many
of the observed moiré patterns originate from such boundaries,
rather than from 2 superposed independent crystals. The spacing
of the screw dislocations in such a boundary is given by $D_1 = b/\vartheta$
where b is the Burgers vector and ϑ the twist angle.

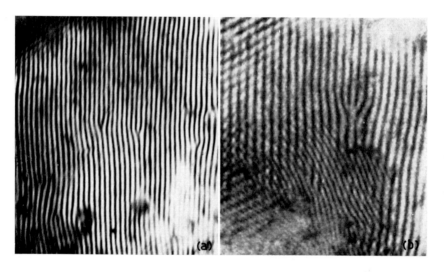

FIG. 223. (a) Moiré dislocation containing a single supplementary half-row.
(b) Moiré dislocation containing 2 supplementary half-rows.

This formula is of exactly the same form as that which gives the
spacing $\varLambda_r$ of moiré fringes, Eq. (12.1). As a result there will be a
simple relationship between the moiré spacing $\varLambda_r$ and the mesh size
D_1. For some reflections $d_{hkl} = b$ and the spacing will be identical.
This makes it sometimes difficult to distinguish dislocation networks
from moiré patterns. It is in fact also a very direct way to prove the
correctness of both formulas for small values of ϑ (or d_r) where a
direct measurement of ϑ is very difficult.

Figure 224a shows a dislocation network in graphite and Fig. 224b
the corresponding moiré pattern; notice that the fringe spacing is
half the mesh size of the network.

(b) Parallel moiré.   Let us consider an array of parallel edge dislocations all of the same sign. Such an array causes a lattice curvature,
but in practice the specimen may be forced to remain relatively flat.
One has now obtained a sandwich in which, say the upper part, contains a number of supplementary half-planes equal to the number of
dislocations. On the average the lattice spacing will be smaller in

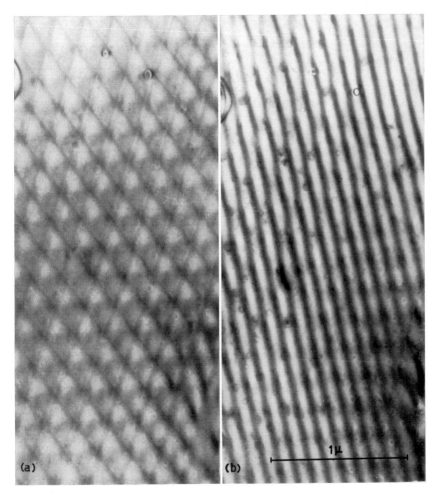

(a)   (b)

FIG. 224.   Relation between moiré pattern and dislocation network. Notice that
the spacing of the fringes is half the mesh size. (a) Network in graphite. (b) Corresponding moiré pattern.

the upper part and parallel moiré fringes will result, the direction of which coincides with that of the dislocations.

The spacing of such fringes will be related in a simple way to the spacing of the dislocation; for $d_{hkl} = b$, the spacing will be equal. In this case also the distinction between fringes and dislocation images may become difficult.

(c) Mixed moiré.   An array of mixed dislocations will in general cause a mixed moiré since both orientations and a difference in lattice parameter are present.

(3) *Influence of a dislocation parallel to the foil plane on the moiré pattern.*   (a) Kinematical theory.   Let the dislocation with Burgers vector $\bar{b}$ in the foil plane lie parallel to the y axis at a depth t_3 below the surface, as in Fig. 225. We will assume that the part of the crystal

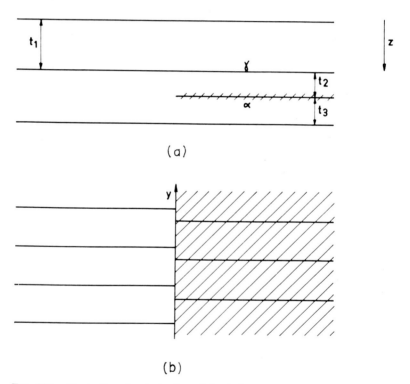

Fɪɢ. 225.   Illustrating the derivation of the influence of a dislocation in the foil plane on the moiré pattern. (a) Side view of sandwich. (b) Top view.

to the right of the y axis has been displaced over a vector $\bar{b}$ as a consequence of the presence of the dislocation. If the dislocation has partial character the region will contain a stacking fault and an additional phase shift α will be introduced. We will now deduce how the fringe pattern will be influenced by partial and perfect dislocations.

The amplitude of a wave diffracted by a column of crystal on the region to the right of the y axis is then given by

$$\pi s S = e^{\pi i s t_1} \sin \pi s t_1 + e^{\pi i s (2t_1 + t_2)} \sin \pi s t_2 \cdot e^{i\gamma}$$

$$+ e^{\pi i s (2t_1 + 2t_2 + t_3)} \sin \pi s t_3 \, e^{i(\gamma + \alpha)} \qquad (12.40)$$

where formula (9.23) was applied; γ is as above $\gamma = (2\pi/\Lambda)x$ and $\alpha = 2\pi\bar{n} \cdot \bar{b}$. This can be transformed into

$$\pi s S = \sin \pi s t_1 + e^{\pi i s (t_1 + t_2 + t_3)} \, e^{i\gamma} \, [\sin \pi s t_2 \, e^{-\pi i s t_3} + \sin \pi s t_3 \, e^{\pi i s t_2 + i\alpha}] \quad (12.41)$$

when leaving out the common phase factor $e^{\pi i s t_1}$. In this form it is clear that only the part between brackets depends on α, i.e., on the presence of the stacking fault. For a column in the region left of the y axis α is zero and the expression between brackets is real and equal to $\sin \pi s (t_2 + t_3)$; the position of the fringes is now entirely determined by the exponential in front of the brackets. We will now see how the position of the fringes will be influenced by the stacking fault. We write the quantity between brackets which is complex, say $A + Bi$ in the form $\rho e^{i\theta}$. One finds:

$$\text{tg } \theta = \frac{B}{A} = \frac{\sin \pi s t_3 \sin (\pi s t_2 + \alpha) - \sin \pi s t_2 \sin \pi s t_3}{\sin \pi s t_3 \cos (\pi s t_2 + \alpha) + \sin \pi s t_2 \cos \pi s t_3}. \qquad (12.42)$$

It is clear that for $\alpha = 0$ $\theta = 0$. If $\theta \neq 0$ it has to be added to γ and hence the fringe pattern is shifted in the faulted region by a quantity determined by θ which is not a simple function of α; in particular it depends also on s. In general the phase shift will be fractional for a partial dislocation and a visible shift will occur. For a perfect dislocation no shift will be observed since it is equal to an integral number of interfringe distances.

It has been deduced erroneously from the optical analog that the phase shift would be given in all cases by $\theta = \alpha$. This value is obtained if we go over to the limit for $s \to 0$ in formula (12.42), demonstrating again that the optical analog is equivalent to a kinematical approximation with $s = 0$.

For the special value of t_2 such that $\sin \pi s t_2 = 0$ one obtains $\mathrm{tg}\theta = \mathrm{tg}\alpha$ and the fractional shift is now also given by α.

On the other hand, if $\sin \pi s t_3 = 0$ one finds $\theta = 0$ and again no visible shift occurs.

(b) Dynamical theory. The dynamical theory for this case has been formulated by Gevers[435] and by Naiki.[436] Expressions for the shift of moiré fringes at a partial dislocation parallel to the surface of one of the foils were derived. The changes in contrast that result from the presence of the stacking fault were also discussed. The dependence of both features: fringe position and contrast on the different parameters involved, i.e., on the foil thicknesses, the position of the stacking fault, and the interference errors, is rather complicated. A discussion is outside the scope of this review.

(4) *Influence of an arbitrary dislocation on the moiré pattern.* The discussion will be limited to geometrical considerations following Basett *et al.*[431] The two parts of the crystal delineated by the glide plane of an edge dislocation are displaced with respect to each other over a distance and in a direction given by $b = [uvw]$. Let the diffraction vector be $\bar{n} = [hkl]$ then the lattice spacing is $d_{hkl} = 1/|\bar{n}|$. The projected length of $\bar{b}$ on the normal to the set of diffracting planes is then $\bar{b} \cdot \bar{n}/|\bar{n}|$. The number of supplementary half-planes of the type (hkl) required to make this edge dislocation, and hence also the number of half fringes N in the moiré pattern is given by

$$N = \frac{b \cdot \bar{n}}{|\bar{n}|} : d_{hkl} = \bar{b} \cdot \bar{n} = hu + kv + lw. \tag{12.43}$$

In Fig. 223b the moiré pattern exhibits, for instance, 2 supplementary half-rows. As shown graphically by Bassett *et al.*[431] this relation does not depend on the character of the dislocation. It also holds for screw dislocations. The observation of an "edge dislocation" in the moiré pattern therefore cannot be taken as evidence that the corresponding lattice dislocation has this same character.

In the face-centered cubic lattice perfect dislocations have a vector $\bar{b} = 1/2 \langle 110 \rangle$; moreover the indices for all reflections are unmixed, i.e., all even or all odd. From this can be concluded that

[435] R. Gevers, *Proc. 5th Intern. Conf. on Electron Microscopy, Philadelphia, 1962* 1, B-12 (1962).

[436] T. Naiki, *J. Phys. Soc. Japan* **17**, 145 (1962).

$N = 1/2 \,(h + h)$ is always an integer. Depending on the operative reflection the number N for $b = 1/2\,[110]$ can be zero ($2\bar{2}0$, $2\bar{2}4$), one ($20\bar{2}$, 202 ...), two (220, 224), or even three (422, 242, $4\bar{2}2$), etc. This was verified experimentally.[431]

For partial dislocations with a vector of the type $(a/6)\,[121]$ the number of supplementary half-lines can be calculated again from formula (12.43). It may now become fractional. It is either 0, $1/3$, $2/3$, 1, $4/3$, or $5/3$. In the stacking fault region the moiré fringes will be shifted over this fraction of the interfringe spacing on both sides of the intersection line of the fault. Different images for ribbon dislocations are shown in Fig. 226. Partials will have individual images provided their separation is larger than the moiré spacing.

(5) *Influence of a dislocation perpendicular to the foil plane; quantitative treatment.* The influence of the moiré pattern of an edge dislocation perpendicular to the foil plane, in one of the components of the sandwich, will now be considered. Use is made of the column approximation introduced by Gevers.[231b] The use of the method is justified because the displacement is constant within each column.

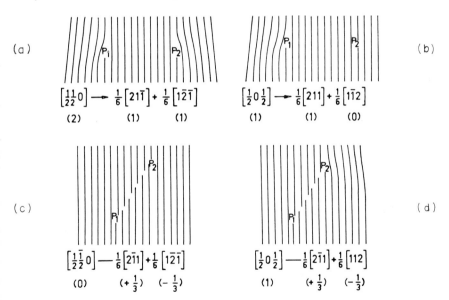

FIG. 226. Images of ribbon dislocations containing a stacking fault, and intersecting the foil surface. [After G. A. Bassett, J. W. Menter, and D. W. Pashley, *Proc. Roy. Soc.* **A246**, 345 (1958).]

However the changes in lattice parameter and the resulting changes in s are neglected. The phase shift between waves diffracted by the two parts of the sandwich is $\alpha = 2\pi\bar{n} \cdot \bar{u}$. If the 2 components were perfect the phase shift $\alpha = 2\pi\bar{n} \cdot \bar{u} = 2\pi\bar{n} \cdot \overline{\Delta r}$ would only be due to either the rotation of the two parts, or their difference in lattice parameter, or both. If a dislocation is present an additional phase shift results from the strain field of the dislocation: $\bar{u} = \overline{\Delta r} + \bar{R}$, where R is the displacement function of the dislocation.

Let the edge dislocation be at the origin of the reference system (x, y) with its Burgers vector along the x axis. It is further assumed that the dislocation is in the exit foil of the sandwich. The strain field is then described by Eq. (9.45). In the case of a rotation moiré the fringes in the perfect region are parallel to the diffraction vector $\bar{n}$ which we will choose along the x axis. An explicit form for α can now be written:

$$\alpha = 2\pi \frac{y - y_0}{\Lambda} + 2\pi\bar{n} \cdot \bar{R}$$

$$= 2\pi \frac{y - y_0}{\Lambda} + \bar{n} \cdot \bar{b} \left(\mathrm{arctg} \frac{y}{x} + \frac{1}{2(1 - \nu)} \frac{xy}{x^2 + y^2} \right).$$

The bright fringes, which are the loci of the points for which $\cos \alpha = 1$, have as an equation:

$$\frac{y - y_0}{\Lambda} + \frac{\bar{n} \cdot \bar{b}}{2\pi} \left(\mathrm{arctg} \frac{y}{x} + \frac{1}{2(1 - \nu)} \frac{xy}{x^2 + y^2} \right) = r$$

where r is an integer. Assuming $(y_0 = 0)$ and transforming into polar coordinates (ρ, ϑ) one obtains (for $\nu = 1/3$) for the rotation moiré:

$$\frac{\rho \sin \vartheta}{\Lambda} = \frac{y}{\Lambda} = r - \frac{n}{2\pi} \left(\vartheta + \frac{3}{8} \sin 2\vartheta \right)$$

where $n = \bar{n} \cdot \bar{b}$ can be positive or negative depending on the signs of $\bar{b}$ and $\bar{n}$. Figure 227 shows the computed patterns for $n = 1$, $n = 2$, and $n = 2/3$. It is clear that n gives the number of supplementary fringes.

For the parallel moiré pattern the diffraction vector is again taken along the x axis. The fringes in the perfect region are now parallel to the y axis. The equation of the fringes becomes:

$$\rho \frac{\cos \vartheta}{\Lambda} = \frac{x}{\Lambda} = r - \frac{n}{2\pi} \left(\vartheta + \frac{3}{8} \sin 2\vartheta \right).$$

If the dislocation is a partial n may become fractional. The dislocation is now the limit of a stacking fault. The presence of the stacking fault is reflected in the discontinuous shift of the fringes in the case of the parallel moiré. Figure 227c shows the computed pattern for $n = 2/3$. The sign of the moiré dislocation follows from the following rules.

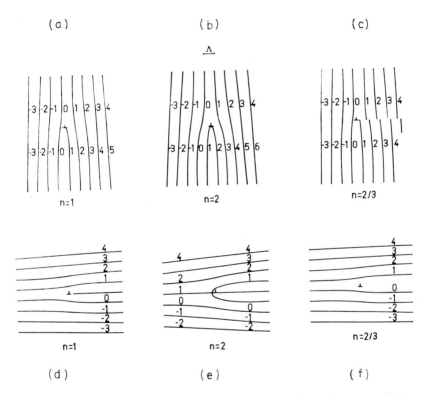

FIG. 227. Calculated moiré fringe patterns for an edge dislocation perpendicular to the foil plane. (a), (b), (c) Parallel moiré. (d), (e), (f) Rotation moiré.

(i) Parallel moiré. If the perfect crystal has a larger lattice parameter than the dislocated one, the moiré dislocation has the same sign as the lattice dislocation. The inverse is true if the lattice parameter of the perfect dislocation is the smaller one.

(ii) Rotation moiré. For a rotation moiré pattern the rule is as follows: If the perfect crystal is rotated in the clockwise sense with respect to the dislocated crystal, when looking along the incident beam, the supplementary "half-plane" is rotated over about $\pi/2$ in the same sense.

In a recent paper Mannami[238b] has derived the intensity distribution in the moiré fringe system in the vicinity of an edge dislocation perpendicular to the foil plane.

i. Applications

(1) *Imaging of dislocations.* A large number of papers are concerned with this subject. Hashimoto and Yueda were the first to demonstrate that a lattice dislocation is pictured as a moiré dislocation.[430] They used cupric sulfide crystals. The first moiré patterns were reported by Mitsuishi *et al.*[437] in graphite. Systematic use of the method was first made by Bassett *et al.*[429,431] Further applications can be found in Gillet[438] and others.[439-447]

Moiré patterns can be used to determine dislocation densities, if one uses at least two different operating reflections; for a single reflection some dislocations will be invisible. Dislocation densities in these evaporated films are of the order of $10^{10} - 10^{11}$ cm². Movement of dislocations has also been observed.[431,441,447a]

In principle the moiré fringes for three different noncoplanar diffraction vectors permit the determination of the direction and magnitude of the Burgers vector. The principle is very simple; the three numbers N_1, N_2, and N_3 corresponding to the three diffraction vectors $\bar{n}_1$, $\bar{n}_2$, and $\bar{n}_3$ give the projections of $\bar{b}$ along three different directions; hence $\bar{b}$ can be deduced.

The separation into partials can also be observed[441] and hence one can in principle measure the stacking fault energy. One must be sure, however, that the partials have their equilibrium separation.

[437] T. Mitsuishi, H. Nagasaki, and R. Uyeda, *Proc. Japan Acad.* **27**, 86 (1951).

[438] M. Gillet, *Bull. Soc. Franç. Mineral. Crist.* **83**, 245 (1960).

[439] H. Hashimoto and T. Naiki, *J. Phys. Soc. Japan* **13**, 764 (1958).

[440] H. Hashimoto and K. Fujita, *Nature* **187**, 51 (1960).

[441] O. Rang and H. Poppa, *Naturw.* **45**, 239 (1958).

[442] F. E. Fujita and K. Izui, *J. Phys. Soc. Japan* **16**, 214 (1961).

[443] G. Möllenstedt and H. Duker, *Physik. Verhandl.* **4**, 98 (1953).

[444] J. F. Goodman, *Nature* **180**, 425 (1957).

[445] K. Izui, *J. Phys. Soc. Japan* **14**, 1829 (1959).

[446] Y. Seki, *J. Phys. Soc. Japan* **6**, 534 (1951).

[447] O. Rang, *Z. Physik* **136**, 465 (1953).

[447a] J. Demny, *Z. Naturforsch.* **15a**, 194 (1960).

(2) *Epitaxy*. Kamiya and Uyeda[448] studied the epitaxial growth of a silver film on a molybdenum sulfide single crystal substrate. They find that dislocations in these films result from the junction of different nuclei.

Bassett[449] studied the same system by evaporation of the silver in the microscope itself. In this way he could follow continuously the growth of the silver film. The difference in lattice spacing (d_{220} of Ag $= 1.44$ A: $d_{11\bar{2}0}$ of $MoS_2 = 1.57$ A) gives rise to a moiré pattern with a spacing of 17 A. As can be deduced from formula (12.4) a small relative rotation of the 2 lattices together with the difference in lattice spacing produces a large rotation of the moiré fringes. This phenomenon allows us to follow to what extent the expitaxially growing film is parallel to the substrate. It is found that many of the silver nuclei have moiré fringes rotated with respect to one another over angles of up to 30° (i.e., about 3° actual lattice rotation). These relative rotation angles change during growth. It is suggested that the model of interfacial dislocations[398,450] is not adequate to account for the large rotations of the nuclei. The relative orientations of film and substrate would correspond to cusps in the curve relating grain boundary energy to orientation difference. The nuclei would rotate bodily from one such position to another while they grow.

Matthews[451] [see also Section 10e(1)] made somewhat similar observations on the system silver on a mica substrate. In this case also deviations from perfect alignment are found especially for small nuclei. As the layer increases in thickness, the deviations decrease. Most of the silver nuclei are dislocation free; only when they coalesce are dislocations formed in accord with the finding of the two previous authors.

[448] Y. Kamiya and R. Uyeda, *Acta Cryst.* **14**, 70 (1961).

[449] G. A. Bassett, *Proc. European Reg. Conf. on Electron Microscopy, Delft 1960* **1**, p. 270 (1961).

[450] N. Cabrera, *in* "Structure and Properties of Thin Films," (C. A. Neugebauer *et al.*, eds.), pp. 42, 529. Wiley, New York, 1959.

[451] J. W. Matthews, *Proc. European Reg. Conf. on Electron Microscopy, Delft, 1960* **1,** 276 (1961).

IV. Field Emission and Field Ion Microscopy

13. Principle of the Method

In field emission electron microscopy the electrons emitted from a fine point of some refractory metal under the influence of an intense radial electrostatic field, cast an image of the point on a fluorescent screen. The point develops facets by evaporation and they can be recognized clearly on the image. The contrast arises from differences in electron emission or electron surface density. The magnification for this extremely simple microscope is only determined by geometry, i.e., by the ratio of the screen radius to the tip radius which may be of the order of 10^6. The resolution is about 25 A and therefore individual atoms are not revealed.

However the use of ions instead of electrons, as the image producing carrier, has improved the resolution considerably. This is done in the field ion microscope. Hydrogen or, better, helium ions are used. The tip end of the microscope is cooled to liquid helium or hydrogen temperature and a residual pressure of 1 to 5 microns is left in the microscope. The helium atoms are ionized when they come into the immediate vicinity of the atoms of the tip which is kept at a positive potential. As they form the ions are accelerated in the high field, which is of the order of 10^8 volts/cm near the tip. Due to their large mass the ions suffer very little deviation from their radial path. Since helium has the largest ionization potential of the residual gases the helium atoms will come closer to the tip than other gas atoms, before being ionized. This is the reason why helium is the best choice as the image producing carrier for achieving optimum resolution. Dot patterns which are strongly suggestive of the atom arrangement in a tungsten tip were obtained by Müller.[452] The observed spacing between dots is somewhat larger than the spacing calculated on the assumption that individual atoms are revealed. This point needs further consideration. However, it is reasonably sure that individual

[452] E. W. Müller, Z. *Physik* **131**, 136 (1951).

439

atoms are indeed revealed. Mostly atoms in "edge" positions, for instance along surface steps, show up preferentially.

14. DISLOCATIONS

In the field ion microscope spiral features have occasionally been observed by Drechsler[453,454] on a tungsten tip subject to field evaporation. The author attributed the features to evaporation spirals centered on screw dislocations. However, this interpretation has been questioned by Müller,[455] who showed that an eccentric succession of "plateaux" may produce a "pseudo-spiral" as a consequence of the bunching of steps, especially since emission at the steps is observed. Müller,[455] on the other hand, produced pictures of the arrangement of atoms around an edge dislocation.

Well-resolved images of the core structure of edge dislocations with Burgers vector $(a/2) \langle 111 \rangle$ in molybdenum were obtained by Brandon and Wald.[456] It is shown that the arrangement of atoms in the core is more like that shown in Fig. 228a than that in Fig. 228b.

In more recent work by Wald[456a] the structure of grain boundaries is studied. He finds that the structure closely ressembles that exhibited by the bubble model.

   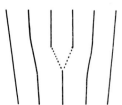

(a)   (b)

FIG. 228.   Arrangement of atoms in the core of a dislocation in molybdenum. The arrangement is more like in (a) than like in (b).

[453] M. Drechsler, G. Pankow, and R. Vanselow, *Z. Physik. Chem. (Frankfurt)* **4**, 17 (1955).

[454] M. Drechsler, *Z. Metallk.* **47**, 305 (1956).

[455] E. Müller, *Acta Met.* **6**, 620 (1958).

[456] G. Brandon and W. Wald, *Phil. Mag.* [8] **6**, 1035 (1961).

[456a] W. Wald, private communication (1963).

Appendix A. Properties of Stacking Fault Fringes Taking into Account the Anomalous Absorption[456b]

1. APPROXIMATE EXPRESSIONS FOR THE AMPLITUDES OF TRANSMITTED AND SCATTERED BEAM CLOSE TO THE FOIL SURFACES

The conclusions of Section 9h(6) which were originally obtained from machine calculations can also be derived analytically. This was shown for some of them in Art *et al.*[457] The procedure consists in evaluating the derivatives $\partial I_I / \partial t_1$ and $\partial I_S / \partial t_1$ at both foil surfaces. The stacking fault is at t_1 from the first surface and at t_2 from the back surface. A positive derivative at the entrance face, i.e., for $t_1 = 0$ evidently means that the first fringe is a bright one, whereas a positive derivative at the exit surface $t_1 = t$ (or $t_2 = 0$) means a last dark fringe. The derivation in Art *et al.*[457] is rigorous. The treatment presented here is based on an approximation introduced by Gevers.[458] For simplicity we consider in detail only the case $s = 0$, however the method can readily be generalized for small s.[459]

a. The bright field image

We start with formula (9.191) which can be rewritten as

$$T = T_{12} - (1 - e^{i\alpha}) S_1 S_2.$$

For $s = 0$ this becomes explicitly (using formulas 9.123 and 9.124):

$$T = \cos \pi \sigma t + (1 - e^{i\alpha}) \sin \pi \sigma t_1 \sin \pi \sigma t_2. \tag{A1}$$

[456b] The attenuation factor describing normal absorption is left out.
[457] A. Art, R. Gevers, and S. Amelinckx, *Phys. Stat. Solidi* **3**, 697 (1963).
[458] R. Gevers, *Phys. Stat. Solidi* **3**, 415 (1963).
[459] R. Gevers, A. Art, and S. Amelinckx, *Phys. Stat. Solidi* **3**, 1563 (1963).

with

$$\sigma = \frac{1}{t_e} + \frac{i}{\tau_e} \; ; \; t = t_1 + t_2$$

After some transformations this can be written as [cf. Eq. (9.210)]

$$T = e^{i\alpha/2} \left[\cos \pi\sigma t \cos \frac{\alpha}{2} - i \sin \frac{\alpha}{2} \cos \pi\sigma(t - 2t_1) \right]. \tag{A2}$$

We now introduce the approximation[460]:

$$\cos \pi\sigma(t - 2t_1) \simeq \left(\tfrac{1}{2}\right) e^{-i\pi\sigma(t-2t_1)} \tag{A3}$$

valid for a value of the real part of the exponent $\pi(t - 2t_1/t_e)$ which is not too small, i.e., for t large enough and t_1 small enough, e.g., $t - 2t_1 < 0.4 \, \tau_e = 4t_e$ if $\tau_e = 10t_e$. One then obtains:

$$\cos \pi\sigma(t - 2t_1) \simeq \tfrac{1}{2} e^{-i\pi\sigma t} \, e^{2\pi i\sigma t_1} \simeq \cos \pi\sigma t \cdot e^{2\pi i\sigma t_1}$$

$$= T_0 \, e^{2\pi i\sigma t_1} = T_0 \, e^{-2\pi(t_1/\tau_e)} \, e^{2\pi i(t_1/t_e)} \tag{A4}$$

where T_0 is the amplitude of the wave transmitted by the perfect crystal of thickness t. This allows to rewrite (A2) as:

$$T(t_1) = T_0 \, e^{i(\alpha/2)} \left[\cos \frac{\alpha}{2} - i \sin \frac{\alpha}{2} e^{-2\pi(t_1/\tau_e)} \, e^{2\pi i(t_1/t_e)} \right]. \tag{A5}$$

This expression lends itself to a convenient graphical representation of the fringe profile. Let us first consider the limiting case $\tau_e \to \infty$:

$$e^{-i(\alpha/2)} \, T = T_0 \left[\cos \frac{\alpha}{2} - i \sin \frac{\alpha}{2} e^{2\pi i(t/t_e)} \right] = A_0 + A_1 \, e^{\pi i(t_1/t_e) - i(\pi/2)} \tag{A6}$$

with $A_0 = T_0 \cos (\alpha/2)$ and $A_1 = T_0 \sin (\alpha/2)$. In the complex plane A_0 is represented by a vector OB. As t_1 increases the end point of vector $A_1 \equiv$ BC describes a circle with radius $|A_1|$ and center B

---

[460] One has

$$\cos (x + iy) = \tfrac{1}{2} \left[e^{i(x+iy)} + e^{-i(x+iy)} \right] = \tfrac{1}{2} \left[e^{-y+ix} + e^{y-ix} \right].$$

To a good approximation $\cos (x + iy) \simeq (\tfrac{1}{2})e^{y-ix}$ provided $e^{-y} \ll e^y$ or $e^{-2y} \ll 1$ say $e^{-2y} < 0,1$ which is true for $y \geqslant 1,2$ and similarly $\sin (x + iy) \simeq (i/2) e^{y-ix}$. If on the contrary y is small $e^y \simeq e^{-y} \simeq 1$ and one has to a reasonable approximation

$$\cos (x + iy) \simeq \cos x ; \qquad \sin (x + iy) \simeq \sin x.$$

(Fig. A1, a). The final amplitude is then given (in absolute value) by the vector $\overline{OC}$, which is seen to fluctuate between two extreme values $|A_0| + |A_1|$ and $|A_0| - |A_1|$.

Taking into account the exponential attenuation factor amounts to writing:

$$e^{-i(\alpha/2)} T = A_0 + A_1 e^{-2\pi(t_1/\tau_e)} e^{2\pi i (t_1/t_e)-i(\pi/2)} \tag{A7}$$

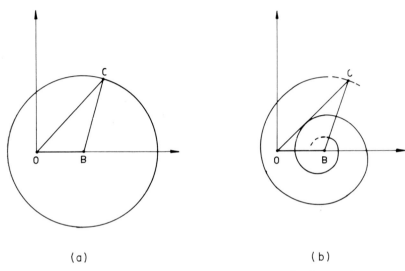

(a)   (b)

FIG. A1.   Construction of the amplitude of the transmitted wave for a crystal containing a stacking fault. The construction is valid only if the fault is close to one of the surfaces. (a) The attenuation factor $\exp(-2\pi t_1/\tau_e)$ is left out; as the distance t_1 increases the point C describes the circle. (b) Taking into account the attenuation factor transforms the circle of (a) into a logarithmic spiral.

As a consequence of the factor $e^{-2\pi(t_1/\tau_e)}$ the point C of Fig. A1,b now describes a logarithmic spiral instead of a circle. This clearly means that successive fringes become less and less pronounced, or that the oscillations are damped. The quasiperiodicity in depth is $\Delta t_1 = t_e$ as opposed to $(\frac{1}{2})t_e$ if absorption is neglected [see Section 9h(4)].

The intensity of the transmitted wave in the faulted region is given by:

$$I_T = I_0 \left[\cos^2 \frac{\alpha}{2} + \sin^2 \frac{\alpha}{2} e^{-4\pi t_1/\tau_e} + \sin \alpha \sin 2\pi(t_1/t_e) e^{-2\pi(t_1/\tau_e)} \right] \tag{A8}$$

which can easily be computed. However, for our purpose we mainly want to know the nature of the first fringe. As shown this can be deduced from the sign of the derivative at the surface:

$$\left(\frac{\partial I_T}{\partial t_1}\right)_{t_1=0} = -I_0 \sin^2 \frac{\alpha}{2} \cdot \frac{4\pi}{\tau_e} + I_0 \sin \alpha \frac{2\pi}{t_e} \tag{A9}$$

with

$$\alpha = \pm \frac{2\pi}{3} \, , \left| \frac{2\pi}{t_e} \sin \alpha \right| \gg \left| \frac{4\pi}{\tau_e} \sin^2 \frac{\alpha}{2} \right|.$$

This shows that the sign of the derivative is only determined by the sign of α, and does not depend on the thickness provided the crystal is thick enough to allow the use of the approximation. For $\alpha = 2\pi/3$ the derivative is positive and consequently the first fringe is bright; the reverse is true for $\alpha = -2\pi/3$. Changing the sign of α consequently means that the bright field fringe pattern changes into the "pseudocomplementary" pattern at least for the first few fringes. This is also directly evident from (A8).

We shall now estimate the position of the first fringe and show that it is sufficiently differentiated from the background to allow an unambiguous determination of its nature.

We first consider the case $\sin (\alpha/2) > 0$. From (A5) we conclude that T reaches maxima for $t_1 \simeq (k + \frac{1}{4})t_e$ and minima for $t_1 = (k + \frac{3}{4})t_e$ where k is an integer. The center of the first bright fringe is therefore at $\frac{1}{4} t_e$ from the surface and the center of the first dark fringe is at $\frac{3}{4} t_e$ from the surface.

The intensity in the maxima is given by:

$$I_T^{\max} = I_0 \left| \frac{1}{2} + \frac{\sqrt{3}}{2} e^{-(\pi/2)(t_e/\tau_e)(1+4k)} \right|^2.$$

For $k = 0$ this becomes $1.54 \, I_0$ and for $k = 1 : 0.81 \, I_0$ if one assumes $\pi t_e/\tau_e = 0.3$. This means that the intensity in the first bright fringe is about 50% above the background intensity, whereas the intensity in the second fringe is already slightly less than the background intensity. The first fringe can therefore be distinguished very clearly from the background. The intensity in the minima is given by

$$I_T^{\min} = I_0 \left| \frac{1}{2} - \frac{\sqrt{3}}{2} e^{-(\pi/2)(t_e/\tau_e)(3+4k)} \right|^2.$$

For $k = 0$ this gives $0.00 \, I_0$ and for $k = 1 : 0.04 \, I_0$.

If on the other hand $\sin(\alpha/2) < 0$ the second terms in these expressions change sign, the minima now occur for $t \simeq (k + \frac{1}{4})t_e$ and the maxima for $t = (k + \frac{3}{4})t_e$. This leads to the following intensities in the maxima: for $k = 0$, $1.10\,I_0$; for $k = 1$, $0.64\,I_0$. The first fringe is again above the background.

One can determine the positions of the fringes more accurately by determining the zeros of the derivative of I_T in (A8). However, this leads to a small correction only and the main conclusion remains valid. It is as a consequence justified to identify the extinction distance t_e to the depth distance between the first two fringes.

b. The dark field image

We now consider the amplitude of the scattered wave. According to formula (9.193) we have

$$S = S_{12} - (1 - e^{-i\alpha})\,T_1 S_2. \tag{A10}$$

For $s = 0$ this becomes explicitly

$$S = ie^{-i(\alpha/2)}\left[\cos\frac{\alpha}{2}\sin\pi\sigma t - i\sin\frac{\alpha}{2}\sin\pi\sigma(t - 2t_1)\right]. \tag{A11}$$

In the vicinity of the entrance face one finds to the same approximation as above:

$$\sin\pi\sigma(t - 2t_1) \simeq \frac{i}{2}e^{-\pi i\sigma(t - 2t_1)} = \frac{i}{2}e^{-\pi i\sigma t}\,e^{2\pi i\sigma t_1} = -iS_0\,e^{2\pi i\sigma t_1} \tag{A12}$$

and finally (A11) becomes:

$$S(t_1) = e^{-i(\alpha/2)}\,S_0\left(\cos\frac{\alpha}{2} - i\sin\frac{\alpha}{2}e^{-2\pi(t_1/\tau_e)}\,e^{2\pi i(t_1/t_e)}\right). \tag{A13}$$

We have further, to the same approximation,

$$S_0 = i\sin\pi\sigma t \simeq i\left(\frac{i}{2}e^{-\pi i\sigma t}\right) = -\frac{1}{2}e^{-\pi i\sigma t} = -T_0. \tag{A14}$$

Combining (A13) and (A5) one concludes that in the vicinity of the entrance face defined as

$$t_1 < \sim \frac{t - 4t_e}{2},$$

$$I_T(t_1) \simeq I_S(t_1) \tag{A15}$$

which means that the *fringe pattern is similar in bright and dark field image near the entrance face.*

Let us now describe the situation at the exit face of the crystal. We first consider the bright field image. The expression (A2) can be rewritten as:

$$T = e^{i\alpha/2} \left[\cos \pi \sigma t \cos \frac{\alpha}{2} - i \sin \frac{\alpha}{2} \cos \pi \sigma (t - 2t_2) \right] \qquad (A16)$$

since

$$t - 2t_1 = t_1 + t_2 - 2t_1 = -t_1 + t_2 = -(t - 2t_2) \qquad (A17)$$

This means that we only have to substitute $t_1 \rightarrow t_2$ in order to obtain the corresponding approximate expressions valid for small t_2, i.e., at the back-surface. By analogy with (A5), we can write

$$T(t_2) = T_0 \, e^{i(\alpha/2)} \left[\cos \frac{\alpha}{2} - i \sin \frac{\alpha}{2} e^{-2\pi(t_2/\tau_e)} e^{2\pi i(t_2/t_e)} \right]. \qquad (A18)$$

One concludes

$$I_T(t_1) = I_T(t_2) \qquad (A19)$$

for $t_1 = t_2$, i.e., at equal distances from the foil surfaces, in accord with *the symmetry of the bright field image* with *respect to the foil center.*

The approximate expression for S at the exit surface, i.e., for t_2 small can similarly be obtained by making the substitution (A17) $[(t - 2t_1) = -(t - 2t_2)]$ in (A11) and transforming. One finds:

$$S(t_2) = e^{-i(\alpha/2)} S_0 \left[\cos \frac{\alpha}{2} + i \sin \frac{\alpha}{2} e^{-2\pi(t_2/\tau_e)} e^{2\pi i(t_2/t_e)} \right]. \qquad (A20)$$

Only the term containing $\sin \alpha/2$ has changed sign.
The expression for the intensity becomes

$$I_S(t_2) = I_0 \left[\cos^2 \frac{\alpha}{2} + \sin^2 \frac{\alpha}{2} e^{-4\pi(t_2/\tau_e)} - \sin \alpha \sin 2\pi \frac{t_2}{t_e} e^{-2\pi(t_2/\tau_e)} \right]. \qquad (A21)$$

The first part of the expressions (A8) and (A21) is slowly decreasing with t_2, while the second part represents a damped oscillation with period t_e. Comparing (A8) with $t_1 \rightarrow t_2$ and (A21), it becomes clear that the oscillations for $I_S(t_2)$ and $I_T(t_2)$ are opposite in phase—i.e., *the bright and dark field images are "pseudocomplementary" at the back surface,* whereas they were in phase at the entrance face, as is clear from (A15). The formulas (A8) and (A21) also show that a change in sign of α results in a change in sign of $\sin \alpha$ and hence in a change in phase of π for the oscillating parts of the intensities. This statement is equivalent to saying that the nature of the first fringe changes with the sign of α whereas it is independent of the thickness provided this is large enough to allow the application of this approximation $(t \geqslant 4t_e)$.

2. The Fringe Pattern Close to the Center of the Foil

a. The bright field image

The approximations used in the previous paragraph were only valid close to the surface. Near the middle of the foil the difference $|t_1 - t_2|$ is small and one can use the approximation:

$$\cos \pi\sigma(t_1 - t_2) \simeq \cos \frac{\pi}{t_e}(t_1 - t_2). \tag{A22}$$

One can therefore write

$$T = e^{i(\alpha/2)} \left[\cos \pi\sigma t \cos \frac{\alpha}{2} - i \sin \frac{\alpha}{2} \cos \frac{\pi}{t_e}(t_1 - t_2) \right]. \tag{A23}$$

On the other hand

$$T_0 = \cos \pi\sigma t \simeq \frac{1}{2} e^{\pi(t/\tau_e)} e^{-i\pi(t/t_e)} \tag{A24}$$

and hence

$$T = e^{i(\alpha/2)} \left[\frac{1}{2} e^{\pi(t/\tau_e)} e^{-\pi i(t/t_e)} \cos \frac{\alpha}{2} - i \sin \frac{\alpha}{2} \cos 2\pi \frac{u}{t_e} \right] \tag{A25}$$

where $2u = t_1 - t_2$; u is the distance of the fault plane from the foil center.

In the center of the foil, i.e., for $u = 0$, $T_0 \cos (\alpha/2) \gg \sin (\alpha/2)$; $T \simeq T_0 \cos (\alpha/2)$ and hence

$$I_T = I_T^{(0)} \cos^2 \frac{\alpha}{2} \simeq \frac{1}{4} I_T^{(0)},$$

i.e., the center of the fault should be significantly darker than the background. Equation (A25), can be written:

$$T = e^{i(\alpha/2)} \left[\bar{A}(t) + i\bar{B}(u) \right] \tag{A26}$$

where

$$\bar{A}(t) = \frac{1}{2} e^{\pi(t/\tau_e)} e^{-\pi i(t/t_e)} \cos \frac{\alpha}{2} = T_0 \cos \frac{\alpha}{2}$$

and

$$\bar{B}(u) = -\sin \frac{\alpha}{2} \cos 2\pi \frac{u}{t_e}.$$

The shape of the fringe profile is most conveniently discussed by means of a graph in the complex plane. Let us consider α as positive. One first represents the vector $\bar{A} = \overline{OM}$ [Fig. A2 (a)] which only depends on t. From the point M as the origin we represent the vector $\bar{B}$. For $u = 0$ one obtains the point D; for $u = \frac{1}{2} t_e$ the point C is reached. An arbitrary value of u is represented by a point P on the segment CD. The amplitude of the transmitted wave is represented by OP. As u changes the point P oscillates between the two extreme points C and D and the amplitude T oscillates between a maximum OC and a minimum OD. The period in u is t_e, i.e., the fault changes t_e in depth, between two fringes.

However the situation of Fig. A2(a) is not the only possible one. An essentially different one is represented in Fig. A2(b); here $MD > ME$. As a consequence of this the amplitude now oscillates between a first maximum OC, a minimum OE and a second maximum OD. The fringe profiles for these two cases are shown in Fig. A3, as curves 1 and 2.

In the special case where $e^{-\pi i(t/t_e)} = \pm 1$, i.e., for $t = kt_e$ the vector $\overline{OM}$ coincides with the real axis and the two maxima have now equal magnitude. One has an apparent halving of the fringe spacing, the depth periodicity being now $(\frac{1}{2}) t_e$. The condition for the occurrence of secondary maxima is evidently $MD > ME$, i.e.,

$$\left| \sin \frac{\alpha}{2} \right| > \frac{1}{2} e^{\pi(t/\tau_e)} \cos \frac{\alpha}{2} \left| \sin \pi \frac{t}{t_e} \right| \left(\mathrm{tg} \frac{\alpha}{2} = \sqrt{3} \right)$$

or

$$e^{\pi(t/\tau_e)} \left| \sin \pi \frac{t}{t_e} \right| < 2 \sqrt{3}. \tag{A27}$$

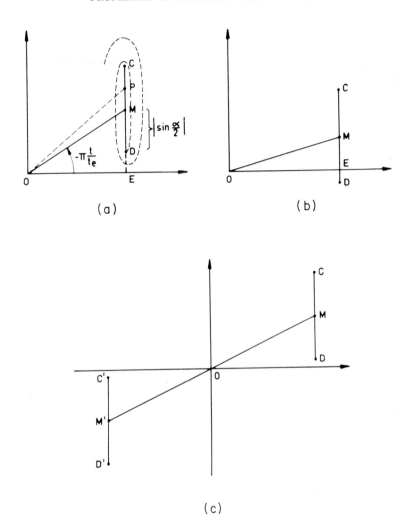

FIG. A2. Construction of the amplitude of the transmitted wave for a crystal containing a stacking fault. This construction is valid for a fault located close to the center of the foil. (a) As the foil position changes the point P describes the segment CD. The point D corresponds to $u = 0$ and the point C to $u = \frac{1}{2}t_e$. The amplitude OP is seen to oscillate between a maximum OC and a minimum OD. (b) For a different thickness the situation may be as shown here. The amplitude now oscillates between a maximum OC, a minimum OE, and a secondary maximum OD. (c) If the thickness increases by t_e the vector OM turns over π and becomes OM'. Minima and maxima are now reversed. If the correct expressions were used the point P would not describe the line segment CD but an elongated elliptical spiral drawn as a dotted line in (a).

The phenomenon will therefore occur provided the thickness is right. If the thickness changes gradually there will periodically be regions where this condition is satisfied. These regions become narrower, however, as the thickness becomes larger.

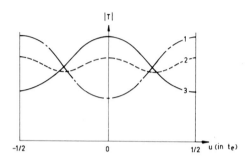

FIG. A3.   Bright field fringe profiles close to the center of the foil for three different thicknesses. Between curves 1 and 3 the crystal differs t_e in thickness. Curve 2 represents the profile at an intermediate thickness.

The effect of increasing thickness is further to make the vector $A \equiv \overline{OM}$ larger. Since the magnitude of $\bar{B}$ does not depend on the thickness, this results in a decrease in contrast as measured by the difference $(OC) - (OD)$ or $(OC) - (OE)$, depending on the type of fringe profile. In thick foils the central fringes should therefore become washed out.

The nature of the central fringe, for a given thickness depends on the sign of $\sin(\alpha/2)$. For $\sin(\alpha/2) > 0$ the point corresponding to $u = 0$ [in Fig. A2(a)] is D, and the central fringe will be dark. However if $\sin(\alpha/2) > 0$, the point corresponding to $u = 0$ is C which means that the central fringe is now bright. This property of the image depends on the thickness. If the thickness increases by t_e the vector $\overline{OM}$ turns over π and the point C' now corresponds to a minimum [Fig. A2(c)] and D' to a maximum. This behavior is reflected in the fringe pattern as fringe reversal. As t changes, the central fringe inverts in character; a black central fringe becomes a bright one and vice versa every time the foil becomes t_e thicker. On "reversal" the vector $\overline{OM}$ has necessarily to pass through the point where it coincides with the horizontal axis and hence where a subsidiary maximum occurs. Reversal proceeds as shown in Figs. A3 and A4. Just prior to reversal the contrast becomes a minimum.

The additional fringes resulting from the increase in thickness are in a sense "born" in the center of the foil, initially as a subsidiary maximum.

The behavior suggested by Fig. A4 can be observed in Fig. A5. This behavior has been observed also by Whelan and Hirsch.[229] However, they explained this feature on the basis of the dynamical theory without absorption. From the discussion given here it becomes clear that it is a typical absorption effect resulting from a change in thickness. Fringe reversal without a change of the number of fringes, also takes place if a second stacking fault in a very nearby lattice plane overlaps partly the first one. In the region of overlap this results in a change of the sign of α, which causes fringe reversal as we have seen. It should be pointed out that the approximation is not as good to describe the fringe behavior close to the center as it is in the vicinity of the surfaces. A numerical estimate shows that the behavior is

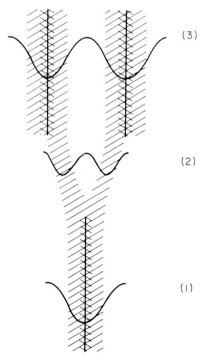

(3)

(2)

(1)

Fig. A4.   The configuration of fringes where reversal takes place. This figure can be compared with Fig. A3 and with Fig. A5.

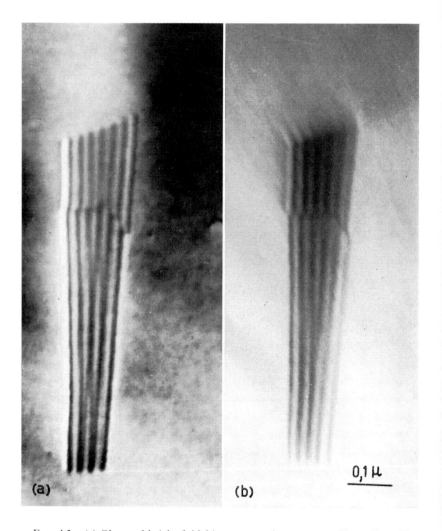

FIG. A5. (a) Observed bright field fringe pattern in a copper-gallium alloy of low stacking fault energy. The phenomenon of fringe reversal is seen to occur everytime the foil thickness increases by t_e. This can be judged from the spacing of the thickness extinction contours. (b) In the dark field image fringe reversal (forking) is observed at a thickness which is precisely the intermediate of the thicknesses where reversal occurs in the bright field image. This behavior is a consequence of anomalous absorption.

correctly described at least qualitatively for the first fringe each side of the center. If use were made of the correct expression, it can be shown that the point P in Fig. A2(a) would not describe a line segment CD but a spiral of elongated elliptical shape, as shown schematically by the dotted line in Fig. A2,a. Provided the distance OM is larger than CD the resulting fringe profile is not much changed as a consequence of this.

b. The dark field image

We start from:

$$S = i e^{-i(\alpha/2)} \left[\cos \frac{\alpha}{2} \sin \pi\sigma t - i \sin \frac{\alpha}{2} \sin \pi\sigma(t_2 - t_1) \right]. \qquad (A28)$$

Using the approximations

$$\sin \pi\sigma t \simeq \frac{i}{2} e^{(\pi t/\tau_e)} e^{-i\pi(t/t_e)}; \sin \pi\sigma(t_2 - t_1) \simeq \sin \frac{\pi}{t_e} (t_2 - t_1) = -\sin \frac{2\pi u}{t_e}$$

with $2u = t_1 - t_2$ one can transform this expression into:

$$S = -e^{-i(\alpha/2)} \left[\frac{1}{2} \cos \frac{\alpha}{2} e^{\pi t/\tau_e} e^{-i\pi(t/t_e)} + \sin \frac{\alpha}{2} \sin \frac{2\pi u}{t_e} \right] \qquad (A29)$$

which can again easily be interpreted graphically in the complex plane. Leaving out the phase factor one can represent S as the sum of two vectors,

$$S = \bar{A} + \bar{B},$$

where $|\bar{A}| = \frac{1}{2} \cos \alpha/2 \ e^{\pi t/\tau_e} e^{-i\pi(t/t_e)}$ and $|\bar{B}| = \sin \alpha/2 \sin (2\pi u/t_e)$; $\bar{A} \equiv OM$ forms an angle $-\pi(t/t_e)$ with the real axis. The vector $\bar{B} \equiv MP$ has to be constructed parallel to the real axis. For $u = 0$ P coincides with M, for $u = t_e/4$ one obtains point C, while for $u = 3t_e/4$ or $-t_e/4$ one obtains point D. The amplitude S oscillates in magnitude between the two extreme values OD and OC. Just like in the bright field image one can expect subsidiary maxima and in particular an apparent halving of the fringe spacing for $e^{-i\pi(t/t_e)} = \pm i$, i.e., for $t = [k + (1/2)]t_e$. In the latter case the line OM coincides with the imaginary axis.

The condition for subsidiary maxima to occur is again $DM > EM$ or explicitly

$$\left| \sin \frac{\alpha}{2} \right| > \frac{1}{2} e^{\pi(t/\tau_e)} \cos \frac{\alpha}{2} \left| \cos \frac{\pi t}{t_e} \right|$$

or

$$e^{\pi(t/\tau_e)}\left|\cos\frac{\pi t}{t_e}\right| < 2\sqrt{3}. \tag{A30}$$

Unlike for the bright field image the center of the foil does *not* correspond to the position of a fringe (bright or dark), except in the particular case $t = [k + (1/2)]t_e$ where $u = 0$ corresponds to a dark fringe.

Changing the sign of α at constant thickness now changes the *slope* of the fringe profile at $u = 0$ into the symmetrical one. A change in thickness of t_e has the same effect.

The change in profile as the crystal becomes thicker is shown schematically in Fig. A6(a). It is clear that one passes necessarily through a position where subsidiary maxima arise.

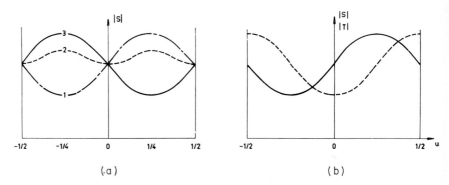

(a)   (b)

Fɪɢ. A6.   Fringe profiles close to the center of the foil. (a) Dark field fringe profile (u is in units t_e). Between curves (1) and (3) the specimen thickness has changed by t_e. Curve (2) corresponds to an intermediate thickness. Notice that $u = 0$ does not correspond to a fringe position in general. (b) Relation between the dark (S) and bright (T) fringe profile. The two curves are $\pi/2$ out of phase.

The relation between dark and bright field image for the same u value can be derived very easily by comparing Fig. A3 and Fig. A6 (a). It turns out that the relation is as shown in Fig. A6(b) for $\sin(\alpha/2) > 0$; i.e., there is a phase shift of $\pi/2$ between the two curves. This is of course in line with the fact that at the front surface the two curves are in phase, at the back surface they are out of phase by π (pseudo-complementary), in the center they are out phase by half this amount. One can therefore state roughly that the two profiles become gradually out of phase as one goes from front to back surface. This is precisely the behavior found by the machine calculation.

The same approximation can be used also if $s \neq 0$. The expressions for T and S become slightly more complicated. One obtains

$$T = e^{i(\alpha/2)} T_0 \left[\cos \frac{\alpha}{2} - i \frac{s}{\sigma} \sin \frac{\alpha}{2} - i \left(1 - \frac{s}{\sigma}\right) \sin \frac{\alpha}{2} e^{-2\pi(t_1/\tau_e)} e^{2\pi i(t_1/t_e)}\right] \quad \text{(A31)}$$

$$= e^{i(\alpha/2)} T_0 \Phi(\alpha, s, t_1) \quad \text{(A32)}$$

$$S = e^{-i(\alpha/2)} S_0 \Phi(\alpha, s, t_1) \quad \text{(A33)}$$

where as an additional approximation we have accepted that only the real part of $1/\sigma$ and $1 - (s/\sigma)$ is conserved. This introduces only a small error. These expressions clearly reduce to the previous ones (A5) and (A13) if s is put equal to zero. It should be pointed out however that for $s \neq 0$ we no longer have $|S_0| = |T_0|$ and hence in general $I_T(t_1) \neq I_S(t_1)$.

For details we refer to Gevers et al.[459]

c. Determination of the nature of stacking faults

As discussed in Section 10a(1), one can consider two kinds of stacking faults in the face-centered cubic lattice. Howie[461] and Hashimoto et al.[462] have proposed the principle of a method to distinguish these two kinds making use of the contrast effects discussed in the previous paragraph. A practical procedure based on this principle was worked out by Art et al.[457] and applied to a copper-18% gallium alloy of low stacking fault energy, and for which it was found that the stacking faults are of the *intrinsic* type. We summarize this procedure here.

The Burgers vector of a partial dislocation generating a stacking fault can be considered as the shear vector $\bar{R}$ of the fault. The shear vector $\bar{R}_i$ for an intrinsic fault on the (111) plane is $\bar{R}_1 = (a/6)[11\bar{2}]$, $\bar{R}_2 = (a/6)[\bar{1}2\bar{1}]$, or $\bar{R}_3 = (a/6)[2\bar{1}\bar{1}]$. The shear vector for the extrinsic fault would then be the opposite of one of these. Such a vector would describe a displacement which brings one atomic layer on "top" of another one. The corresponding displacement

[461] A. Howie, *Met. Rev.* **6**, 467 (1961).
[462] H. Hashimoto, A. Howie, and M. J. Whelan, *Proc. Roy. Soc.* **A269**, 80 (1962).

therefore requires two motions along "valleys," but between successive lattice planes as described by the letter sequences :

$$abc \mid abcabca \cdots \rightarrow$$
$$abc \mid bcabcab \cdots \rightarrow$$
$$abcb \mid abcabc \cdots$$
$$\uparrow$$

From these considerations it is clear that $\alpha = 2\pi\bar{n} \cdot \bar{R}$ has opposite sign for intrinsic and extrinsic faults when the same diffraction vector is operating. The problem now is to relate the sign of α to a particular geometrical situation observed in a thin foil. To make this correlation it is easier to use the displacement vector $\bar{D}$, based on Frank's original definition of the two types of faults, given in Section 10.9(1). We consider the front part of the foil as being at rest and we displace the part below the stacking fault over the vector $\bar{D}$, describing the displacement resulting from the removal or the insertion of one layer. This convention is in agreement with that adopted in writing the system of equations (9.195) on which the derivation of all the results of Section 10h(2) is based. When writing this system it is assumed that the exit part of the column of crystal is displaced with respect to the entrance part of the column kept fixed. The vectors $\bar{D}_i$ and $\bar{D}_e$ for intrinsic and extrinsic faults respectively are shown in Fig. A7. The three vectors $\bar{R}_1$, $\bar{R}_2$, and $\bar{R}_3$ are now replaced by a single vector $\bar{D}_i$, and $-\bar{R}_1$, $-\bar{R}_2$, and $-\bar{R}_3$ are replaced by $\bar{D}_e = -\bar{D}_i$. It is clear that $\bar{D}_i = (a/3) [111]$. The use of the vector $\bar{D}_i$ is equivalent to the use of $\bar{R}_i$ because both vectors only differ by a lattice vector, for instance:

$$\frac{a}{6} [1\bar{1}2] = \frac{a}{3} [111] + \frac{a}{2} [1\bar{1}0].$$

The same remark applies of course to $\bar{D}_e$. The lattice vector only contributes a term 2π to α and hence gives rise to a factor unity for the phase factor $e^{i\alpha}$.

The procedure now consists in deducing the angle β between $\bar{n}$ and $\bar{D}$ and hence determine the sense of $\bar{D}$. One can always look at the fault in such a way that it is sloping upwards to the right as in Fig. A7; this face is called (111). The diffraction vector $\bar{n}$ can then be either to the left or to the right of the line AB which passes through the center C of the diffraction pattern and which is parallel to the intersection line of the fault and the surface. We will say that $\bar{n}$ is either left (L) or right (R). It is necessary of course, in order to establish this relationship, that the diffraction pattern is correctly oriented with respect to the image. The four possible situations are shown in Fig. A7; the sign of $\cos \beta$ is immediately evident. It now remains to be shown how the knowledge of the sign of α allows to deduce the sign of $\cos \beta$. One can always write $\alpha = 2\pi \mid \bar{n} \mid \mid \bar{D} \mid \cos \beta$.

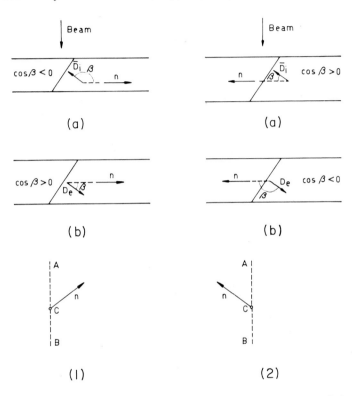

(a)   (a)

(b)   (b)

(1)   (2)

Fig. A7. Different relative orientations of the diffraction vector $\bar{n}$ and the shear vector D_i (intrinsic fault) and D_e (extrinsic fault). The line AB is parallel to the intersection of the fault plane and the surface: (1) $\bar{n}$ lies at the right side of AB (a) intrinsic stacking fault, (b) extrinsic stacking fault; (2) $\bar{n}$ lies at the left side of AB (a) intrinsic stacking fault, (b) extrinsic stacking fault.

For a given diffraction vector $\bar{n}$ and noting that the magnitude of $\bar{D}$ is always a $\sqrt{3}/3$ one can deduce the possible values of $\cos \beta$ compatible with the observed sign of α and with the known direction and sense of the diffraction vector. For all the simple diffraction vectors encountered in practice it is possible to assign an unambiguous value to $\cos \beta$

30 *

and hence determine the sense of $\bar{D}$. Table AI summarizes the different possibilities.

TABLE AI

$\langle n \rangle$	$\lvert n \rvert$ in units $1/a$	Calculated value for $\alpha \pmod{2\pi}$ $\alpha_{\text{calc}} = 2\pi \lvert n \rvert \lvert D \rvert \cos \beta$	$\cos \beta$ for $x_{\text{obs}} = \dfrac{2\pi}{3}$	$\alpha_{\text{obs}} = -\dfrac{2\pi}{3}$
200	2	$\dfrac{4\pi}{3}\sqrt{3}\cos\beta$	$-\dfrac{1}{\sqrt{3}}$	$+\dfrac{1}{\sqrt{3}}$
400	4	$\dfrac{8\pi}{3}\sqrt{3}\cos\beta$	$\dfrac{1}{\sqrt{3}}$	$-\dfrac{1}{\sqrt{3}}$
220	$2\sqrt{2}$	$\dfrac{4\pi}{3}\sqrt{6}\cos\beta$	$\dfrac{2}{\sqrt{6}}$	$-\dfrac{2}{\sqrt{6}}$
111	$\sqrt{3}$	$2\pi\cos\beta$	$1/3$	$-1/3$
222	$2\sqrt{3}$	$4\pi\cos\beta$	$-1/3$	$+1/3$

The use of the table is as follows. Suppose that the diffraction vector is of the type $\langle 200 \rangle$. The only possible values for $\cos \beta$ are $\pm 1/\sqrt{3}$ because these are the only ones that can occur for the angles between $\langle 111 \rangle$ and $\langle 100 \rangle$ directions. If the observed value of α is $+ 2\pi/3$ it is clear from this table that $\cos \beta$ has to be $-1/\sqrt{3}$ and hence β must be obtuse. Suppose that the diffraction vector is to the right as in (1) of Fig. A7 then β obtuse means $D = D_i$, i.e., the fault is intrinsic as in Fig. A7(a). Should the diffraction vector be at the left then β obtuse means $D = D_e$ and the fault would be extrinsic.

From the considerations given above it is clear that the following information is required: (1) the sense of sloping of the fault plane; (2) the sign of α; (3) the diffraction vector in the correct orientation with respect to the slope of the fault plane.

A convenient "check list" for determining the nature of the fault is now as follows.

(1) Take a bright and dark field image of the fault without changing the orientation of the foil. The value of s has to be close to zero. Be sure that you have, as close as possible, a "two-beam" case, i.e., that only one diffracted beam is strong, the one which is also used for making the dark field image.

(2) Take a diffraction pattern and identify the operating diffraction vector.

(3) Orient the diffraction pattern with respect to the image taking into account all electron optical rotations.

(4) By comparing the bright and dark field images it is possible to identify the top and bottom end of the fault. The bottom part is pseudocomplementary in the two images. The bright field image is now oriented so that the fault is sloping upwards towards the right.

(5) Determine whether the diffraction vector is left or right, i.e., whether you are in situation (1) or (2) of Fig. A7 by combining the conclusions form (2), (3), and (4) above.

(6) Determine the nature of the first fringe in the bright field image.

(7) It is now sufficient to consult Table AII to determine the character of the fault.

Using the method outlined here, the type of several stacking faults occurring in foils of Cu + Ga alloy (15% Ga) has been determined: they were all intrinsic.

TABLE AII

Diffraction vector: n	Orientation of n	First fringe in bright field image	
		bright	dark
200	L	E	I
	R	I	E
400	L	I	E
	R	E	I
220	L	I	E
	R	E	I
111	L	I	E
	R	E	I
222	L	E	I
	R	I	E

L: left   R: right
E: extrinsic   I: intrinsic

Hashimoto *et al.*[462] have reported that their preliminary results for Cu + Al alloys indicate that also for this alloy the stacking faults are of the intrinsic type.

If one is only interested in the nature of the fault one can obtain this information from the dark field image alone.[459] In fact it is sufficient to reformulate the results of Tables AI and AII. In this way we can construct the table in Fig. A8 in a different manner in which all possible geometrical cases are considered. In each of them as well the dark field as the bright field image is shown schematically (although the dark field image alone is needed). It is assumed that the diffraction vector is always pointing to the right. This means no loss

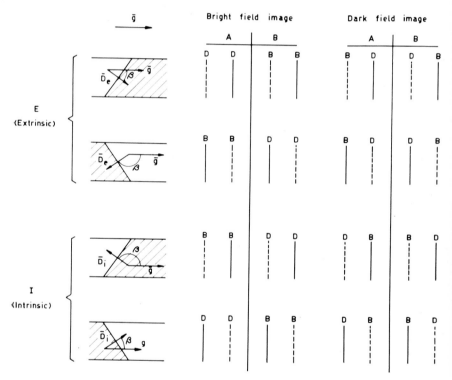

FIG. A8. Schematic representation of the bright and dark field images for the four possible geometrical situations. A full line represents the first fringe at the top surface while a dotted line represents the first fringe at the back surface. The nature of the fringes is indicated by *B* (for bright) or *D* (for dark). In all cases the diffraction vector (noted *g* on this drawing) is oriented towards the right.

in generality since it is always possible to orient the image in such a way that the operating diffraction vector points to the right of the intersection line of the fault plane and the foil plane. We have to consider two classes of reflections: class A (200, 222) and class B (400, 200, 111). The reasoning used in constructing the table is illustrated by means of one example. We consider: (1) the intrinsic fault E (2) with β_p obtuse (β_p is the angle between the displacement vector $\bar{D}$ and the projection n_p of the diffraction vector on the plane normal to the intersection line of fault plane and foil plane) and (3) for a diffraction vector $\bar{n}$ of the type $\langle 200 \rangle$ (class A). The sign of α is now positive since

$$\alpha = 2\pi \,|\, n_p \,|\,|\, D \,| \cos \beta_p$$

is equivalent to

$$\alpha = 2\pi \,|\, \bar{n} \,|\,|\, \bar{D} \,| \cos \beta = 2\pi \frac{2}{a} \cdot a \frac{\sqrt{3}}{3} \cdot -\frac{1}{\sqrt{3}} = -\frac{4\pi}{3} = \frac{2\pi}{3} \,(\mathrm{mod}\, 2\pi)$$

where β is now the real angle between $\bar{n}$ and $\bar{D}$. The identity of both expressions is evident if one writes $\bar{n} = \bar{n}_p + \bar{n}_0$ where $\bar{n}_0$ is the component perpendicular to the plane considered and one notes that $\bar{n} \cdot \bar{D} = (\bar{n}_p + \bar{n}_0) \cdot D = \bar{n}_p \cdot \bar{D}$ since $n_0 \cdot \bar{D} = 0$. When $\alpha = 2\pi/3$ the top fringe (represented by a full line in Fig. A8) is bright (B) in the dark field image and the bottom fringe (represented by a dotted line) is dark (D). Making the same reasoning for the different geometrical situations and for the different diffraction vectors leads to the table in Fig. A8.

The result of this table can be summarized in one rule: orient the image in such a way that the diffraction vector points to the right of the intersection line of the fault plane and of the foil plane. If the reflection belongs to class A, and if the diffraction vector points towards the bright fringe in the dark field image, the fault is intrinsic. The reverse is true for reflections belonging to class B. This rule becomes evident when looking at Fig. A8.

Appendix B. Symmetry of Dislocation Images for $s = 0$, in the Absence of Absorption

If absorption is neglected and taking $s = 0$ the system of differential Eq. (9.150) reduces to:

$$\frac{dT}{dz} = \frac{i\pi}{t_e} S \, e^{+i\alpha}; \quad \frac{dS}{dz} = \frac{i\pi}{t_e} T \, e^{-i\alpha}. \tag{B1}$$

For a screw dislocation the phase angle is $\alpha = n\phi$, where $\phi = \text{arctg} \, [(z - z_0)/x]$, with z_0, the distance of the dislocation below the entrance face. This system of equations has to be integrated with the initial conditions $T(0) = 1$; $S(0) = 0$. It describes the behavior of a column say at $x = x_0 (x_0 > 0)$. For a column at $x = -x_0$ the phase angle becomes $n(\pi - \phi)$. We call the amplitudes emerging from this column $T_{(-)}$ and $S_{(-)}$. They satisfy the system

$$\frac{dT_{(-)}}{dz} = \frac{i\pi}{t_e} S_{(-)} \, e^{in\pi} \, e^{-i\alpha}; \quad \frac{dS_{(-)}}{dz} = \frac{i\pi}{t_e} T_{(-)} \, e^{-in\pi} \, e^{i\alpha} \tag{B2}$$

with the same boundary conditions as above: $T_{(-)}(0) = 1$; $S_{(-)}(0) = 0$.

The complex conjugate quantities T_-^* and S_-^* satisfy the system of equations: ($e^{in\pi} = (-1)^n!$)

$$\frac{dT_{(-)}^*}{dz} = \frac{i\pi}{t_e^*} (-S_{(-)}^*) (-1)^n \, e^{i\alpha}; \quad \frac{d}{dz} [-S_{(-)}^*(-1)^n] = \frac{i\pi}{t_e^*} T_{(-)}^* \, e^{-i\alpha} \tag{B3}$$

with the boundary conditions $T_{(-)}^* = 1$; $S_{(-)}^*(0) = 0$.

By comparing Eqs. (B3) and (B1) it becomes evident that T and S and $T_{(-)}^*$ and $-(-1)^n S_{(-)}^*$ satisfy the same system of equations, and are subject to the same boundary conditions, since

$$t_e = t_e^*. \tag{B4}$$

One can conclude that $T_{(-)}^* = T$ and $-(-1)^n S_-^* = S$ and hence that $I_T^{(-)} = I_T$ and $I_S^{(-)} = I_S$ or that the image is symmetrical. The condition (B4) implies that anomalous absorption is neglected.

For an edge dislocation with slip plane parallel to the foil plane and assuming that the diffraction vector is in the foil plane one has $\alpha = n\phi + \{n/[4(1-\nu)]\} \sin 2\phi$. Since $\sin(2\pi - 2\phi) = -\sin 2\phi$, the reasoning therefore remains valid for such a dislocation. It is also valid for a mixed dislocation with Burgers vector parallel to the foil plane, if the strain field is considered as the superposition of the edge and the screw components separately.

Appendix C

The properties of image profiles for $s = 0$ as found by means of machine calculations[241] using the 2-beam column approximation can also be deduced analytically.[458] A simplified model is introduced for a foil containing an arbitrary dislocation parallel to the foil plane with Burgers vector also parallel to this plane. The model, although approximate, exhibits most of the properties found by the exact machine calculations. Some of the symmetry properties which we will discuss are only approximately true for the exact model; they become exact, however, for the approximate model. This somewhat peculiar situation follows from the fact that with the exact equations these "approximate" symmetry properties cannot be deduced without actually solving the equation. The special merits of the approximate model are precisely to make the "approximate" symmetry properties exact ones, so that it becomes possible to deduce them from the equations without solving the latter. The model also provides some physical insight and we will therefore discuss it in some detail.

The columns are dissected into three parts:

(1) the entrance part or part I (thickness t_1);

(2) the central part, called part II (thickness t_2);

(3) the exit part, denoted by III (thickness t_3).

Parts I and III are perfect, while part II contains the dislocation at t_2' from the front of part II and at t_2'' from its back face Fig. C1. The origin of the reference system is chosen at the dislocation. For the calculation of the bright and dark field image use will be made of the matrix method [Section 9i(5)].

(a) Derivation of the Scattering Matrix

For $s = 0$ the system (9.157) reduces to

$$\frac{dT}{dz} = \frac{\pi i}{\tau} S e^{i\alpha}$$

$$\frac{dS}{dz} = \frac{\pi i}{\tau} T e^{-i\alpha} \qquad \text{(C1)}$$

with $(1/\tau) = (1/t_e) + (i/\tau_e)$ and where, for a dislocation of the kind postulated, α is given by Eq. (9.34). In the vicinity of the dislocation and for $t_1 \gg |x|$ one has to a good approximation $\vartheta \simeq - (\pi/2)$ [or $\alpha \simeq - n(\pi/2)$] in part I and $\vartheta \simeq + (\pi/2)$ [or $\alpha \simeq n(\pi/2)$] in part III. Since the width of that part of the image where the intensity changes rapidly is only $0.2\, t_e$ one can to a good approximation put $\vartheta = - (\pi/2)$ and $+ (\pi/2)$ in parts I and III provided t_2 is taken to be at least one extinction distance, i.e., $t_2 \geqslant t_e$. Apart from these constant shifts parts I and III are perfect. The scattering matrices for such perfect slabs, taking into account the phase shifts, are given by Eq. (9.202). For part I one has:

$$_1M = \begin{pmatrix} T_1 & S_1\, e^{-in(\pi/2)} \\ S_1\, e^{in(\pi/2)} & T_1 \end{pmatrix} \tag{C2}$$

and for part III

$$_3M = \begin{pmatrix} T_3 & S_1\, e^{in(\pi/2)} \\ S_3\, e^{-in(\pi/2)} & T_3 \end{pmatrix} \tag{C3}$$

where [according to Eqs. (9.123) and (9.124), in which $s = 0$]

$$T_j = \cos \pi(t_j/\tau), \quad S_j = i \sin \pi(t_j/\tau)\ (j = 1, 3). \tag{C4}$$

The scattering matrix for the central part is now required. In order to obtain it, we first specify the significance of the elements of a scattering matrix for the perfect crystal.

The elements of the first column M_{11} and M_{21} represent the amplitudes of the transmitted and the scattered waves if the boundary conditions at the entrance face are $T = 1, S = 0$. They can be obtained by integrating the system (C1) with these boundary conditions. The matrix elements of the second column, M_{12} and M_{22}, represent respectively the amplitudes of the transmitted and the scattered wave if the boundary conditions at the entrance face are $T = 0$ and $S = 1$. This can again easily be derived by integrating the system (C1) with these boundary conditions. It can be deduced also very simply by interchanging S and T and replacing $\alpha \rightarrow - \alpha$. With this substitution the system (C1) remains unaltered but the boundary conditions reduce to the previous ones. The second column therefore contains the same elements but in a different succession and with a different sign for α. We now apply the same method for deducing the scattering matrix of the central part, which is no longer perfect.

We call $T_c(z, x)$ and $S_c(z, x)$ the elements of the first column; they ought to be calculated from the system (C1) with the boundary conditions $T_c(-t'_2, x) \equiv 1$ and $S_c(-t'_2, x) \equiv 0$ at the entrance face. The elements of the second column $T'_c(z, x)$ and $S'_c(z, x)$ could be derived in the same way as for the perfect crystal, by integrating the system (C1), but now with the boundary conditions $T'_c(-t'_2, x) \equiv 0$ and $S'_c(-t'_2, x) \equiv 1$ at the entrance face.

Explicit calculations can be avoided in the following way. Interchanging S and T in Eq. (C1) and at the same time changing x into $-x$ or, what is equivalent, changing ϑ into $\pi - \vartheta$ and taking into account that $\alpha(z, -x) = n\pi - \alpha(z, x)$ leads to the following system for $T'_c(z_1 - x)$ and $S'_c(z_1 - x)$

$$\frac{dS'_c(z, -x)}{dz} = \frac{i\pi}{\tau} e^{-in\pi} T'_c(z, -x) e^{i\alpha(z, -x)}$$

$$e^{-i\pi n} \frac{dT'_c(z, -x)}{dz} = \frac{i\pi}{\tau} S'_c(z, -x) e^{-i\alpha(z, -x)}$$

(C5)

where now the boundary conditions have become again $T'_c = 1$ and $S'_c = 0$ at the entrance face. From these considerations it is clear that $S'_c(z, -x)$ and $e^{-in\pi} T'_c(z, -x)$ satisfy the same system as $T_c(z, x)$ and $S_c(z, x)$ with the same boundary conditions, and hence we can conclude that, for a slab of thickness t_2,

$$T'_c(t_2, x) = e^{i\pi n} S_c(t_2, -x)$$

and

(C6)

$$S'_c(t_2, x) = T_c(t_2, -x).$$

The transmission matrix for the central part III is therefore:

$$_2M(t_2, x) = \begin{pmatrix} T_c(t_2, x) & e^{i\pi n} S_c(t_2, -x) \\ S_c(t_2, x) & T_c(t_2, -x) \end{pmatrix}.$$

(C7)

Explicit expression for T_c and S_c are not known, but are not required for our purpose. The scattering matrix for the complete column is now of course:

$$M = {}_3M\,{}_2M\,{}_1M$$

(C8)

and the amplitudes of the transmitted and scattered wave for the complete crystal are then given by

$$\begin{pmatrix} \mathscr{T} \\ \mathscr{S} \end{pmatrix} = M \begin{pmatrix} 1 \\ 0 \end{pmatrix} = {}_3M_2M \begin{pmatrix} T_1 \\ S_1 e^{in(\pi/2)} \end{pmatrix}. \tag{C9}$$

(b) Variations of the Image Profile with Depth

An increase in depth of a dislocation over a distance $t_e \varDelta$ where $\varDelta \leqslant 1$ can be represented now as an increase in thickness $t_e \varDelta$ of parts I and a decrease in thickness by the same amount of part III. The central part does not change. We now compare the profiles for the same dislocation at the two positions differing by $t_e \varDelta$ in depth.

Let us consider first the dislocation in a slab consisting of the following parts: $t_1 = 0$; t_2, $t_3 = t_e$; we call this combination A; the change in depth is then represented by taking $t_1 = \varDelta t_e$, t_2, $t_3 = (1 - \varDelta)t_e$; we call this combination B. At this point we make the additional assumption that anomalous absorption can be neglected for slabs of a thickness equal to a or smaller than t_e, i.e., $(1/\tau) \simeq (1/t_e)$.

We then have the following scattering matrices.

(1) for combination A:

$$\begin{pmatrix} \mathscr{T}_A \\ \mathscr{S}_A \end{pmatrix} = \begin{pmatrix} 1 & 0 \\ 0 & 1 \end{pmatrix} {}_2M(t_2, x) \begin{pmatrix} 1 & 0 \\ 0 & 1 \end{pmatrix} \begin{pmatrix} 1 \\ 0 \end{pmatrix} = \begin{pmatrix} T_c(x) \\ S_c(x) \end{pmatrix} \tag{C10}$$

(2) for combination B:

$$\begin{pmatrix} \mathscr{T}_B \\ \mathscr{S}_B \end{pmatrix} = \begin{pmatrix} -\cos \pi\varDelta & i\, e^{in(\pi/2)} \sin \pi\varDelta \\ i\, e^{-in(\pi/2)} \sin \pi\varDelta & -\cos \pi\varDelta \end{pmatrix} {}_2M(t_2, x) \begin{pmatrix} \cos \pi\varDelta \\ e^{in(\pi/2)} \sin \pi\varDelta \end{pmatrix}. \tag{C11}$$

We now consider two special cases:

(a) $\varDelta = 1$. Substituting this value into (C11) and working out the product of matrices gives

$$\begin{pmatrix} \mathscr{T}_B \\ \mathscr{S}_B \end{pmatrix} = - \begin{pmatrix} T_c(x) \\ S_c(x) \end{pmatrix} \quad \text{and hence} \quad \begin{matrix} {}_AI_t(x) = {}_BI_t(x) \\ {}_AI_s(x) = {}_BI_s(x). \end{matrix} \tag{C12}$$

This evidently means that the dark as well as the bright field image remains unchanged by the displacement over a distance t_e. This property is very approximately correct as can be deduced from the machine calculations.[241]

(b) $\Delta = 1/2$. Substituting this value into relation (C11) and again working out the product of matrices yields:

$$\begin{pmatrix} \mathcal{T}_B \\ \mathcal{S}_B \end{pmatrix} = -e^{-in\pi} \begin{pmatrix} T_c(-x) \\ S_c(-x) \end{pmatrix} \tag{C13}$$

which means that

$$\mathcal{T}_B(x) = -e^{in\pi} \mathcal{T}_A(-x)$$
$$\mathcal{S}_B(x) = e^{-in\pi} \mathcal{S}_A(-x) \tag{C14}$$

or finally

$$_B I_t(x) = {_A}I_t(-x) \qquad {_B}I_s(x) = {_A}I_s(-x). \tag{C15}$$

These relations show that image profiles for dislocations that differ $1/2\, t_e$ in depth are mirror images with respect to $x = 0$. As a consequence of this the depth period is apparently halved and becomes $(1/2)t_e$ for symmetrical images.

(c) *Change of the Profiles with Total Thickness of the Foil*

Consider now a dislocation in the combination of slabs A and the same dislocation at the same distance from the entrance face but in a foil which has a thickness larger by $\frac{1}{2}\, t_e$, i.e., in combination B (Fig. C1).

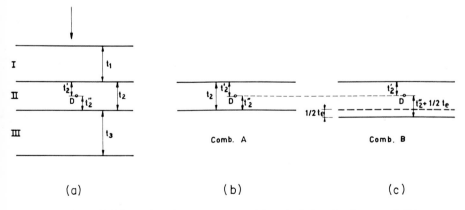

(a)   (b)   (c)

FIG. C1.   (a) Illustrating the geometry used in calculating the images of dislocations. The foil consists of three parts I, II, and III. Parts I and III are perfect, while part II contains the dislocation (D). (b) Combination A. (c) Combination B.

One has

$$\begin{pmatrix} \mathcal{T}_B \\ \mathcal{S}_B \end{pmatrix} = M \begin{pmatrix} \mathcal{T}_A \\ \mathcal{S}_A \end{pmatrix} \tag{C16}$$

where M is the scattering matrix for a perfect foil of thickness $\frac{1}{2} t_e$
which is

$$M = \begin{pmatrix} 0 & 1 \\ 1 & 0 \end{pmatrix},$$

and consequently

$$\mathcal{T}_B = \mathcal{S}_A \quad \text{and} \quad \mathcal{S}_B = \mathcal{T}_A \tag{C17}$$

which evidently means that dark and bright field images are interchanged. Should the added layer have a thickness of t_e then the matrix
$M = \begin{pmatrix} 1 & 0 \\ 0 & 1 \end{pmatrix}$ and both bright and dark field images are equal in combinations A and B.

All properties discussed so for are only valid for dislocations not
too close to the surface. Near the surface the dislocation images have
properties which are very similar to those of stacking fault fringes; at
the front surface dark and bright field images are similar; whereas at the
back surface they are "pseudocomplementary." Therefore it is possible
to deduce which end of the dislocation belongs to the front surface.

Most properties which follow from the machine calculations by
Howie and Whelan[241] could be deduced in a similar way by Gevers.[458]
We refer to this paper for further details.

(d) Applications

The properties just discussed are very useful in determining the
geometry of dislocations in space; they provide a depth scale, which is
completely lacking in the electron microscope as a consequence of the
considerable depth of focus. It is possible, for instance, to estimate
the depth of a particular segment of an inclined dislocation in a plane
parallel foil. If a dislocation is parallel to 1 plane of a wedge, the
oscillating contrast allows us to deduce its distance to the other wedge
plane. Whether this condition is satisfied or not can be deduced from
the comparison of the lateral periodicity of the wedge fringes and of
the oscillating contrast. The method has been used to estimate the
distance from the surface of dislocation ribbons.[252a]

Author Index

Numbers in parentheses are footnote numbers and indicate that an author's work is referred to although his name is not cited in the text.

Subject Index

Index of Substances

Printed in Belgium